19.50₂₅

BIOLOGY OF NORMAL AND ABNORMAL MELANOCYTES

BIOLOGY OF NORMAL AND ABNORMAL MELANOCYTES

Edited by
Taro KAWAMURA,
Thomas B. FITZPATRICK, and
Makoto SEIJI

UNIVERSITY PARK PRESS
Baltimore · London · Tokyo

© UNIVERSITY OF TOKYO PRESS, 1971
UTP 3047–67389–5149
Printed in Japan.

Originally published in 1971 by
UNIVERSITY OF TOKYO PRESS

UNIVERSITY PARK PRESS
Baltimore • London • Tokyo
ISBN 0–8391–0586–X
Library of Congress Catalog Card Number 70–150605

FOREWORD

During the period since the U.S.-Japan Cooperative Science Program was formally initiated there have been more than 180 seminars and 200 cooperative research projects involving the participation of several thousand American and Japanese scientists. These scientific meetings and research projects have no doubt made a great contribution to the promotion of science, to scientific cooperation, and especially to mutual understanding and friendly relations among scientists of both countries, as well as among scientists of other lands.

It is indeed meaningful for both the U.S. National Science Foundation and the Japan Society for the Promotion of Science to support seminars and cooperative research projects to promote further the benefits mentioned above. It is our great pleasure to know that the excellent papers presented at the Seminar on the Biology of the Normal and Abnormal Melanocyte are appearing in book form, which we believe will be of great use and value to many scientists in this field.

On behalf of the implementing agencies of the U.S.-Japan Cooperative Science Program, we wish to extend our thanks and grateful appreciation to all who helped make this seminar successful. In particular, we would like to extend special thanks to Professor Thomas B. Fitzpatrick, of Harvard University, Professor Taro Kawamura, of the University of Tokyo, and Professor Makoto Seiji, of Tohoku University. Without their strenuous efforts and the sacrifice of their precious time for the organization of the seminar it would have been difficult to have obtained these very fruitful results.

Walter H. HODGE
Head
National Science Foundation Tokyo Office

Masao YOSHIKI
Chief Director
Japan Society for the Promotion of Science

PREFACE

Twenty-two years have elapsed since the late Dr. Myron Gordon, in his foreword to *The Biology of Melanomas*, stressed the "urgent need for a concerted attack upon the nature of melanomas" and "for a broader research program to achieve the earliest possible solution of the melanoma problem." International congresses such as this one are of immense value in furthering this objective and Dr. Gordon's request is being heeded now by an intensive approach to the study of melanocytes on three planes: the multidisciplinary distribution of scientists, involving biologists, biochemists, chemists, and clinicians, comprises two horizontal planes, and the depth of penetration of their studies—extending from grossly visible color phenomena through findings made with light and electron microscopes down to information concerning molecular structure—comprises the third, which is perpendicular to the first two.

The idea for this joint meeting between Japanese and American scientists was conceived by Professor Makoto Seiji. We are obliged to Dr. Walter H. Hodge, of the National Science Foundation of the United States and to Mr. H. Amatsuchi, of the Japan Society for the Promotion of Science, for the help that they extended to us and for their addresses. We are indebted to Professor K. Kitamura, who kindly attended and addressed the seminar, to Professor M. Ito, Doctors Y. Sato and T. Irimagiri, of the Department of Dermatology, Tokyo Medical and Dental University, and

Dr. M. Miyamoto, Department of Dermatology, University of Tokyo, who provided invaluable help in preparing the seminar and in arranging the discussion manuscripts.

The seminar moved at a steady but leisurely pace with time for discussions not only at the regular sessions but at "tea breaks," receptions, and casual gatherings. These informal, spontaneous discussions are possibly the most rewarding periods of any seminar. The interaction of two or more scientists produces new ideas which then become the seeds of future research. During this period of drastic restriction of funds for foreign travel, it seems appropriate to recall the words of the American philosopher William James on his return from the International Congress on Psychology held in Paris in 1896:

> The open results were, however (as always happens at such gatherings), secondary in real importance to the latent ones—the friendships made, the intimacies deepened, and the encouragement and inspiration which come to everyone from seeing before them in flesh and blood so large a portion of that little army of fellow students from whom and for whom all contemporary psychology exists. The individual worker feels much less isolated in the world after such an experience.
> Cited in *William James,* by G. W. Allen (New York: The Viking Press, 1967), p. 310.

It is hoped that the publication of this book will promote further interaction among scientists interested in pigment cells.

Taro KAWAMURA
Department of Dermatology, Faculty of Medicine, University of Tokyo, Tokyo

Thomas B. FITZPATRICK
Department of Dermatology, Harvard Medical School, Boston, Massachusetts

Makoto SEIJI
Department of Dermatology, Tohoku University School of Medicine, Sendai

OPENING ADDRESS

Colleagues from America and Japan:
Having welcomed our colleagues from America to this United States-Japan Seminar, I shall take this opportunity to talk to you about the pigment studies that have been done in the past in this country, and about our friendly relationship with American fellow dermatologists during these studies.

In my previously published memoirs covering forty years of dermatological studies, I wrote, "Pigmentary disorders of the skin or problems of skin pigment in general indeed constitute an extremely fascinating subject for every dermatologist. Those who have entered the study of dermatology, without exception, are enchanted by it at least once."

In the prewar years, before the development of electron microscopes and when optical microscopes were the only means available for histological studies, many Japanese dermatologists, including myself, devoted themselves to the study of all sorts of pigmentary disorders of the skin as well as to more fundamental research on skin pigment itself. Among our senior dermatologists Keizo and Shoji Dohi, Toyama, Matsumoto, Ota, Minami, and Ito deserve to be called forerunners in these studies. Dr. Toyama, my teacher, once stated that because of their moderate pigmentation—neither too heavy nor too light—among the Japanese some forms of hyper- and depigmentation that are likely or easy to be overlooked in white and colored people can be found. This probably explains why some pigment anomalies, such as *dyschromatosis hereditaria, leucoderma pseudosyphiliticum,* and *acropigmentatio reticularis* were first described by Japanese dermatologists and were regarded as indigenous to the Japanese.

Since the end of the Word War II, skin pigment and related problems have again become a matter of concern for us, and studies using new electron-microscopy techniques, which are incomparably more precise than the old techniques have been carried out. Among the pigmentary disorders we have observed clinically in recent years, the following three in particular should be mentioned. First, Riehl's melanose, which is noteworthy because it appeared in the early postwar years and occurred as a result of using some kinds of black-market cosmetics which may have contained some photosensitizing substance. Second, *incontinentia pigmenti* has been the subject of frequent case reports suggesting its probable high incidence among the Japanese. Finally, Peutz-Jeghers' syndrome deserves mention because of its atavistic genesis, deduced from the histological similarity of its palmoplantar pigmentations to those of apes. The so-called *guanofuracin leucoderma* of eyelids caused by the use of eye lotion containing guanofuracin is also a new form of depigmentation observed in the postwar years.

In regard to our pigment studies since the end of the war, we should also mention the friendly associations we have enjoyed with American dermatologists. The very instructive lectures given by Dr. Fitzpatrick, our friend for many years, in Sendai in 1956 and again in April of this year in Tokyo, and the many valuable pigment research projects which our young Japanese colleagues in Portland, Boston, New Haven, Detroit, and elsewhere have completed under their American teachers are results of this association. Needless to say, all these have encouraged and promoted, to no small degree, our pigment studies.

The present United States-Japan Seminar is a place for cooperation among all of you, the participants from both countries. I hope that you will strive for definitive results at the highest level of investigation of the structures, functions, and behavior of melanin and melanocytes, which play an indispensable part in the formation and disappearance of skin pigment. I am sure that the seminar will contribute to the further development of melanin research, not only in America and Japan, but also throughout the world.

Kanehiko KITAMURA
President
Tokyo Medical College
Tokyo

LIST OF CONTRIBUTORS

BAGNARA, Joseph T.
Department of Biological Sciences, University of Arizona, Tucson, Arizona, U.S.A.

BLOIS, Marsden S., Jr.
Department of Dermatology, University of California Medical Center, San Francisco, California, and
Department of Dermatology, Stanford University School of Medicine, Stanford, California, U.S.A.

CHAVIN, Walter
Department of Biology, Wayne State University, Detroit, Michigan, U.S.A.

COOPER, Michael
Wayne State University School of Medicine, Detroit, Michigan, and Veterans Administration Hospital, Allen Park, Michigan, U.S.A.

FERRIS, Wayne
Department of Biological Sciences, University of Arizona, Tucson, Arizona, U.S.A.

FITZPATRICK, Thomas B.
Department of Dermatology, Harvard Medical School, and the Massachusetts General Hospital, Boston, Massachusetts, U.S.A.

Fujii, Ryozo
Department of Biology, Sapporo Medical College, Sapporo, Japan

Fukushiro, Ryoichi
Department of Dermatology, Kanazawa University School of Medicine, Kanazawa, Japan

Hirone, Takae
Department of Dermatology, Kanazawa University School of Medicine, Kanazawa, Japan

Hori, Yoshiaki
Department of Dermatology, Faculty of Medicine, University of Tokyo, Tokyo, Japan

Ikeda, Shigeo
Department of Dermatology, Faculty of Medicine, University of Tokyo, Tokyo, Japan

Imai, Katsutoshi
Faculty of General Studies, Gunma University, Maebashi, Japan

Itakura, Hideko
Department of Dermatology, Tokyo Medical and Dental University School of Medicine, Tokyo, Japan

Jimbow, Kowichi
Department of Dermatology, Sapporo Medical College, Sapporo, Japan

Kasuga, Tsutomu
Section of Pathology, Division of Physiology and Pathology, National Institute of Radiological Sciences, Chiba, Japan

Kawamura, Taro
Department of Dermatology, Faculty of Medicine, University of Tokyo, Tokyo, Japan

Kikuyama, Sakae
Department of Biology, Waseda University School of Education, Tokyo, Japan

Kinebuchi, Shoichi
Department of Dermatology, Tokyo Teishin Hospital, Tokyo, Japan

Kobori, Tatsuji
Department of Dermatology, Tokyo Teishin Hospital, Tokyo, Japan

Kukita, Atsushi
Department of Dermatology, Sapporo Medical College, Sapporo, Japan

Lerner, Aaron B.
Section of Dermatology, Department of Medicine, Yale University School of Medicine, New Haven, Connecticut, U.S.A.

MATSUBARA, Tameaki
Department of Dermatology, Kanazawa University School of Medicine, Kanazawa, Japan

MISHIMA, Yutaka
Department of Dermatology, Wakayama Medical University, Wakayama, Japan

MIYAZAKI, Kazuhiko
Department of Dermatology, Tokyo Medical and Dental University, School of Medicine, Tokyo, Japan

MORIYA, Tsuneo
Department of Biology, Waseda University School of Education, Tokyo, Japan

NAGAI, Tadashi
Department of Dermatology, Kanazawa University School of Medicine, Kanazawa, Japan

NAKAYASU Michie
Biochemistry Division, National Cancer Center Research Institute, Tokyo, Japan

NIIMURA, Michihito
Department of Dermatology, Faculty of Medicine, University of Tokyo, Tokyo, Japan

OBATA, Hiroko
Department of Dermatology, Faculty of Medicine, University of Tokyo, Tokyo, Japan

OIKAWA, Atsushi
Biochemistry Division, National Cancer Center Research Institute, Tokyo, Japan

PATHÁK, Madhukar A.
Department of Dermatology, Harvard Medical School, and the Massachusetts General Hospital, Boston, Massachusetts, U.S.A.

QUEVEDO, Walter C., Jr.
Division of Biological and Medical Sciences, Brown University, Providence, Rhode Island, U.S.A.

SEIJI, Makoto
Department of Dermatology, Tohoku University School of Medicine, Sendai, Japan

SZABÓ, George
Department of Anatomy, Harvard Dental School, and the Massachusetts General Hospital, Boston, Massachusetts, U.S.A.

TAKAHASHI, Ichi
Section of Pathology, Division of Physiology and Pathology, National Institute of Radiological Sciences, Chiba, Japan

TAKAHASHI, Makoto
Department of Dermatology, Sapporo Medical University, Sapporo, Japan

TAKEUCHI, Takuji
Department of Biology, Faculty of Sciences, Tohoku University, Sendai, Japan

TODA, Kiyoshi
Department of Dermatology, Harvard Medical School, and the Massachusetts General Hospital, Boston, Massachusetts, U.S.A.

TSUCHIYA, Eiko
Section of Pathology, Division of Physiology and Pathology, National Institute of Radiological Sciences, Chiba, Japan

YASUMASU, Ikuo
Department of Biology, Waseda University School of Education, Tokyo, Japan

CONTENTS

xvi

BIOLOGY OF NORMAL AND ABNORMAL MELANOCYTES

Neural Control of Pigment Cells*

Aaron B. Lerner

Section of Dermatology, Department of Medicine, Yale University
School of Medicine, New Haven, Connecticut, U.S.A.

I. INTRODUCTION

Hormonal and neural mechanisms are two of the major regulatory forces
in pigment cell function. Both have a long history, but knowledge of
neural control goes back much further—approximately 100 years compared with 50 for hormonal. However, hormonal mechanisms have
received much more attention from investigators in recent decades. There
are reasons for the difference in emphasis. The dramatic color changes
produced in tadpoles and frogs following hypophysectomy or after injection of pituitary extracts were discovered at the end of the First World
War and laid the basis for the entire field of modern endocrinology. This
opening inspired work on numerous projects related to pituitary and
adrenal regulation of metabolic processes. Furthermore, in the last
fifteen years application of hormonal mechanisms to the control of mammalian pigment cells, particularly human, has been strikingly successful.
By contrast the modern phase of neural control began in the 1930s and
even now it is not yet applicable to mammalian pigment systems. Nevertheless, neural mechanisms probably loom large in the regulation of
human pigment cells and thus deserve intensive study.

The leader in the field of neural control of pigment cells, George H.

* This research was supported by U.S. Public Health Service Grant No. CA 04679 and
by the American Cancer Society Grant P 168 C.

Parker,[1] carried out his investigations from 1930 to 1945 and was followed by his students Abramowitz,[2] Kleinholz,[3] and Scott.[4-6] During the past ten years the subject has been extended by the studies of Scott,[4-6] Hadley,[7-12] Novales,[13-16] Bagnara,[12] Goldman,[9-11] Fujii,[15,16] Abe,[17,18] members of our own laboratory,[19-27] and others.[28-32]

Pigment cells, i.e., melanocytes or melanophores, are nerve cells. Embryologically they are derived from the neural crest, and it is appropriate to ask whether pigment cells are under neural control. If so, in what way does such control differ from hormonal regulation?

II. PIGMENT CELLS WITH DIRECT INNERVATION

In 1876 Pouchet[33] reported that electrical stimulation of cutaneous nerves in a flatfish produced blanching of the skin supplied by the nerves. Subsequently a variety of experiments involving either cutting or electrical stimulation of nerves showed that, in many kinds of fish, pigment cells are under direct nerve control.[2,4-6] These conclusions were given additional support by the recent experiments of Bikle, Tilney, and Porter,[28] who found unmyelinated nerve fibers in direct apposition to pigment cells in *Fundulus* and by Jacobowitz and Laties[29] who observed catecholamine-containing fibers in anatomic proximity to conjunctival and dermal melanocytes of a *Teleost* fish.

In all fish having directly innervated pigment cells, electrical stimulation of a nerve to the skin produces localized blanching and cutting the nerve produces immediate darkening.

III. PIGMENT CELLS WITHOUT INNERVATION

Rana pipiens, the experimental hero of research on hormonal control of pigmentation, has two major populations of pigment cells—dermal and epidermal—neither of which is innervated directly.[26] Yet these cells, like the lizard's[9, 10] which also have no direct nerve control, respond to pharmacologically active adrenergic and cholinergic chemicals as well as to their blocking agents. In the frog the epidermal and dermal pigment cells show separate properties of response. For example, *in vitro* both types darken with MSH, ACTH, or caffeine and lighten with their removal.[8, 25] Darkened dermal melanophores will lighten in response to adrenaline or melatonin but darkened epidermal cells will not. Yet in some frogs both kinds of cells lighten on exposure to acetylcholine.[8] In addition to the difference in the qualitative behavior of epidermal and dermal cells, quantitative variations in response occur within each group of cells depending largely on their anatomical location. Differences in the qualitative and quantitative response of pigment cells within a single animal are undoubtedly not limited to frogs but probably occur in all animals. In man the melanocytes of the skin at the epidermal-dermal junction, around hair bulbs, and in the dermis, oral mucosa, eyes, and central nervous system most likely have overlapping as well as distinctive features. And for each type of melanocyte there would be variations in responsiveness depending on location. For example, the melanocytes at the epidermal-dermal junction in the eyelids may react to hormones more readily than the melanocytes of the cheeks, which in turn are more reactive than those of the thighs.

IV. ADRENERGIC RECEPTORS OF MELANOCYTES

In 1906 Dale,[34] studying the adrenergic blocking action of ergot on different effector systems, found that in some tissues ergot blocking was pronounced, e.g., pressor action of adrenaline, whereas in others it was not, e.g., gut muscle relaxation by adrenaline. The concept favoring the existence of two types of adrenergic receptors was sharpened further by Ahlquist in 1948,[35] when he introduced the classification of alpha- and beta-adrenergic receptor sites based on investigation of the potencies of adrenergic substances and their blocking agents. Extension of his studies led to the view that alpha receptors mediate responses for which adrenaline $>$ noradrenaline $>$ phenylephrine $>$ isoproterenol[36] (see Table 1). These responses are blocked by ergotanine, dibenamine, priscoline and phentolamine. For beta receptors the reaction order is isoproterenal $>$ adrenaline $>$ noradrenaline $>$ phenylephrine. The responses are blocked by propranalol, pronethalol and dichloroisoproterenol.

Table 1. Natural and synthetic agents that stimulate alpha- and beta-adrenergic receptor sites are listed in order of decreasing potency. Substances that can block the action of the stimulators are also listed.

α-Stimulators	α-Blockers	β-Stimulator	β-Blockers
Adrenaline	Ergotamine	Isoproterenol	Propranalol
Noradrenaline	Dibenamine		Pronethalol
Phenylephrine	Priscoline		Dichloroisoproterenol
	Phentolamine		

Thirteen years after Ahlquist's report, Graham[30] concluded from his own experiments that catecholamine activated beta-adrenergic receptors of *Xenopus laevis* to produce skin darkening. Subsequent investigations by Hadley and Goldman,[9-11] Scott,[4-6] Gupta and Bhide,[31] and McGuire[27] showed clearly that melanocytes of all the animals they studied—fish, frogs, toads, and lizards—have alpha- and/or beta-adrenergic receptors. A constant finding is that stimulation of alpha receptors leads to aggregation of type IV melanosomes and skin lightening. Activation of beta

5

receptors brings about dispersion of pigment granules and skin darkening. A single agent, e.g., adrenaline, may stimulate both alpha- and beta-receptors but seldom equally. Generally the alpha receptors respond more vigorously than the beta.

The above findings allow one to clarify a number of paradoxical states. The dual lightening and darkening actions of adrenaline on frog skin can be explained as follows: MSH-darkened skin treated with adrenaline will lighten because alpha-adrenergic receptors of dermal pigment cells will be activated. However, if adrenaline is added to skin washed free of MSH and consequently light in color, the skin darkens. In the light state the alpha receptors cannot be stimulated further; and, because adrenaline can trigger beta receptors, the skin darkens a little.

In the lizard *Anolis carolinensis* excitement induces generalized pallor which is followed rapidly by the appearance of scattered dark spots that make the skin look stippled. In *Anolis* most pigment cells possess both alpha- and beta-adrenergic receptors, but some cells have only beta receptors. Stimulation of the alpha sites by catecholamines produces the excitement pallor. At about the same time dark islands result from activation of the melanophores possessing only beta receptors.[9, 10]

The spadefoot toad *Scaphiopus couchi* has only beta-adrenergic receptors, the alpha receptors being absent or physiologically inactive.[11] Therefore, stimulation *in vitro* by adrenaline or noradrenaline produces a further increase in darkening of MSH-darkened skin or a darkening of skin maintained in Ringer's solution free of MSH.

From these as well as other studies a consistent pattern emerges. Pigment cells of fish, amphibians, and reptiles have alpha- and/or beta-adrenergic receptors. Stimulation of alpha sites produces an aggregation of melanosomes and subsequent lightening. Stimulation of beta sites brings about dispersion of pigment granules followed by darkening. When both kinds of receptors function within a single melanophore, the alpha receptor's role overrides the beta's.

Stimulation by faradic current of a cut nerve in a fish produces immediate blanching through activation of alpha receptors. If one could find a fish having melanophores with only beta receptors, then faradic stimulation of a cut nerve would produce either no change or further darkening.

In mammals the only evidence linking catecholamines to lightening is that injection of adrenaline into rat skin produces gray hair.[37, 38] Also in two human subjects with cervical sympathectomies, the development of gray hair with age was retarded on the denervated sides.[22, 39]

V. CHOLINERGIC RECEPTORS OF MELANOCYTES

In contrast to the numerous papers on the effects of adrenergic substances and their blocking agents on pigment cells, only a handful of experiments have been reported on the action of acetylcholine.[8,19,25] According to Parker et al.[40] acetylcholine can darken many fishes by dispersing the pigment granules within melanophores. In our laboratory acetylcholine was found to be an effective but inconsistent lightening agent of dermal melanocytes in *Rana pipiens*. A detailed investigation by Möller and Lerner[25] explained the inconsistency: the skin from only one out of three frogs reacted to acetylcholine; that is, approximately two out of three animals were nonreactors. If one piece of skin from a frog lightened upon addition of acetylcholine, all other pieces of skin from that animal would lighten. In addition only skin from reactor animals responded to acetylcholine-related chemicals, and all reactions were blocked by atropine. We could not alter the ratio of reactors to nonreactors. Genetic differences may exist in the cholinergic receptor sites of dermal melanocytes in the two populations of frogs. MSH will not darken skin previously lightened by acetylcholine.

Dermal cells of frogs darkened by MSH lighten on exposure to catecholamines or melatonin, but darkened epidermal cells do not. However, in some specimens of skin acetylcholine will lighten both the dermal and epidermal pigment cells.

It has been suggested that in man some populations of melanocytes in skin and hair may lighten in response to acetylcholine but not to catecholamines or melatonin.[8,21] The entire subject of acetylcholine regulation of pigment cells of fish, amphibians, reptiles, and mammals must be studied thoroughly if we are to achieve a clearer picture of neural control of pigmentation.

VI. MELATONIN CONTROL OF PIGMENTATION

Melatonin, a methoxyindole unrelated chemically to either catecholamines or acetylcholine, is produced in the pineal gland. The output of melatonin is determined by the quantity of light to which the animal is exposed.[41] On dermal melanocytes of *Rana pipiens* melatonin causes reversible lightening in concentrations less than those required for adrenaline or acetylcholine, but the degree of lightening is less with melatonin than with adrenaline. Alpha- and beta-adrenergic blockers and atropine cannot prevent the action of melatonin.

In contrast to the activity of highly diluted melatonin on frog dermal pigment cells, this methoxyindole has no direct effect on mammalian melanocytes. However, the possibility of melatonin's acting centrally to prevent release of MSH from the pituitary has been suggested by Rust and Meyer's experiments on the weasel[42] and by Kastin and Schally's work on the rat.[43]

Melatonin seems to control the circadian color changes of the pencil fish.[32] Whether the control is direct or indirect on the melanophores of the fish is not known.

VII. MECHANISM OF HORMONAL AND NEURAL CONTROL

Cyclic AMP (3′,5′-adenosine monophosphate) has been labeled a "second messenger" because several hormones exert their metabolic control by regulating the intracellular levels of this nucleotide which in turn monitors the reaction. Cyclic AMP is formed from ATP through the action of adenyl cyclase and is converted to inactive 5′-AMP by a phosphodiesterase. Therefore increasing cyclase and/or decreasing phosphodiesterase activities brings about an increase in cyclic AMP.

Early experiments disclosed that ATP without any other agent could disperse melanin granules in isolated pieces of frog skin.[19] If frog skin were made to darken and lighten several times by immersion in Ringer's solution with and without MSH, after several hours the melanocytes failed to respond to MSH. Addition of ATP restored the MSH-dispersing reaction. More direct evidence that cyclic AMP could regulate the dispersion of melanosomes came when Bitensky and Burstein[44] and later Novales and Davis[13] found that cyclic AMP darkened frog skin. Also caffeine, an inhibitor of phosphodiesterase, caused dispersion of pigment granules. Before the end of 1969 the diverse and elegant investigations by Bitensky and Demopoulous,[45] Abe et al.[17,18] and Goldman and Hadley[11] left little doubt that cyclic AMP was the critical biological agent mediating the action of both darkening and lightening agents (Fig. 1). Bitensky and Demopoulous found that MSH and ACTH increased adenyl cyclase activity *in vitro* in fish and mouse melanomas that were either melanotic or amelanotic. Abe et al. found that in skin from *Rana pipiens* α-MSH, β-MSH, and ACTH increased cyclic AMP levels with the same relative potencies at which they stimulated darkening. On the other hand nor-

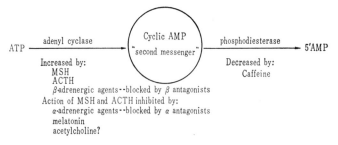

Fig. 1. The intracellular level of cyclic AMP is determined by the balance between factors favoring its enzymic synthesis and its breakdown.

adrenaline and melatonin inhibited the MSH-induced increase in cyclic AMP. Neither substance influenced the darkening reaction of exogenous cyclic AMP. The inhibitory action of noradrenaline but not that of melatonin could be stopped by alpha-adrenergic blocking agents. Beta blockers were ineffective. Theophylline, a phosphodiesterase inhibitor, led to accumulation of cyclic AMP as well as to darkening of frog skin *in vitro*. Goldman and Hadley showed that in both the toad *Scaphiopus couchi* and the lizard *Anolis carolinenis* stimulation of beta-adrenergic receptors darkened skin probably by increasing intracellular concentrations of cyclic AMP.

The easily demonstrable dispersion and aggregation of melanosomes through relatively fixed channels in pigment cells of fish, amphibians, and reptiles—in response to MSH, caffeine, catecholamine, acetylcholine, and melatonin—has not been shown in mammalian systems. Movement of granules *in vivo* was looked for but not found in bats[46] and mice[47] treated with MSH. Also cultured guinea pig melanocytes showed no movement of pigment granules.[48] Nevertheless, MSH has a pronounced darkening effect on the skin or hair of man, guinea pigs, hamsters, and mice. In each case the evidence is clear that darkening is associated with an increase in tyrosinase activity.[21,49–52] The increase in tyrosinase most likely is due to increased enzyme synthesis and not simply to activation of tyrosinase already present in the cell.

At present it can be said that MSH darkens the skin of fish, amphibians, and reptiles by causing dispersion of pigment granules within intracellular channels surrounded by microtubules;[28,53] and MSH darkens the skin and/or hair of mammals by stimulating tyrosinase synthesis. It is not known whether the microtubules have a part in melanosome movement.[54] The role of cyclic AMP in the induction of tyrosinase is still unknown, but since MSH increases cyclic AMP in melanomas it is possible that the nucleotide is responsible for initiating tyrosinase synthesis.

What is the relation between dispersion of melanosomes and synthesis of tyrosinase? Perhaps, as has been suggested, dispersion leads to increased synthesis of melanin.[55,56] However no one has shown that increased synthesis of tyrosinase leads to dispersion of melanosomes.

VIII. SUMMARY

In all animals melanin-producing pigment cells are derived from neural tissue, and melanin is made by the oxidation of tyrosine in the presence of tyrosinase. The pigment cells or melanophores of fish, amphibians, and reptiles contain alpha- and/or beta-receptor sites and melanosomes that move within cytoplasmic channels. MSH, ACTH, and caffeine cause dispersion of granules by increasing the level of intracellular cyclic AMP

MELANIN DISPERSION

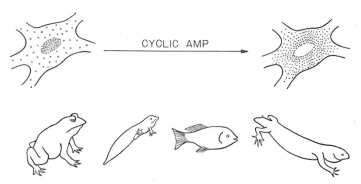

Fig. 2. α- and β-MSH, ACTH, and caffeine raise the intracellular levels of cyclic AMP in melanophores of amphibians, fish and reptiles to initiate dispersion of melanosomes

MELANIN SYNTHESIS

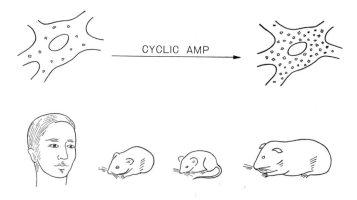

Fig. 3. Injection of MSH into man, hamsters, mice and guinea pigs increases tyrosinase activity, which in turn results in melanin formation. It is likely that these events are triggered by MSH, which raises cellular levels of cyclic AMP.

(Fig. 2). Stimulation of alpha sites produces aggregation of granules leading to lightening of skin, whereas beta stimulation yields melanosome dispersion and darkening of skin. Lightening by catecholamine and melatonin is associated with a decrease in cyclic AMP but each neurohormone acts at different receptor sites. Acetylcholine darkens some fish but lightens about one out of three *Rana pipiens*.

In mammalian pigment cells or melanocytes MSH increases cyclic AMP, induces tyrosinase synthesis, and darkens skin (Fig. 3). These steps may be related sequentially. Alpha- and beta-receptor sites have not been found in the mammalian cells, but it is likely that the melanocytes are under neural control. Acetylcholine may be a lightening agent for mammalian melanocytes.

We do not know if one can classify pigment cells as melanophores or melanocytes depending on the presence or absence of alpha- and beta-adrenergic receptor sites. Also separating melanophores and melanocytes on the basis of darkening due to melanosome dispersion through channels or as a result of tyrosinase synthesis is not possible at the present time.

The relationship between movement of pigment granules and tyrosinase synthesis and the action of neural agents other than catecholamines are important current problems that require solution.

REFERENCES

1. Parker, G. H., *Animal colour changes and their neurohumours*. Cambridge University Press, Cambridge (1948).
2. Abramowitz, A. A., Physiology of the melanophore system in the catfish, Ameiurus. *Biol. Bull.*, **71**, 259 (1936).
3. Kleinholz, L. H., Studies in reptilian colour changes. *J. Exp. Biol.*, **15**, 492 (1938).
4. Scott, G. T., Physiology and pharmacology of color change in the sand flounder, *Scopthalmus aquosus. Limnol. Oceanog.* vol. 10, suppl., p. R230 (1965).
5. Scott, G. T. and Wong, K. K., Evidence against the presence of functional pigment-dispersing nerve fibers in the sand flounder, *Scophthalmus aquosus, Biol. Bull.*, **131**, 406 (1966).
6. Scott, G. T., The action of phenothiozines on pigment cells of the flatfish, *Scopthalmus aquosus*, Second International Symposium on Action Mechanism and Metabolism of Psychoactive Drugs Derived from Phenothiozine and Structurally Related Compounds, Oct. 17–19, 1967. Paris.
7. Hadley, M. E. and Quevedo, W. C., The role of epidermal melanocytes in adaptive color changes in amphibians. In *Advances in Biology of Skin*, vol. 8, *The Pigmentary System* (W. Montagna and F. Hu, eds.), p. 337, Pergamon Press, Oxford (1967).
8. Goldman, I. P. and Hadley, M. E., Acetylcholine-induced Aggregation of melanin granules within epidermal (frog) melanocytes, *J. Invest. Derm.*, 50, 59 (1968).
9. Goldman, J. M. and Hadley, M. E., *In vitro* demonstration of adrenergic receptors controlling melanophore responses of the lizard, *Anolis carolinensis. J. Pharmacol. Exp. Ther.*, **166**, 1 (1969).
10. Hadley, M. E. and Goldman, J. M., Physiological color changes in reptiles. *Amer. Zool.*, **9**, 489 (1969).
11. Goldman, J. M. and Hadley, M. E., The beta adrenergic receptor and cyclic 3′, 5′-adenosine monophosphate: Possible roles in the regulation of melanophore responses of the spadefoot toad, *Scaphiopus couchi. Gen. Comp. Endocr.*, **13**, 151 (1969).
12. Bagnara, J. T. and Hadley, M. E., The control of bright-colored pigment cells of fishes and amphibians, *Amer. Zool.*, **9**, 465 (1969).
13. Novales, R. R. and Davis, W. J., Melanin-dispersing effect of adenosine 3′, 5′ monophosphate on amphibian melanophores. *Endocrinology*, **81**, 283 (1967).
14. Novales, R. R. and Davis, W. J., Cellular aspects of the control of physiological color in amphibians. *Amer. Zool.*, **9**, 479 (1969).

15. Fujii, R. and Novales, R. R., Cellular aspects of the control of physiological color changes in fishes. *Amer. Zool.*, **9,** 453 (1969).

16. Fujii, R. and Novales, R. R., The nervous mechanism controlling pigment aggregation in Fundulus melanophores. *Comp. Biochem. Physiol.*, **29,** 109 (1969).

17. Abe, K., Butcher, R. W., Nicholson, W. E., Baird, C. E., Liddle, R. A., and Liddle, G. W., Adenosine 3', 5'-monophosphate (cyclic AMP) as the mediator of the actions of melanocyte-stimulating hormone (MSH) and norepinephrine on the frog skin. *Endocrinology,* **84,** 362 (1969).

18. Abe, K., Gobison, G. A., Liddle, G. W., Butcher, R. W., Nicholson, W. E., and Baird, C. E., Role of cyclic AMP in mediating the effects of MSH, norepinephrine and melatonin on frog skin color. *Endocrinology,* **85,** 674 (1969).

19. Lerner, A. B. and Takahashi, Y., Hormonal control of melanin pigmentation. *Recent Progr. Hormone Res.*, **12,** 303 (1956).

20. Chanco-Turner, M. L. and Lerner, A. B., Physiologic changes in vitiligo. *Arch. Derm.*, **91,** 390 (1965).

21. Lerner, A. B., Snell, R. S., Chanco-Turner, M., and McGuire, J. S., Vitiligo and sympathectomy. *Arch. Derm.*, **94,** 269 (1966).

22. Lerner, A. B., Gray hair and sympathectomy. *Arch. Derm.*, **93,** 235 (1966).

23. McGuire, J. and Möller, H., Response of melanocytes of dermis and epidermis to lightening agents. *Nature,* **208,** 493 (1965).

24. McGuire, J. and Möller, H., Differential responsiveness of dermal and epidermal melanocytes of *Rana pipiens* to hormones. *Endocrinology,* **78,** 367 (1966).

25. Möller, H. and Lerner, A. B., Melanocyte-stimulating hormone inhibition by acetylcholine and noradrenaline in the frog skin bioassay. *Acta Endocr.*, **51,** 149 (1966).

26. Snell, R. S. and Kulovich, S., Nerve stimulation and the movement of melanin granules in the pigment cells of the frog's web. *J. Invest. Derm.*, **48,** 438 (1967).

27. McGuire, J. S., Adrenergic control of melanocytes. *Arch. Derm.* (in press).

28. Bikle, D., Tilney, L. G., and Porter, K. R., Microtubules and pigment migration in the melanophores of *Fundulus heteroclitus* L. *Protoplasma,* **61,** 322 (1966).

29. Jacobowitz, D. M. and Laties, A. M., Direct adrenergic innervation of a teleost melanophore. *Anat. Rec.*, **162,** 501 (1968).

30. Graham, J. D. P., The response to calecholamines of the melanophores of *Xenopus laevis. J. Physiol.* (London), **158,** 5P–6P (1961).

31. Gupta, I. and Bhide, N. K., Nature of adrenergic receptors on the skin melanophores of *Rana tigrina. J. Pharm. Pharmacol.*, **19,** 768 (1867).

32. Reed, B. L., The control of circadian pigment changes in the pencil fish: A proposed role for melatonin. *Life Sci.,* vol. 7, pt. 2, p. 961 (1968).

33. Pouchet, G., Des changements de coloration sous l'influence des nerfs. *J. Anat., J. Physiol.* (Paris), **12,** 1 (1876).

34. Dale, H. H., On some physiological actions of ergot. *J. Physiol.* (London), **34,** 163 (1906).

35. Ahlquist, R. P., A study of the adrenotropic receptors. *Amer. J. Physiol.*, **153,** 586 (1948).

36. Weyer, E. M. (ed.), New adrenergic blocking drugs: Their pharmacological, biochemical and clinical actions. *Ann. N. Y. Acad. Sci.,* vol. 139, art. 3 (1967).

37. Munan, L. P., Depigmentation of hair by oxidized adrenaline solutions. *Lancet,* **1,** 600 (1953).

38. Shelley, W. B. and Öhman, S., Epinephrine induction of white hair in ACI rats. *J. Invest. Derm.*, **53,** 155 (1969).

39. Lerner, A. B., unpublished observations.

40. Parker, G. H., Welsh, J. H., and Hyde, J. E., The amounts of acetylcholine in the dark skin and in the pale skin of the catfish. *Proc. Nat. Acad. Sci. U.S.A.*, **31,** 1 (1945).

41. Wurtman, R. J., Axelrod, J., and Kelly, D. E., *The pineal,* Academic Press, New York (1968).

42. Rust, C. C. and Meyer, R. K., Hair color, molt, and testis size in male, short-tailed weasels treated with melatonin. *Science,* **165,** 921 (1969).

43. Kastin, A. J. and Schally, A. V., Autoregulation of release of melanocyte stimulating hormone from the rat pituitary. *Nature,* **213,** 1238 (1967).

44. Bitensky, M. W. and Burstein, S. R., Effects of cyclic adenosine monophosphate and melanocyte stimulating hormone on frog skin *in vitro. Nature,* **208,** 1282 (1965).

45. Bitensky, M. W. and Demopoulos, H. B., Activation of melanoma adenyl cyclase by MSH. Abstract from the Seventh International Pigment Cell Conference, Sept. 2–6, 1969, *J. Invest. Derm.* (in press), Seattle, Washington (1970).

46. McGuire, J. S., Examination of melanocytes *in vivo*: A new approach. *J. Invest. Derm.,* **46,** 311 (1966).

47. Snell, R. S., Silver, A. F., and Chase, H. B., A new method for studying mammalian melanocytes *in vivo. Nature,* **210,** 219 (1966).

48. Klaus, S. N. and Snell, R. S., The response of mammalian epidermal melanocytes in culture to hormones. *J. Invest. Derm.,* **48,** 352 (1967).

49. Snell, R. S., Effect of the alpha melanocyte stimulating hormone of the pituitary on mammalian epidermal melanocytes. *J. Invest. Derm.,* **42,** 337 (1964).

50. Pomerantz, S. H., L-tyrosine-3, 5-³H assay for tyrosinase development in skin of newborn hamsters. *Science,* **164,** 838 (1969).

51. Pomerantz, S. H. and Chuang, L., Tyrosinase in the skin of newborn hamsters and mice: Effects of β-melanocyte-stimulating hormone, cortisol and adrenocorticotropin (forthcoming).

52. Geschwind, I. I., Hormonal modification of coat color in the laboratory mouse, Abstract from the Seventh International Pigment Cell Conference, Sept. 2–6, 1969. *J. Invest. Derm.* (in press), Seattle, Washington (1970).

53. Green, L., Mechanism of movements of granules in melanocytes of Fundulus heteroclitus. *Proc. Nat. Acad. Sci. U.S.A.,* **59,** 1179 (1968).

54. Wise, G. E., Ultrastructure of amphibian melanophores after light-dark adaption and hormonal treatment. *J. Ultrastruct. Res.,* **27,** 472 (1969).

55. Foster, M., Physiological studies of melanogenesis. *Pigment Cell Biology* (M. Gordon, ed.), p. 301, Academic Press, New York (1959).

56. Lee, T. H. and Lee, M. S. Studies on melanocyte stimulating hormone induced melanogenesis in frog skin. *Fed. Proc.* (abstract), **28,** 871 (1969).

DISCUSSION

DR. BAGNARA: Has anyone determined the level of cholinesterase activity in vitiligo skin as opposed to normal skin?

DR. LERNER: No. Cholinesterase levels should be determined. It is more important to carry out microscopic localization of cholinesterase than simply to make gross assays. There is a lot of cholinesterase in the skin, but the amount in the melanocyte should be ascertained.

DR. BAGNARA: I wonder if one could extend vitiligo to amphibians, considering that the interspot area of *Rana pipiens* might be an example of vitiligo.

DR. LERNER: Yes. I suppose one can look at the frog pigment pattern in a reverse way.

DR. FITZPATRICK: It is worthwhile, perhaps, to mention that the nerve and hormonal control mechanisms act on the melanin-dispersion and melanin-aggregating structure apparatus rather than on the melanosome itself. That is, pituitary hormones can also promote aggregation and dispersion of pigments other than melanin, such as pterin-containing organelles (pterinosomes), present in erythrophores. The nerve and hormonal control mechanisms thus probably act on the microtubules that have been observed in electron micrographs of fish melanophores and have also been observed recently in mammalian melanocytes by Dr. Toda in my laboratory.

DR. LERNER: With knowledge of cyclic AMP, we advance one more step; no one knows what happens after the formation of the cyclic AMP, or why the dispersion then takes place. But that is an interesting problem in its own right. Microtubules may take an important part especially since colchicine causes damage to microtubules and the cells and the granules remain in the dispersed state.

DR. BAGNARA: In response to Dr. Fitzpatrick's point, Hadley and I have demonstrated that amphibian iridophores are sensitive to cyclic AMP.

DR. MISHIMA: Since you are also discussing vitiligo, I would like to point out that in vitiligo lesions there is a disappearance of melanocytes and a discontinuance of synthesis of premelanosomes. The genesis of this process is one of the most direct problems in vitiligo research. I would hope you could look not only for dispersion or aggregation but also for the cellular change of the epidermal melanocytes in the frog skin in your experiments.

DR. LERNER: Yes, it is important to determine whether or not melanosome dispersion occurs in the mammalian melanocyte.

DR. BLOIS: What is the time required to demonstrate the effect of cyclic AMP in increasing tyrosinase activity in mammalian skin? We have studied the rate of melanin synthesis in mouse melanoma *in vivo* by using the rate of ^{14}C-dopa incorporation, and we have found no stimulating effect.

Dr. Lerner: The increased tyrosinase activity is demonstrated at the end of 7 days, but gross darkening appears to begin after 24 hours.

Dr. Pathak: In measuring the skin reflectance of man, I have noticed significant variation. If the skin reflectance of a given area of the body is measured, it shows detectable changes; sometimes one observes darkening of the skin and sometimes the same area shows lightening. Do you think that the variation in skin reflectance at different hours of the day could be explained by variation in the cyclic AMP activity? Is there any evidence of *diurnal* variation in the cyclic AMP levels?

Dr. Lerner: More studies must be carried out to decide whether or not melanosome dispersion occurs in human pigment cells.

Dr. Fujii: According to the conventional concept of the autonomic nervous system, acetylcholine and catecholamines act antagonistically. This idea might be applied to the innervated effector cells only. However, in fish melanophores where dual nervous mechanisms probably exist, acetylcholine has only a very weak effect or more often no effect on the cells. How can you explain this phenomenon? By the way, we have in Japan a good subject for the study of the beta-receptor mechanism in the fish melanophore system. The melanophores of the Japanese catfish, *Parasilurus asotus,* respond to catecholamines by dispersion of melanosomes. We are now planning to do an experiment on this line.

Dr. Lerner: I have no explanation at present. I believe one can say that granules in the human melanocyte do not respond beautifully to catecholamines. The receptor groups may vary in the different melanocytes. We have the same situation with the frog. Some frogs respond to acetylcholine, while most of them do not.

Dr. Oikawa: Could you demonstrate the darkening of skin by administration of cyclic AMP or butyryl-cyclic AMP?

Dr. Lerner: Isolated pieces of frog skin can be darkened with cyclic AMP or butyryl-cyclic AMP. To my knowledge none of these agents have been injected into mammalian skin.

Dr. Fujii: In relation to Dr. Oikawa's question to Dr. Lerner, I would like to refer to our recent findings on the melanin-dispersing effect of cyclic AMP in fish melanophores. Novales and Fujii are the authors of the paper concerned (*J. Cell. Physiol.,* **75**, in press), and an outline of the results will be given later in my paper in this session.

DR. BAGNARA: With regard to concentration effects of cyclic AMP, what is your feeling about this? One has to use such high concentrations of cyclic AMP to elicit any response, especially in reference to the dibutyryl-cyclic AMP which worked at lower concentrations.

DR. LERNER: I don't know. Of course, it has always been said that the difference is that one substance does not readily get into the cell whereas the other does. That explanation may be good enough. Everything appears to fit so wonderfully that it's hard not to accept this concept.

DR. CHAVIN: A reason for the lack of long-term fish experiments utilizing denervation may well be due to the fact that fish show remarkable abilities to regenerate even nerves.

Color Change and Pituitary Function in Xenopus laevis

Katsutoshi Imai

Faculty of General Studies, Gunma University, Maebashi, Japan

I. THE GENERAL IDEA OF COLOR CHANGE AND PITUITARY FUNCTION IN AMPHIBIANS

Since Hogben and Slome[1,28] have performed extensive experiments on amphibian color change, the endocrine mechanism involved in this biological phenomenon has been gradually elucidated. Like most lower vertebrates, *Xenopus* can change its own color according to the light condition in which the animal is placed. Under illuminated conditions, *Xenopus* becomes darker if in a black container (black adaptation), and lighter if in a white container (white adaptation). It is well known that in this sort of color change, i.e., physiological color change, dermal and epidermal melanophores play a most important role. Microscopic observation reveals that in a dark animal the melanin granules in the melanophores are dispersed all over the cells, while in a white one the pigment granules are concentrated near the central part of the melanophores.

Dispersion and concentration of melanin granules in melanophores are under neural and/or humoral control in many fishes and reptiles. Melanophores of amphibians, however, are not known to be linked to any controlling nerve. Melanin movements in amphibian melanophores are for most part regulated by endocrine factors. A bihumoral mechanism possibly involved in the color change, proposed by Hogben and Slome,[1,28] was recently reviewed by Bagnara,[2] who suggested the pineal gland. Hogben and Slome suggested that the pars tuberalis of the pituitary may secrete

17

"W substance" which is antagonistic to "B substance" (MSH). Since a highly active melanophore-concentrating agent (melatonin) was extracted from the mammalian epiphysis,[3] efforts were concentrated on elucidating the pineal function in the controlling mechanism of color change in amphibians. Bagnara[2] found a so-called body-lightening reaction in *Xenopus* tadpole which turn very pale as a result of maximal concentration of body melanophores when it was put into complete darkness. Recently Charlton[4] insisted that the function of the pineal gland in determination of color could not be neglected in the tadpole or in the young adult. Under illuminated conditions, however, the pineal gland is apparently inert, and the pituitary gland seems to be most important in regulating the color.

Generally speaking, top illumination without reflected light (black adaptation) accelerates release of MSH from the pars intermedia. MSH released from the pars intermedia into general circulation reaches melanophores to cause dispersion of melanin granules. Top illumination plus reflected light (white adaptation), on the other hand, inhibits release from the pars intermedia, which causes the melanin granules to concentrate. Thus, the color of amphibians is a direct indication of secretory activity in the pars intermedia of the pituitary gland.

II.   MSH CONTENT IN THE PARS INTERMEDIA OF *XENOPUS LAEVIS* IN RELATION TO COLOR

Several investigators have reported MSH activity in the pars intermedia under differently illuminated conditions. Koller and Rodewald[5,6] reported that MSH activity in the pituitary gland of *Rana temporaria* decreased rapidly in complete darkness and finally disappeared when the animals were kept in it longer. Burgers et al.,[7] however, could not confirm these results in the same animals. Masselin[8] and Stoppani[9] observed that MSH activity was higher in light than in darkness. American investigators,[10] on the contrary, could not find any difference in MSH activity in darkness and under illumination.

Apart from experiments with the presence or absence of light, Ortman[11,12] seems to be the first to test MSH activity of amphibian pituitary placed on different backgrounds under constant illumination. According to his experiment, MSH content in the pars intermedia/posterior lobe complex of *Rana pipiens* did not show any significant difference between white, black, and neutral adaptations, when bioassay was performed on the hypophysectomized frogs *(Rana pipiens)*. Using the African clawed toad, *Xenopus laevis*, in different background conditions, Burgers et al.,[7,13] Takahashi and Imai,[14] and Imai[15] tested MSH activity in the pars intermedia/posterior lobe complex. According to them, MSH content in the pars intermedia/posterior lobe complex was much higher in the white-adapted animal than in the black-adapted one. The experiment was performed as follows.

Xenopus laevis were first adapted to a black background to make the pituitary MSH condition constant throughout. After two days, half of the animals were transferred to a white container, while the rest were kept in the black container for another two days. During the adaptation period, animals were illuminated from the top with a fluorescent lamp. At the end of the adaptation period, the animals were autopsied and the anterior lobe and the pars intermedia/posterior lobe complex were separately assayed by the Anolis test.[7] At the end of the adaptation period, the mean melanophore index of white-adapted animals was 2.6, and that of black-adapted ones 4.9. According to the results of the assay, the level of MSH activity in the anterior lobe was not different for white- and black-adapted *Xenopus*, while the levels of activity in the pars intermedia/posterior lobe complex were significantly different from each other. The activity was 150–600 (350 on an average) *Anolis* units in white-adapted animals and 10–60 (38.3) *Anolis* units in black-adapted ones. From these data, the following explanation of the secretory activity of the pars intermedia, at least in *Xenopus laevis*, may be possible. When the animals are kept in the black container, stored MSH is released into general circulation, which causes pigment dispersion in the melanophores. At the same time, synthesis of MSH starts in order to replace the lost hormone. When the animal is transferred into the white container, release of the hormone ceases, but production proceeds for a while until hormone content reaches certain level, then stops. Since as long as the animals are kept on white background, they are pale, hormone release and synthesis must not be taking place. Similar results were also reported by Ito,[16] using the Japanese frog, *Rana nigromaculata*. MSH content is an indication of the counterbalance of production and release. Differences between the results found with *Rana pipiens, Xenopus laevis,* and *Rana nigromaculata* were not always unexpected.

III. HISTOLOGY OF THE PARS INTERMEDIA IN RELATION TO COLOR CHANGE

A. *Ordinary microscopy*

As was shown in the biological assay, the content of MSH and the secretory activity of the pars intermedia in *Xenopus laevis* differ greatly between white- and black-adapted animals. An investigation of the histology of the tissue in the different experimental conditions mentioned above should be very valuable. According to Ortman,[11, 12] in the pars intermedia of *Rana pipiens,* there are two types of granules distinguished by histochemical methods. Granules of one type (A granules) are numerous, small in size, and show positive reactions for carbohydrates and protein; the other granules (B granules) are less abundant, of larger size, and can be stained by the method for demonstrating phospholipids. Ortman changed the background but did not find any remarkable change in the cytology of the pars intermedia.

The brains and pituitary glands of white- and black-adapted *Xenopus* were fixed with Bouin's fixative and embedded in paraffin. Serial 6 μ sections of the tissue were stained with aldehyde fuchsin, PAS-orange G, or Mallory's triple stains. In the pars intermedia of *Xenopus laevis,* no secretory granules comparable to the A granules described by Ortman can be observed by the ordinary histological methods mentioned above. Larger granules (which may be identical to Ortman's B granules), how-ever, can be seen in *Xenopus laevis.* These granules are found in the cells situated near the blood vessels or the border of the pars intermedia and the nervosa. Some small difference was found in histology of the pars intermedia between white- and black-background adapted *Xenopus.* In the black-adapted animals in which MSH is synthesized and released

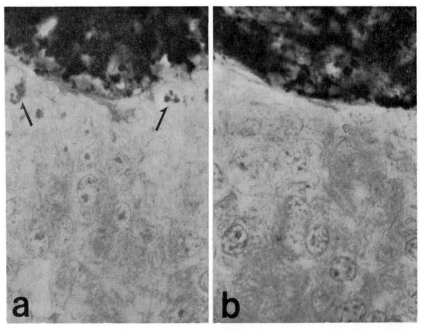

Fig. 1. The pars intermedia of the pituitary gland of *Xenopus laevis* adapted to black background (a) and white background (b). Arrows show colloidal bodies in the pars intermedia of the black-adapted *Xenopus.* There are no colloidal bodies in the white-adapted animal. Aldehyde fuchsin stain. Oil immersion.

from the tissue, the cytoplasm of the cells is a little lighter than that in the white-adapted ones; a more remarkable difference is that the B granules described by Ortman do not appear in the white-adapted animals (Fig. 1a,b). Recently, Iturriza,[17] and Iturriza and Koch[18] mentioned that this colloidal substance may be a reservoir of MSH. However, measurement of MSH content in *Xenopus laevis* denies this possibility, because the MSH content is higher in white-adapted animals in which the colloidal substance is absent than in black-adapted ones where the substance is abundant. For the present, it is tentatively proposed that the substance may be involved in some physiological events relating to synthesis or release of MSH, but discovering the real physiological meaning of the colloidal substance requires further investigation.

B. *Electron microscopy*

Recent electron microscopic studies revealed that cytological pictures of the amphibian pars intermedia differ remarkably according to the secretory activity of the tissue.[15,16,19,20] A similar kind of cytological difference was found in cells of the pars intermedia of *Xenopus* adapted on white and black backgrounds. *Xenopus laevis* were adapted to a white or black background as described before. After adaptation for 2, 4, and 7 days, they were killed by decapitation, and the pituitary glands were immersed in buffered OsO_4 for one and a half hours and embedded in EPON 812 through routine alcoholic dehydration. Sections were made using Porter's ultramicrotome MT-1, stained with uranyl acetate and lead citrate, and observed with an electron microscope, Hitachi UM-7.

Black adaptation

In the black-adapted animals, cells of the pars intermedia are characterized by the presence of cell organelles known to be related to protein synthesis. In the cytoplasm, development of rough-surfaced endoplasmic reticulum (rough ER), arranged as a lamellated structure, is remarkable· Such an arrangement of rough ER has already been reported by Ito[16] in *Rana nigromaculata* with hypothalamic lesion, and by Saland[19] in dark-adapted *Rana pipiens*. Golgi elements which usually appear near the nuclei are well developed. A number of secretory granules ranging from 200 to 240 mμ in size are seen in various parts of the cytoplasm, frequently associated with Golgi membrane (Fig. 2b). In short, the pars intermedia in black-adapted animals shows a picture characteristic of tissue which produces protein or related substances. In addition to the secretory granules, large, dark bodies are seen in the cells near the blood vessels and the border between the pars intermedia and the nervosa (Fig. 2c). These dark bodies are the same substance seen under the ordinary microscope in the black-adapted *Xenopus*. It should be mentioned that the colloidal bodies are completely different from the secretory granules, because the

21

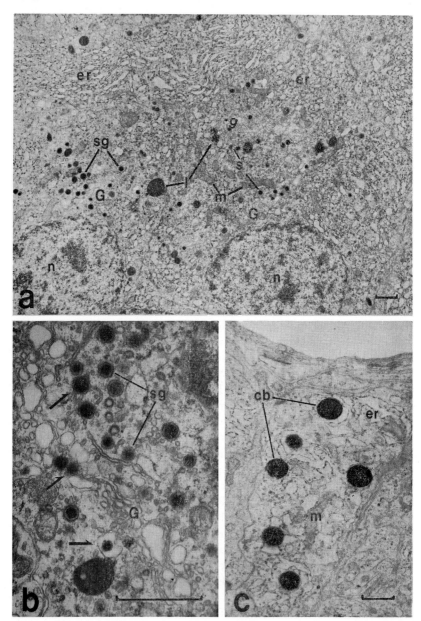

Fig. 2. Electron micrographs of the pars intermedia of the pituitary of *Xenopus laevis* adapted to black background for 7 days. (a) A picture of low-power magnification show-ing developed cell organelles. (b) Golgi element. Arrows show the possible condensation of the secretory material in the Golgi membrane. (c) Colloidal bodies in the cells of the pars intermedia. cb: colloidal bodies; er: rough ER; G: Golgi elements; l: lysosomes; m: mitochondria; n: nucleus; sg: secretory granules.

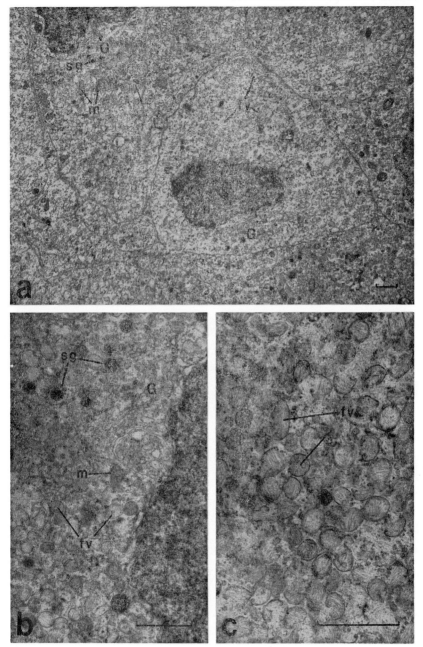

Fig. 3. Electron micrographs of the pars intermedia of the pituitary of *Xenopus laevis* adapted to white background for 7 days. (a) A picture of low-power magnification showing underdevelopment of cell organelles. (b) A picture showing transformation of secretory granules into filamentous vesicles. (c) Filamentous vesicles. fv: filamentous vesicles; other abbreviations are the same as in Fig. 2.

former are surrounded by rough ER, while the latter arise in the Golgi membrane. The significance of the large colloidal bodies was discussed above, but the possible explanation proposed by Saland[19] that the bodies are an aggregation of hormone at elevated synthetic activity level is not ruled out, because the cells of the pars distalis, which is thought to be active in synthesis of the hormone,[21] have a similar appearance.

White adaptation

In the white-adapted animals, the cells of the pars intermedia have several distinct characteristics (Fig. 3a). The first is a decrease of rough ER, Golgi elements, and secretory granules. The tendency toward under-development of these cell organelles is remarkable in animals adapted for 7 days on a white background. Instead of reduction of the secretory granules, light and slightly larger vesicles are found even after only 2 days of adaptation, and at the end of 7 days of adaptation a large area of the cytoplasm is occupied by these vesicles. In these vesicles, fine filamentous structures can be seen (Fig. 3c). The filamentous vesicles are quite similar to the vesicles or light granules observed in the rat's pars intermedia.[22] Furthermore, these vesicles may be identical to the dark granules found

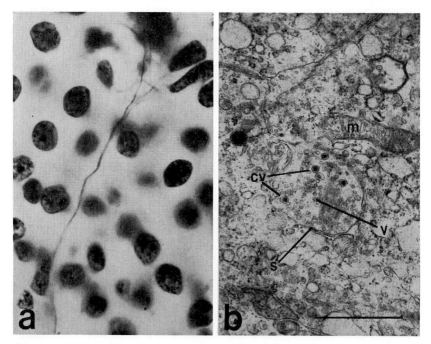

Fig. 4. (a) Nerve fibers in the pars intermedia of the pituitary of *Xenopus laevis* demonstrated by Bodian's protargol stain. Oil immersion. (b) Synapsis on the cell of the pars intermedia of the pituitary. cv: monoamine-containing synaptic vesicles; s: synapsis; v: cholinergic synaptic vesicles; other abbreviations are the same as in Fig. 2.

by several investigators[18,19] who fixed the material at first with glutaraldehyde and then with buffered OsO_4. Sometimes, a transitional form of vesicles from secretory granules is observed (Fig. 3b). Therefore, it is highly possible that the vesicles found in white-adapted *Xenopus* are a morphological indication of stored MSH.

It must be added that many typical and atypical synapses are observed at the surface of the cells of the pars intermedia in *Xenopus laevis*. Many authors have insisted that secretory activity is regulated by the hypothalamus through inhibition.[16–18, 23–25] Enemar et al.[24] and Iturriza[25] demonstrated that the nerve is an adrenergic one. In accordance with their results, a number of ordinary nerve fibers are demonstrated and electron microscopic observation reveals that the synapses contain so-called monoamine-containing synaptic vesicles, i.e., cored synaptic vesicles (Fig. 4). It is highly probable that the pars intermedia is controlled by adrenergic nerves.

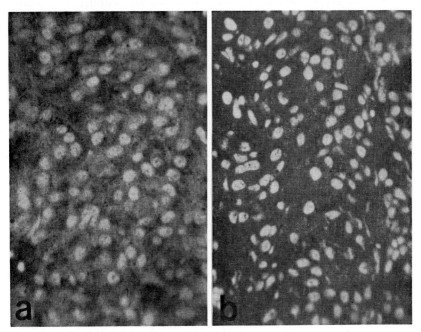

Fig. 5. Monochrome photographs of the pars intermedia of the pituitary of *Xenopus laevis* stained with acridine orange and observed with a fluorescent microscope. White dots are nuclei which are yellow under the fluorescent microscope. Grey areas around nuclei in the pictures which are vermilion red under the fluorescent microscope represent the presence of RNA. (a) Dark-adapted *Xenopus*. (b) White-adapted *Xenopus*. Grey area is much more abundant in the dark-adapted *Xenopus* than in the white-adapted one. × ca. 200.

C. *Fluorescent microscopy*

According to recent biochemical information, synthesis of protein is carried out by ribosomal RNA in the cytoplasm. Electron microscopic observation revealed clearly that in dark animals a large number of ribosomes are found at the surface of rough ER, while in white-adapted *Xenopus* it is rare. Therefore, a cytochemical method was applied to the *Xenopus* pars intermedia in order to demonstrate the presence of RNA by means of a fluorescent microscopic technique.[26] Fresh frozen sections of pituitary gland of *Xenopus laevis* adapted on a white or black background as described before were fixed with alcoholic fixative and stained with buffered acridine orange (pH 6.25). The sections were then observed with a fluorescent microscope.

Nuclei of the cells of the pars intermedia in both animals display similar secondary fluorescent yellow color. On the other hand, cytoplasm of the pars intermedia of the dark-adapted animal (black animal, increased hormone synthesis) is characteristically vermilion red, resulting from induced ultraviolet activation, while in the white-adapted one (white animal, decreased synthesis) such characteristic secondary color is not so prominent (Fig. 5). According to Pearse,[27] the fluorescent method, as well as the methyl green and pyronine methods, are highly reliable. The results obtained in the present observation indicate the presence of a larger amount of RNA in the dark-adapted animal than in the white-adapted one.

IV. SUMMARY

The pituitary gland of *Xenopus laevis* shows different secretory activity according to the background adaptation under constant illumination. MSH content in the pars intermedia is much higher in white *Xenopus* adapted to white background than in black animals adapted to black background, as a result of the counterbalance of production and release of MSH. The difference in secretory activity was demonstrated also by histological methods, notably by use of electron microscopic and fluorescent microscopic techniques. In dark *Xenopus*, in which MSH is released and synthesized and hormone content is low, cell organelles related to the synthesis of proteinaceous substance such as rough ER Golgi elements, secretory granules, and so on, are well developed. In these cells, cytoplasmic RNA is abundant. These electron microscopic and fluorescent microscopic pictures represent clearly the active synthesis of the hormone (MSH). In a white animal, in which release and synthesis of the hormone do not take place and the hormone content is high, development of cytoplasmic organelles is very poor. According to the electron microscopic pictures, RNA is less abundant. Instead of poorly developed organelles, the cytoplasm is occupied by a number of characteristic (filamentous) vesicles. It was suggested that the filamentous vesicles may be a morphological indication of stored MSH at electron microscopic level.

REFERENCES

1. Hogben, L. and Slome, D., The pigmentary effector system. VIII. The dual receptive mechanism of the amphibian background response. *Proc. Roy. Soc.* Ser. B, **120,** 158 (1936).

2. Bagnara, J. T., Pineal regulation of body blanching in amphibian larvae. *Progr. Brain Res.,* **10,** 489 (1965).

3. Lerner, A. B., Shizume, K., and Bunding, I., Endocrine control of melanin pigmentation. *J. Clin. Endocr.,* **14,** 1463 (1954).

4. Charlton, H. M., The pineal gland and color change in *Xenopus laevis* Daudin. *Gen. Comp. Endocr.,* **7,** 384 (1966).

5. Koller, G. and Rodewald, W., Über den Einfluss des Lichtes auf die Hypophysentätigkeit des Frosches. *Arch. ges. Physiol.,* **232,** 637 (1933).

6. Rodewald, W., Der Einfluss der Dunkelheit auf den das Melanophorenhormon bindenden Stoff im Froschblut. *Z. vergleich. Physiol.,* **22,** 431 (1935).

7. Burgers, A. C. J., Imai, K., and van Oordt, G. J., The amount of melanophore-stimulating hormone in single pituitary glands of *Xenopus laevis* kept under various conditions. *Gen. Comp. Endocr.,* **3,** 53 (1963).

8. Masselin, J. N., Influence de la lumière et de l'obscurité sur l'action mélanophore dilatatrice de l'hypophyse. *Compt. Rend. Soc. Biol.,* **91,** 511 (1939).

9. Stoppani, A. O. M., Neuroendocrine mechanism of color change in *Bufo arenarum* (Hensel). *Endocrinology,* **30,** 782 (1942).

10. Kleinholz, L. H. and Rahn, H., The distribution of intermedin: A new biological method of assay and results of tests under normal and experimental conditions. *Anat. Rec.,* **76,** 157 (1940).

11. Ortman, R., Cytochemical study of the physiological activities in the pars intermedia of *Rana pipiens. Anat. Rec.,* **119,** 393 (1954).

12. Ortman, R., A study of the effect of several experimental conditions on the intermedin content and cytochemical reactions of the intermediate lobe of the frog, *Rana pipiens. Acta Endocr.,* **23,** 437 (1956).

13. Burgers, A. C. J. and Imai, K., The melanophore-stimulating potency of single pituitary glands of d-lysergic acid diethylamide (D-LSD) treated *Xenopus laevis. Gen. Comp. Endocr.,* **2,** 603 (1962).

14. Takahashi, H. and Imai, K., Color change and pituitary gland in blind *Xenopus. J. Exp. Morphol.* (Tokyo), **19,** 89 (1965) (in Japanese).

15. Imai, K., Light and electron microscopic studies on the pars intermedia of the pituitary of *Xenopus laevis* under different experimental conditions. *Gumma Symposia on Endocrinology,* **6,** 89 (1969).

16. Ito, T., Experimental studies on the hypothalamic control of the pars intermedia activity of the frog, *Rana nigromaculata. Neuroendocrinology,* **3,** 25 (1968).

17. Iturriza, F. C., Electron microscopic study of the pars intermedia of the pituitary of the toad *Bufo arenarum. Gen. Comp. Endocr.,* **4,** 492 (1964).

18. Iturriza, F. C. and Koch, O. R., Histochemical localization of some alpha MSH amino acid constituents in the pars intermedia of the toad pituitary. *J. Histochem. Cytochem.,* **12,** 45 (1964).

19. Saland, L. C., Ultrastructure of the frog pars intermedia in relation to hypothalamic control of hormone release. *Neuroendocrinology*, **3,** 72 (1968).

20. Imai, K. and Oota, Y., Ultrastructure of the pars intermedia of the pituitary of *Xenopus laevis* under experimental conditions. *J. Exp. Morphol.* (Tokyo), **19,** 88 (1965) (in Japanese).

21. Kurosumi, K., Functional classification of cell types of the anterior pituitary gland accomplished by electron microscopy. *Arch. Histol. Jap.*, **29,** 329 (1968).

22. Kobayashi, Y., Functional morphology of the pars intermedia of the rat hypophysis as revealed with the electron microscope. II. Correlation of the pars intermedia with the hypophseo-adrenal axis. *Z. Zellforsch.*, **68,** 155 (1965).

23. Jørgensen, V. B. and Lasen, L. O., Nature of the nervous control of the pars intermedia function in amphibians: Rate of functional recovery after denervation. *Gen. Comp. Endocr.*, **3,** 468 (1963).

24. Enemar, A., Falck, B., and Iturriza, F. C., Adrenergic nerves in the pars intermedia of the pituitary in the toad, *Bufo arenarum*, *Z. Zellforsch.*, **77,** 325 (1967).

25. Iturriza, F. C., Monoamines and control of pars intermedia of the toad pituitary. *Gen. Comp. Endocr.*, **6,** 19 (1966).

26. Bertalanffy, L. von and Bickis, I. Identification of cytoplasmic basophilia (ribonucleic acid by fluorescence microscopy), **4,** 481 (1956).

27. Pearse, A. G. E., "Histochemistry," theoretical and applied (2nd ed.), Little, Brown, Boston (1961).

28. Hogben, L. and Slome, D., The pigmentary effector system VI. The dual character of endocrine co-ordination in amphibian color change, *Proc. Roy. Soc.* Ser. B, **755,** 10 (1931).

DISCUSSION

DR. FITZPATRICK: I would like to suggest that you attempt to localize the site of MSH using radio-immunoassay for MSH combined with electron microscopy. Also it would be interesting to examine with electron microscopy the pituitary glands of (1) patients with MSH-secreting tumors, such as Nelson's syndrome, where functioning chromophobe adenomas are observed, and (2) patients with Addison's disease.

DR. LERNER: Does melatonin have any influence on the light adaptation or the concentration of MSH in the pituitary? Because recent works have reported that melatonin apparently has a direct effect on the MSH content of the pituitary gland.

DR. IMAI: I don't know, but I think that melatonin may have some influence on the amount of MSH.

DR. BAGNARA: I think that it is very important to establish the role of melatonin in determining the amount of MSH in the pituitary gland. It very well may be that melatonin has an effect on pituitary MSH in mammals, but that it is not a primary method of MSH control in the frog.

DR. KIKUYAMA: Have you observed that large colloidal granules and small secretory granules are present within the same cell?

DR. IMAI: I am not sure, since I have not observed that carefully. But I think that the colloidal granules and secretory granules may be present in the same cell.

DR. KIKUYAMA: We would like to comment on the granules in the pars intermedia of *Rana*. We separated two kinds of granules and confirmed that only the small-granule fraction has a strong MSH activity.

DR. IMAI: I think Dr. Kikuyama's work is excellent. He extracted the colloidal bodies from *Rana* and injected them into frogs; then the body weight of the frog increased. However, I cannot explain the physiological significance of this phenomenon.

The Physiology of Fish Melanophores*

Ryozo Fujii**

Division of Biology, National Institute of Radiological Sciences, Chiba, Japan

I. INTRODUCTION

Recent studies on the physiology of fish melanophores will be briefly reviewed. Special emphasis is laid on the motile mechanism of the cells, since their regulatory systems have recently been discussed by us in a few review articles.[1-3] Current studies of the pigment melanin and of problems concerning morphological color changes have been dealt with in one of those articles,[1] where chromatophores other than melanophores were also touched upon.

II. RECORDING OF MELANOPHORE RESPONSES

It is now well established that intracellular melanosome displacement is the substance of melanophore activities: when the melanosomes are aggregated at the center of the cell, the skin appears pale, and when they are dispersed throughout the cytoplasm, the skin becomes dark. Several attempts to describe quantitatively the state of melanosome distribution have been made.[1] Among them, the "melanophore index" originally devised by Hogben and Slome[4] to describe stages of amphibian melanophores has been widely used and has proved convenient. In this system,

* Aided by a grant from the Ministry of Education.
** Present Address: Department of Biology, Sapporo Medical College, Sapporo, Japan.

numerical figures represent the stages of melanophores, *1* corresponding to maximal pigment aggregation, and *5* to maximal dispersion. The intermediate figures with their decimal fractions, if they have any, indicate only arbitrarily the melanophore stages. Therefore, mathematical treatments such as averaging figures have only little meaning, sometimes leading to an erroneous conclusion. Adopting proper physical parameters, attempts to describe without subjectivity the state of a melanophore have also been made for many years. Several times, changes in the diameter of a melanophore or in the length of its cellular projection have been measured. In some cases, the velocity of migrating melanosomes has been measured directly under a high-magnification microscope.

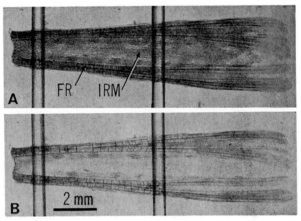

Fig. 1. Split tail fin preparation from the goby, *Chasmichthys gulosus*. The piece is held spread out under the cover slip by means of glass needles. (A) A piece equilibrated in physiological saline solution. All melanophores are in a totally dispersed state. (B) The same piece, but all melanophores are now in fully aggregated state after a 5-minute treatment with 5×10^{-6} M epinephrine. FR: fin ray; IRM: interradial membrane or web.

The photoelectric method was first introduced by Hill et al.,[5] who measured the changes in reflected light from the dorsal skin of an intact killifish *Fundulus*. Smith,[6] on the other hand, employed a photocell to measure the light transmittance through an isolated scale of *Tautoga*. Fujii[7] applied this technique to a split fin preparation of the goby, *Chasmichthys gulosus*. This preparation is made by splitting into two symmetrical halves an isolated long strip of tail fin consisting of two fin rays and an intervening web (Fig. 1). In this preparation, chemicals or drugs can reach the pigment cell layer more easily than in the scale, since there is no need for them to penetrate the epidermis. Thus, all the melanophores respond to chemical stimuli very rapidly and almost simultaneously. A

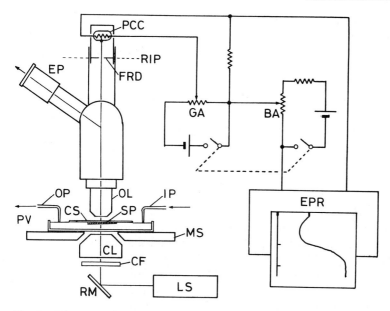

Fig. 2. Diagrammatic representation of a typical photoelectric method for recording melanophore responses *in vitro*. BA: bias potential adjustment resistor; CF: color filter; CL: condenser lens; CS: cover slip; EP: eyepiece; EPR: electronic paper recorder; FRD: diaphragm for field restriction; GA: gain adjustment resistor; IP: inlet pipette for experimental solution; LS: light source; MS: microscope stage; OL: objective lens; OP: outlet pipette for experimental solution; PCC: photoconductive cell; PV: perfusion vessel; RIP: plane of real image; RM: reflecting mirror; SP: split tail fin preparation.

certain area where several melanophores are found is usually focused for measuring its light transmittance, although sometimes the recording of the activities of a single melanophore is necessary, e.g., to determine the pattern of innervation to each melanophore.[8] The activities of other kinds of chromatophores, i.e., xanthophores or erythrophores, can be minimized by using a proper color filter. For instance, Fujii and Taguchi[9] have employed a red filter which eliminates light shorter than 630 mμ in wavelength. A fundamental experimental set-up is shown in Fig. 2, which has been employed by the author. Modifications can be made for electrophysiological application and so on. Sometimes an inverted microscope is conveniently adopted.[8]

III. THE MECHANISM OF MELANOSOME MOVEMENT

It is now generally believed that melanophore activities are caused, not by the amoeboid movement of the cell, but by the intracellular aggregation or dispersion of melanosomes within a fixed shape. Actually, recent electron microscopic studies of the aggregated melanophores have shown the presence of radiating cellular processes without pigment granules.[10-13] Through both light and electron microscopes, on the other hand, it has become apparent that the thickness of the central body of a melanophore becomes greater when the melanosomes are aggregated there (Figs. 3, 4). Submicroscopically, it is also obvious that the thickness of the cellular processes is smaller when the melanosomes are absent (Fig. 4). Thus, the idea that the outline of the cell is "fixed" cannot be applied in the strict sense of the word. Whether or not overall cellular volume changes during the aggregation-dispersion cycle is unknown. But there may not be so much retrograde cytoplasmic flow during melanosome movements to keep the cellular shape constant.

A number of hypotheses have been proposed concerning the possible mechanism of melanosome movements. Early literature can be referred to in the contributions of Franz[14] and Lerner and Takahashi.[15] Among recently presented ideas, the interpretation by Falk and Rhodin[16] of the ultrastructural configuration of *Lebistes* melanophores was plausible and became widely known. They suggested that a melanophore is surrounded by two layers of membranous structures, an outer thick membrane that is fixed, and an inner thin sac containing the usual cellular organelles which

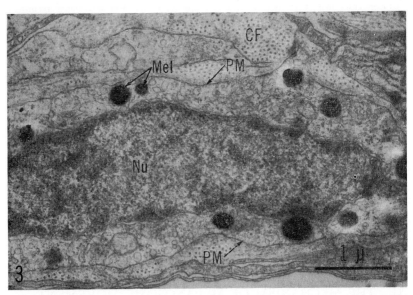

Fig. 3.   Electron micrograph of the central part of a fully dispersed *Chasmichthys* melanophore, fixed with OsO$_4$ after equilibration in physiological saline for a few minutes. Melanosomes are sparsely scattered in the cytoplasm. Thickness of the cell is about 3 μ. CF: collagenous fibrils; Mel: melanosomes; Nu: nucleus; PM: plasma membrane. Stained with uranyl acetate.

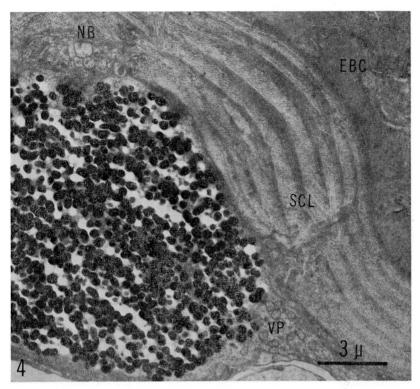

Fig. 4. Electron micrograph of a *Chasmichthys* melanophore in fully aggregated state. A 5-minute treatment with 5×10^{-6} *M* epinephrine was followed by osmium fixation. All melanosomes are compactly gathered in the central body, the thickness of which is somewhat more than 10 μ. EBC: basal cell of epidermis; NB: nerve bundle; SCL: subepidermal collagenous lamella; VP: process of the melanophore now devoid of melanosomes. Stained with uranyl.

is elastic. Between these membranes was a mesh of fibrils, the contraction or the relaxation of which might cause the aggregation or dispersion of melanosomes in the inner bag. However, later electron microscopic studies, including observations on the same species of fish[12] have agreed with the view that melanophores are enclosed by only one membrane, as are usual cells.[10,11,13] Thus, the unique explanation of Falk and Rhodin had to be abandoned.

Working with the melanophores in the scale of *Oryzias latipes*, Kinosita[17,18] recently proposed an entirely different mechanism. According to him, melanosomes are negatively charged and may be moved by the intracellular current flow caused by a local difference in membrane potentials between central and peripheral parts of the cell. Our recent electron microscope observations also seemed to be consistent with this electrophoresis

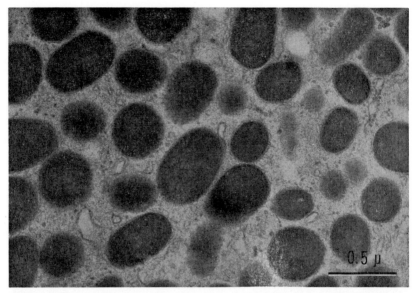

Fig. 5. Electron micrograph of the cytoplasm of a half-aggregated *Chasmichthys* melanophore at high magnification. Glutaraldehyde fixation. Section cut with a diamond
knife and stained with uranyl acetate. Melanosomes are enclosed by a limiting membrane.
The dark matrix of the melanosomes appears to be not entirely homogeneous, but some
swirling pattern is recognizable around the center of the organelle. Free ribosomes and
elements of vesicular smooth-surfaced endoplasmic reticulum are usual inclusions. Rough
endoplasmic reticulum seen on righthand side of the figure is a less common element.

hypothesis:[11, 12] melanophores are surrounded mainly by extracellular
collagenous space, where the specific resistance may be comparatively low,
and there are numerous vesicular elements of endoplasmic reticulum,
which are expected to increase the resistance of the cytoplasm. Each melanosome is also enclosed by a limiting membrane which should be very
lowly conductive (Fig. 5). These structural features seemed to be advanta-
geous for establishing effectively the possible intracellular potential
gradient along the processes of the melanophores. Ultrastructural observations also suggest that the movements of cytoplasmic structures are con-
fined to melanosomes only.[11, 12]

The electrophoresis theory must first meet the argument that equally
charged particles repel each other, thus causing difficulty in forming an
aggregated mass of melanosomes. Our recent results on membrane potentials of *Fundulus* melanophores[2] seem to be at variance also with this unique
interpretation by Kinosita: there was no significant difference between
the mean membrane potential measured in the physiological saline solution and that in the saline containing one of the known melanin-aggregating or -dispersing substances. Using the goby *Chasmichthys*, Fujii and Ta-

guchi[9] recently demonstrated that, even in a potassium-rich medium, the melanophores respond to directly acting melanin-aggregating or -dispersing drugs quite normally when transmitter release from the nerve-endings was blocked by withdrawing alkaline-earth ions from the medium or when denervated preparations were employed. In K-rich solutions, the membrane potentials of *Fundulus* melanophores are very low, being less than 10 mV inside negative, regardless of the presence or absence of melanin-aggregating or -dispersing agencies.[2] These investigations tend to minimize the importance of electrical events in the membrane, raising objections to the electrophoresis hypothesis. Thus, we have to give consideration to another kind of motive force for pigment displacement.

Lerner and Takahashi[15] suggested that pigment movement is associated with ionic exchanges across the cell membrane. Actually, Novales and Gratzer[19] reported that sodium entry into the goldfish melanophore may be involved in the melanin-dispersing action of MSH (melanocyte-stimulating hormone), thus extending Novales' earlier finding on amphibian melanophores.[20] In consideration of the unexceptional role of calcium ions in the activities of motile cells in general, an experiment carried out by Ishibashi[21] deserves to be mentioned. He injected a small amount of solution of sodium oxalate, a calcium precipitant, into an *Oryzias* melanophore and observed melanosome aggregation followed by alternating dispersion and aggregation. Immersing a split tail fin preparation into a divalent cation-free saline solution to which EDTA (ethylenediamine tetra-acetic acid, neutralized) was added up to more than 2 mM in final concentration, the author has also observed a gradual aggregation of melanosomes in *Chasmichthys* melanophores (Fig. 6). These results suggest that the intracellular calcium level may be an important factor in melanosome movements, and that its decrease may be coupled with melanosome aggrega-

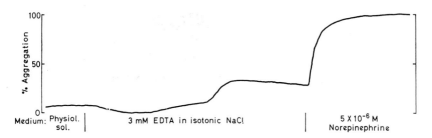

Fig. 6.   Photoelectric recording of the response of *Chasmichthys* melanophores to EDTA-containing, divalent cation-free solution. Following the equilibration in physiological saline, 3 mM EDTA in buffered isotonic NaCl solution was added. A transient small dispersion was gradually reversed by a moderate aggregation of melanosomes. The maximal level of aggregation was finally attained by the addition of norepinephrine to the same medium.

37

tion, while the opposite change in the cell may induce their dispersion. These observations are also in accordance with those on an amphibian form, *Ambystoma*, by Novales and Novales,[22] who reported that calcium withdrawal from the external medium caused the actual contraction of tissue-cultured melanophores.

By inserting a microneedle into a *Fundulus* scale melanophore, S.A. Matthews[23] observed that the cytoplasm of the aggregated cell was in a gelle dstate, while that of the dispersed cell was more serous. Similar observations were made by Marsland,[24] and Marsland and Meisner,[25] who assumed a colloidal framework in which melanosomes are embedded. The gelling or contraction of the framework may produce aggregation of the melanophores, while its solation or relaxation induces the dispersion of melanosomes. In their review, Lerner and Takahashi[15] also expressed that the sol-gel transformation may be the main factor responsible for cellular motility. However, there remains a possibility that these changes in the cytoplasmic elasticity originate merely in the physical distribution of melanosomes which was brought about by some other mechanism.

Even through electron microscopes, neither the cytoplasmic peristaltic canaliculi as described by Ballowitz[26] nor the cytoskeletal system described by Franz[14] has been found. In melanophores described by Ballowitz and Franz and in *Fundulus* scale melanophores, which quite a few researchers have employed, melanosomes move in files or rows. Apparently some sub-cellular structures define the pathway of melanosomes through melano-phore projections, and the recent discovery of microtubules in *Fundulus* melanophores by electron microscopy[2, 10, 13] may have partly explained this characteristic pattern of melanosome motion. These elements were found in melanophore processes regardless of the presence or absence of melanosomes. Indeed, Bikle et al.[10] thought that the microtubules actually play an active role in pigment displacement. The early light microscope observation by Franz[14] of a radiating endoskeletal system in melanophores of a larval bony fish *Atherina* might also be relevant to the presence of microtubules or a colloidal framework in the central body of a melano-phore. However, all the electron microscope studies to date have agreed with the view that these microtubules are not attached to melanosomes. Further, the microtubules are generally considered to be noncontractile, and it should be emphasized here that these melanophore microtubules have not been found in cells other than those of *Fundulus*. Observing the melanophores of a few species of freshwater fish, Kamada and Kinosita[27] noted that melanosomes could alter their location from one process to another during the course of their alternate aggregation and dispersion. Perhaps microtubules were not present in those melanophores. It follows as a logical consequence, therefore, that the tubules do not play an active part in driving melanosomes, and that in some melanophores they are present only as a cytoskeletal element which helps maintain the extended contour of the cell, and at the same time define the paths of melanosomes.

In a branch of a melanophore cut either at its root or at the distal part or at both ends with a microneedle, melanosomes still moved centrally or peripherally in response to a certain stimulus which normally induces the aggregation or dispersion of melanin in intact cells.[27] Thus, we may be able to conclude that the receptors for melanin-aggregating or -dispersing stimuli are distributed along the cellular processes. Even if we assume the positive role of microtubules, their connection with the central part of the cell, possibly with centrioles, or with the tip of a cellular process is not required for pigment displacement. Therefore, interaction between the surface of melanosomes and that of microtubules is considered to be an important factor for pigment movements. Based on these observations, we have recently devised a schematic representation of the possible mode of microtubule participation (Fig. 7). There, a kind of sliding mechanism is assumed as a motive force for driving melanosomes. The direction of particle movements would be determined either by direct action of transmitter or hormonal substances incorporated by micropinocytosis, or indirectly by a second messenger which is produced by receptors for the controlling principles.

Wikswo and Novales,[28] on the other hand, found that the pretreatment of *Fundulus* scale melanophores with colchicine produces a decrease in the rate of melanosome aggregation but not the complete blocking of move-

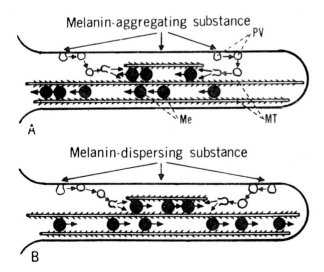

Fig. 7.   Schematic representation of the possible involvement of microtubules in centripetal (A) or centrifugal (B) melanosome migration through a melanophore projection. Me: melanosomes; MT: microtubules; PV: pinocytotic vesicles.[2]

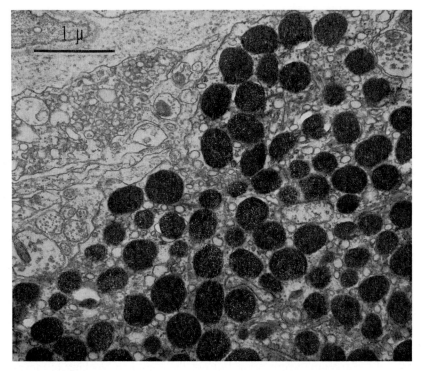

Fig. 8. Electron micrograph of part of an aggregated melanophore of the guppy, *Lebistes reticulatus*. Note filamentous structures which are present in the cytoplasm, mostly in bundles. Osmium fixation; stained with lead citrate.[2]

ments, and that pigment dispersion is somewhat accelerated after colchicine treatment. Their observations are difficult to interpret in the light of the widespread idea that colchicine causes a disassembly of microtubules. Perhaps this alkaloid affects the viscosity of the cytoplasm, bringing about its solation, which may favor a widely scattered distribution of melanosomes within the cytoplasm.

Recently, Rebhun[29] observed saltatory, jumplike movements of melanosomes in embryonic *Fundulus* melanophores, and suggested the presence of contractile filaments closely associated with the microtubules. He thought that these force-generating elements are responsible for cellular activities. On some electron micrographs of *Lebistes* melanophores, Fujii and Novales[2] actually detected fine filaments inside the cells (Fig. 8). They are not associated with microtubules, and sometimes they seem to stick to melanosomes. Novales and Novales (unpublished) also found similar filaments around the melanosomes in a dogfish melanophore. Although these filaments might be responsible for melanosome movements, a comparative survey of this kind of possible force-generating elements should be carried out using elaborate electron microscopic techniques. In an attempt along this line to find crucial elements responsible for pigment motility, possibly fibrillar contractile ones, we have tried to observe the localization of adenosine triphosphatase (ATPase) activity inside *Chasmichthys* melanophores at the level of fine structure.[30] Unfortunately, it was not possible until now to find the reaction precipitates inside the cells, although outside the cell membrane many lead particles, probably formed by membrane ATPase activity, were found. As suggested by the criticisms of ATPase studies on various muscular tissues, it is probable that some of the components in the incubation medium may not have entered the cells, thus causing no reaction inside the cells. The technique should therefore be improved in future trials.

Another approach to this problem is through activation of the glycerinated melanophore model by ATP. However, our preliminary trial of this in the presence of magnesium ions was a failure. The material used was the goby, *Chasmichthys*. With regard to the difficulties inevitably involved in this technique, however, there still remains the possibility that the model can be activated with the addition of the nucleotide under different conditions.

Very recently, the involvement of another kind of adenine nucleotide has been suggested in the melanophores of amphibians.[31] Based mainly on their observation that cyclic 3′, 5′-adenosine monophosphate (cyclic AMP) has a melanin-dispersing effect in frog and tissue-cultured salamander melanophores, Novales and Davis[31] concluded that this kind of nucleotide functions as the second messenger of the MSH action. Quite recently, we did detect the same effect of cyclic AMP in *Fundulus* melanophores.[32] Since MSH may not be involved in the control of *Fundulus* melanophores,[1] we interpret these results to mean that nervous transmitters may act by controlling the intracellular levels of cyclic AMP. Changes in the nucleotide and Ca^{++} levels may be inseparably coupled to control pigment movements in the cell. However, these problems are open for future research.

IV. CONTROL OF MELANOPHORES

Cyclostome and elasmobranch melanophores are thought to be primarily under the control of hypophyseal hormones. On the other hand, melanophores of teleost fish are regulated hormonally or nervously, or more commonly in both ways. Usually, however, the nervous mechanisms seem to dominate, and *Fundulus heteroclitus* and *Oryzias latipes* are examples where the melanophores are controlled exclusively by nerves. Nervous control may have evolved for the rapid background adaptation of animals, while chromatic changes due to endocrine systems are usually slower. For a gradual and sustained adaptation the latter kind of coordination may be efficient. Excellent pieces have appeared in which pituitary involvement in chromatophore control was extensively reviewed.[33-35] Further, current articles including those by Fingerman,[36] Fujii[1] and Fujii and Novales[2] have also dealt with relevant subjects.

Dual systems have been proposed to explain the neural mechanisms controlling melanophores, monoaminergic (sympathetic) nerve fibers which induce melanin aggregation in cells, and the antagonistic melanin-dispersing fibers which are assumed to be parasympathetic. In addition to the classic monographs by Parker[33] and Waring[35] we have current review articles where precise descriptions and some new interpretations of chromatophore activities due to neural mechanisms have been left on record.[1,2] Further, an article by Fujii and Novales[3] deals exclusively with the neural controlling mechanisms of melanophores. Thus, recent advances in these problems can be referred to in these documents.

V. SUMMARY

Current studies on the physiology of melanophores of fish are reviewed. First, methods used up to now for recording melanophore responses are summarized. A photoelectric technique applied on a split fin preparation is put forward as a suitable method for quantitative analysis of cellular responses. Theories about the mechanisms of melanosome movements in melanophores are then enumerated. Microtubules may not be involved, since these elongated elements are not generally found in the melanophores. Melanin movements are found to be not directly correlated with changes in the electric polarization of the melanophore membrane. Thus, the electrophoresis hypothesis may not explain the motility of pigment either. The alternative interpretations include the sol-gel transformation of the cytoplasm and the theory in which the importance of intracellular microfilaments is stressed. Correlation of the Ca^{++} and cyclic AMP levels with the state of melanophores is indicated. This may partly explain steps in a chain of biochemical sequences between the reception of informational molecules and the final manifestation of the effect, pigment displacement. However, none of the current ideas has yet been able to interpret properly the mechanism of pigment movements. Finally, the regulatory mechanisms of pigment cells are mentioned briefly, precise descriptions being left to other reviews.

VI. ACKNOWLEDGMENTS

The author is grateful to Professors N. Egami and H. Kinosita of the University of Tokyo for their frequent encouragement and invaluable advice. Many thanks are also due to Dr. R. R. Novales of Northwestern University with whom the author performed some of the experimental work described in this article.

REFERENCES

1. Fujii, R., Chromatophores and pigments. In *Fish Physiology,* vol. 3 (W. S. Hoar and D. J. Randall, eds.), p. 307, Academic Press, New York (1969).
2. Fujii, R. and Novales, R. R., Cellular aspects of the control of physiological color changes in fishes. *Amer. Zool.,* **9,** 453 (1969).
3. Fujii, R. and Novales, R. R., Nervous control of melanosome movements in vertebrate melanophores. In *Pigmentation: Its Genesis and Biologic Control* (V. Riley, ed.), Appleton-Century-Crofts, New York (in press) (1970).
4. Hogben, L. T. and Slome, D., The pigment effector system. VII. The dual character of endocrine coordination in amphibian colour change. *Proc. Roy. Soc.* [Biol] Ser. B, **108,** 10 (1931).
5. Hill, A. V., Parkinson, J. L., and Solandt, D. Y., Photoelectric records of the colour change in *Fundulus heteroclitus. J. Exp. Biol.,* **12,** 397 (1935).
6. Smith, D. C., A method for recording chromatophore pulsations in isolated fish scales by means of a photo-electric cell. *J. Cell. Physiol.,* **8,** 83 (1936).
7. Fujii, R., Mechanism of ionic action in the melanophore system of fish. I. Melanophore-concentrating action of potassium and some other ions. *Ann. Zool. Jap.,* **32,** 47 (1959).
8. Fujii, R. and Novales, R. R., The nervous mechanism controlling pigment aggregation in *Fundulus* melanophores. *Comp. Biochem. Physiol.,* **29,** 109 (1969).
9. Fujii, R. and Taguchi, S., The responses of fish melanophores to some melanin-aggregating and -dispersing agents in potassium-rich medium. *Ann. Zool. Jap.,* **42,** 176 (1969).
10. Bikle, D., Tilney, L. G., and Porter, K. R., Microtubules and pigment migration in the melanophores of *Fundulus heteroclitus* L. *Protoplasma,* **61,** 322 (1966).
11. Fujii, R., Correlation between fine structure and activity in fish melanophore. In *Structure and Control of the Melanocyte* (G. Della Porta and O. Mühlbock, eds.), p. 114, Springer-Verlag, Berlin (1966).
12. Fujii, R., A functional interpretation of the fine structure in the melanophore of the guppy, *Lebistes reticulatus. Ann. Zool. Jap.,* **39,** 185 (1966).
13. Green, L., Mechanism of movements of granules in melanocytes of *Fundulus heteroclitus. Proc. Nat. Acad. Sci. U.S.A.,* **59,** 1179 (1968).
14. Franz, V., Struktur und Mechanismus der Melanophoren. Teil II. Das Endoskelett. *Z. Zellforsch.,* **30,** 194 (1940).
15. Lerner, A. B. and Takahashi, Y., Hormonal control of melanin pigmentation. *Recent Progr. Hormone Res.,* **12,** 303 (1956).
16. Falk, S. and Rhodin, J., Mechanism of pigment migration within teleost melanophores. In *Electron Microscopy: Proceedings Stockholm Conference* (F. S. Sjostrand and J. Rhodin, eds.), p. 213, Academic Press, New York (1957).
17. Kinosita, H., Studies on the mechanism of pigment migration within fish melanophores with special reference to their electric potentials. *Ann. Zool. Jap.,* **26,** 115 (1953).
18. Kinosita, H., Electrophoretic theory of pigment migration within fish melanophore. *Ann. N. Y. Acad. Sci.,* **100,** 992 (1963).

19. Novales, R. R. and Gratzer, B. M., Effect of sodium concentration on the response of goldfish scale melanophores to MSH. *Anat. Rec.,* **137,** 385 (1960).
20. Novales, R. R., The effects of osmotic pressure and sodium concentration on the response of melanophores to intermedin. *Physiol. Zool.,* **32,** 15 (1959).
21. Ishibashi, T., Effect of the intracellular injection of inorganic salts on fish scale melanophore. *J. Fac. Sci. Hokkaido Univ.* ser. 6, **13,** 449 (1957).
22. Novales, R. R. and Novales, B. J., The effects of osmotic pressure and calcium deficiency on the response of tissue-cultured melanophores to melanocyte-stimulating hormone. *Gen. Comp. Endocr.,* **5,** 568 (1965).
23. Matthews, S. A., Observation on pigment migration within fish melanophore. *J. Exp. Zool.,* **58,** 471 (1931).
24. Marsland, D. A., Mechanism of pigment displacement in unicellular chromatophores. *Biol. Bull.,* **87,** 252 (1944).
25. Marsland, D. A. and Meisner, D., Effects of D_2O on the mechanism of pigment dispersal in the melanocytes of *Fundulus heteroclitus. J. Cell. Physiol.,* **70,** 209 (1967).
26. Ballowitz, E., Ueber die Pigmentströmung in den Farbstoffzellen und die Kanälchenstruktur des Chromatophoren-Protoplasmas. *Pflüg. Arch. Ges. Physiol.,* **157,** 165 (1914).
27. Kamada, T. and Kinosita, H., Movement of granules in fish melanophores. *Proc. Jap. Acad.,* **20,** 484 (1944).
28. Wikswo, M. A. and Novales, R. R., The effect of colchicine on migration of pigment granules in the melanophores of *Fundulus heteroclitus. Biol. Bull.,* **137,** 228 (1969).
29. Rebhun, L. I., Structural aspects of saltatory particle movement. *J. Gen. Physiol.,* **50,** (suppl.), 223 (1967).
30. Fujii, R., Correlation of fine structure to activities in guppy and goby melanophores. *Zool. Mag.* (Tokyo), **75,** 336 (1966).
31. Novales, R. R. and Davis, W. J., Melanin-dispersing effect of adenosine 3', 5'-monophosphate on amphibian melanophores. *Endocrinology,* **81,** 283 (1967).
32. Novales, R. R. and Fujii, R., A melanin-dispersing effect of cyclic adenosine monophosphate on *Fundulus* melanophores. *J. Cell. Physiol.,* **75,** 133 (1970).
33. Parker, G. H., *Animal Colour Changes and their Neurohumours.* Cambridge University Press, Cambridge (1948).
34. Pickford, G. E. and Atz, J. W., The Physiology of the Pituitary Gland of Fishes. *N. Y. Zool Soc.,* New York (1957).
35. Waring, H., *Color Change Mechanisms of Cold-Blooded Vertebrates.* Academic Press, New York (1963).
36. Fingerman, M., Chromatophores. *Physiol. Rev.,* **45,** 296 (1965).

DISCUSSION

DR. CHAVIN: First, do you consider the melanophores in the split tail fin preparations to be ultrastructurally normal after "incubation" in saline during the test procedures? Second, one of my students, Mr. J. P. Stone, has removed the epidermis from goldfish scales so that the dermis is exposed to the environment. He has found at least a tenfold increase in melanophore sensitivity to epinephrine. The epidermal barrier must be considered to be of importance in play serological studies, and thus remained of the epidermis of split fin preparations are essential.

DR. FUJII: Melanophores in a split tail fin preparation respond to various stimuli quite normally even after the prolonged immersion in various experimental solutions. Electron microscopic observations of those melanophores also revealed a normal configuration of cells. The physiological experiments, including the electrical stimulation of the proximal part of the preparation, indicated that the function of the controlling nervous system is also quite normal. For these reasons, I would like to recommend the use of this kind of preparation for physiological experiments on fish melanophores. In the chemical stimulation of scale melanophores, responses proceed at first among the cells located around the peripheral part of the scale, while the cells near the center respond much later. For quantitative experiments, therefore, scales are not suitable.

In reply to your second comment, by treating scales of *Oryzias* of *Carassius* with calcium precipitants or EDTA, Dr. Nagahama (1953) and I (unpublished observations) tried to remove epidermal cells in order to eliminate the barrier function of the epidermal layer overlying the pigment cells. An increased sensitivity to potassium ions or to catecholamines has been demonstrated. This kind of preparation is therefore useful for physiological or pharmacological analysis of melanophore systems.

DR. MISHIMA:
1. Do you think MSH enter into the cytoplasm of melanophores or melanocytes primarily via pinocytotic vesicles?
2. May I ask your thoughts on the effect of MSH on the melanocytes rather than melanophores in reference to melanosome migration?
3. Did you observe synthetic processes of premelanosomes in melanophores which appeared to contain tyrosinase?

DR. FUJII:
1. At this moment, it is too early to assert so. In the electron micrographs of melanophores, however, many pinocytotic vesicles are found, especially just under the membrane where nerve fibers are passing by. Thus, at least the transmitter molecules might be incorporated into the cell by such a mechanism.
2. We have practically no idea about this problem. But, we suspect there is some contractile system even in mammalian melanocytes.
3. In fish melanophores, we have not been able to find premelanosomes. The only case where premelanosomes were described in fish melanophores was the *Fundulus* melanophores described by Bikle, Tilney, and Porter (1966). Our unpublished observations with Dr. Kikuyama on melanophores of hypophysectomized toad tadpole (*Bufo vulgaris*) which was treated with thyroxine indicated the presence of early stages of melanosomes. At least in some cases where morphological color changes are going

on (i.e., on dark backgrounds, or after injection of pituitary preparations), a net melanin synthesis occurs in each melanophore. By observing such a cell through an electron microscope, we may be able to find premelanosomes.

Possible Control Mechanisms of Release and Synthesis of Melanophore-Stimulating Hormone

Tsuneo Moriya, Sakae Kikuyama, and Ikuo Yasumasu

Department of Biology, Waseda University School of Education, Tokyo, Japan

I. INTRODUCTION

In amphibians, melanophore-stimulating hormone (MSH) secreted from the intermediate lobe of the pituitary is believed to be the principal factor for skin darkening. The release of MSH from the pars intermedia is regulated by the hypothalamus through its inhibitory control.[1] The presence of the hypothalamic factor which inhibits the release of MSH from the pars intermedia was demonstrated in frog[2,3] and mammals.[2]

In the present paper, we will report that there may be two factors in frog hypothalamus: MSH-release-inhibiting factor and MSH-synthesis-inhibiting factor: evidence will also be presented that the synthesis and release of MSH are closely related to the synthesis of RNA in the intermediate lobe which seems to be regulated by hypothalamic factors.

II. METHOD OF STUDY

A. *Extraction of hypothalamic tissue*

Three hundred hypothalami were obtained from the brain of bullfrogs, *Rana catesbeiana*. The tissue was homogenized in 4 ml of 0.25 M sucrose. The homogenate was mixed with the same volume of 2 N acetic acid. After 1 hour at room temperature, the mixture was centrifuged at 30,000 $\times g$ for 30 minutes. The supernatant freed of precipitate was neutralized with 1 N sodium hydroxyde and diluted with distilled water so that 0.1 ml

47

contained the extractable substances of 1.5 pieces of hypothalamus. A comparable amount of brain tissue (optic lobe) was obtained from the same materials and the extraction procedure was as described above.

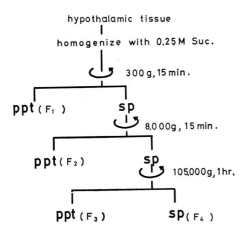

Fig. 1.   Scheme for fractionation of hypothalamic tissue.

B.   *Fractionation of hypothalamic tissue*
The scheme for fractionation of hypothalamic tissue is shown in Fig. 1. The precipitates (F_1, F_2, and F_3) were rehomogenized with 0.25 M sucrose. Extraction was carried out in each of the homogenates as described above. The final volume of each of the extracts was adjusted with distilled water so that 0.1 ml contained the extractable substances of 2 hypothalami.

C.   *Culture of intermediate lobe*
Frogs, *Rana nigromaculata*, were used. Unless otherwise specified, they were kept for a week in darkness before decapitation. The intermediate lobe, including the neural lobe, was separated from the anterior lobe and placed in a dish (2 cm in diameter) with a fitted cover containing 1 ml of modified Gey's (G) solution.[4] To the culture media of every 4–5 dishes, 0.1 ml of any one of the following substances was added: hypothalamic extracts, brain extracts, extracts of fraction 1, fraction 2, or fraction 3, fraction 4, 15 μg of actinomycin D and 500 μg of chloramphenicol. As control, the intermediate lobes were cultured in dishes containing 1.1 ml of G solution. The intermediate lobes were cultured for 48 or 72 hours at 22°C under sterile conditions. Every 24 hours, the media were renewed. The media of the first day of culture were discarded. The media of the second and/or third day of culture were stored at a low temperature. At the termination of the culture, each of the partes intermedia was homogenized with 0.5

ml of 1 N acetic acid. After 1 hour at room temperature, the homogenate was centrifuged at 30,000 $\times g$ for 30 minutes. The supernatant freed of precipitate was neutralized with 1 N sodium hydroxide and stored at a low temperature.

D.   Assay of MSH activity

The skin of the ventral region of a premetamorphic bullfrog tadpole which had been adapted on a white background for a few days was dissected off into G solution and cut into pieces about 4 mm square. The samples were diluted 10 times in a stepwise fashion with G solution. As a standard solution, 0.01 IU/ml of ACTH (Schering) was prepared and diluted in a similar manner. One piece of the skin was immersed in 1 ml of each dilution, contained in a test tube. After 40–60 minutes, the piece was taken out, placed on a glass slide, and observed under a light microscope for reading melanophore index.[5] In each sample and the standard solution, the highest dilution which showed a melanophore index of 3 was obtained. To be used in the assay, the piece of skin had to exhibit a melanophore index of 5, 3, and 1 in the 1-, 10-, and 100-fold dilutions of the standard solution respectively. Total MSH activity of each sample was calculated and expressed as the equivalent dose of standard ACTH. In every culture medium and extract of the cultured intermediate lobe, MSH activity was assayed and the ratio of MSH released from the treated intermedia and MSH content in the treated intermedia to controls was calculated.

E.   Incorporation of ^3H-uridine into RNA and of ^{14}C-proline into protein and acid extractable peptides in the intermediate lobe

The partes intermedia cultured in G solution (control) and the media containing fraction 4, extracts of fraction 3, brain extracts, and actinomycin D for 24 and 48 hours were incubated in the serum-free medium containing ^3H-uridine (1 μCi/ml) and ^{14}C-proline (0.1 μCi/ml) for 2 hours at 25°C. The intermediate lobe was transferred into 5% TCA, homogenized with glass homogenizer, washed 3 times with 5% TCA, once with ethanol, and 2 times with ethanol-ether (3:1). After precipitates were hydrolized with 1 ml of 0.5 N KOH for 16 hours, 1 ml of 10% PCA was added. The supernatant fractions thus obtained were neutralized with 1 N KOH and radioactivities of these fractions were measured by means of a liquid scintillation counter (Aloka, LSC-106). Radioactivities of precipitates hydrolyzed again with 5% PCA at 90°C for 15 minutes were measured by the same liquid scintillation counter.

From the neutralized washings of TCA and the culture medium, peptides into which ^{14}C-proline was incorporated were separated from free ^{14}C-proline by gel filtration on a Sephadex G-25 column (0.5 cm in diameter, 15 cm in height). Each of the radioactivities was measured by liquid scintillation counter. The total of both radioactivities represents the activity of acid-extractable peptides synthesis in the intermedia, and the ratio of the radioactivity of the culture medium to the total radioactivity indicates percent of peptides released from the intermediate lobe.

III. RESULTS AND DISCUSSION

A. *Effects of hypothalamic extracts on the release and synthesis of MSH*
Table 1 shows the effect of hypothalamic extracts on the release of MSH. The results indicate that hypothalamic extracts inhibited the release of MSH from the intermediate lobes following the first 24 hours of culture. Moreover, it revealed that the content of MSH was lowered by the hypothalamic extracts, suggesting that the hypothalamic factor suppressed the synthesis of MSH. Kastin and Schally[2] reported that frog hypothalamic extracts, which lighten the previously darkened frogs, increased the MSH content of rat pituitary glands 20 minutes after injection, and they concluded that the increase in rat pituitary MSH levels by the frog hypo-

Table 1. Effects of hypothalamic extracts on MSH release and MSH content in cultured intermediate lobes.

Treatment	Ratio of MSH released* from the treated intermedia to controls (%)	Ratio of MSH content** in the treated inter-media to controls (%)
Brain extracts	100	95
Hypothalamic extracts	30	60
Hypothalamic extracts for the 1st 24 hours and none for the 2nd 24 hours	71	48

* MSH released during the 24-hour period from 24 to 48 hours of culture.
** MSH content at 48 hours of culture.

Table 2. Effects of extracts of fractionated hypothalamic tissue on MSH release and MSH content in cultured intermediate lobes.

Treatment	Ratio of MSH released* from the treated intermedia to controls (%)	Ratio of MSH content** in the treated inter-media to controls (%)
Fraction 1	78	66
Fraction 2	76	50
Fraction 3	85	10
Fraction 4	42	68

* MSH released during the 24-hour period from 48 to 72 hours of culture.
** MSH content at 72 hours of culture.

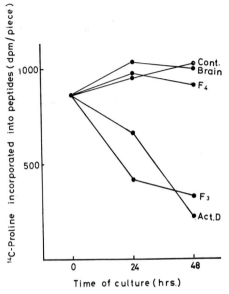

Fig. 2. Effects of extracts of fractionated hypothalamic tissue on acid-extractable peptides synthesis in cultured intermediate lobes.

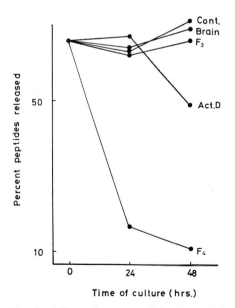

Fig. 3. Effects of extracts of fractionated hypothalamic tissue on acid-extractable peptides release from cultured intermediate lobes.

51

thalamic extracts was due to the inhibition of MSH release. In the present experiments, the frog hypothalamic extracts which inhibited the release of MSH decreased, at the same time, the content of MSH of the frog pituitary glands. It is likely that the observed decrease in frog pituitary MSH levels was due to the long-term effects of the hypothalamic extracts.

Table 2 shows the effects of extracts of fractionated hypothalamus on the release of MSH. Fraction 4 strongly inhibited the release of MSH from the intermedia, while it was fraction 3 that was most effective in lowering the MSH levels.

This was indirectly supported by the results of the experiments carried out to see the effects of fraction 4, extracts of fraction 3, brain extracts, and actinomycin D on the synthesis of acid-extractable peptides, in which MSH is expected to be included, in the frog intermediate lobes. Extracts of fraction 3 suppressed the synthesis of acid-extractable peptides remarkably. Actinomycin D also inhibited the incorporation of ^{14}C-proline into the acid-extractable peptides fraction (Fig. 2). As shown in Fig. 3, fraction 4 strongly inhibited the release of peptides into the culture media from the intermedia. The nature of the active substances in fractions 3 and 4 is under investigation.

B. *Effects of actinomycin D and chloramphenicol on the release and synthesis of MSH*

In the previous experiment, actinomycin D was effective in suppressing the synthesis of acid-extractable peptides in the intermediate lobe. The experiments were carried out to study the effect of actinomycin D and chloramphenicol on the release and synthesis of MSH in the cultured intermediate lobe supplied from the frogs kept in various conditions. As shown in Table 3, chloramphenicol invariably suppressed the release of MSH and decreased the content of MSH. Actinomycin D inhibited the release of MSH from the intermediate lobes supplied from frogs that had been adapted on a white background or kept in darkness, and decreased the MSH content of the intermediate lobe from the frogs kept in darkness. These results are interpreted as follows. For both synthesis and release of MSH, RNA synthesis may be required. RNA synthesis necessary for the release of MSH was taking place in the intermediate lobe of frogs adapted on a black background. As a result, when the intermediate lobe was transferred to the culture medium, the suppression of RNA synthesis by actinomycin D did not bring about the inhibition of the MSH release. RNA synthesis necessary for the MSH synthesis was also taking place in the intermediate lobes of the frogs adapted on an illuminated black or white background. Consequently, actinomycin D hardly suppressed the synthesis of MSH, even after the intermediate lobes were cultured. Suppression of MSH release by the administration of chloramphenicol may

Table 3. Effects of actinomycin D and chloramphenicol on MSH release and MSH content in intermediate lobes.

Type of adaptation of donor frogs of intermediate lobes	Ratio of MSH released from the treated intermedia to controls (%)				Ratio of MSH content in the treated intermedia to controls (%)	
	24–48 hours		48–72 hours		72 hours	
	Act.D	Chlor.	Act. D	Chlor.	Act. D	Chlor.
Black background	80	8	63	4	121	20
White background	26	32	6	6	150	4
Darkness	20	16	4	8	32	20

Table 4. RNA and protein synthesis in intermediate lobes from frogs adapted to various conditions (mean value of 3-4 specimens ±S.E.M.).

Type of adaptation of donor frogs of intermediate lobes	^{3}H-uridine into RNA (dpm/piece)		^{14}C-proline into protein (dpm/piece)	
	0 hours	24 hours	0 hours	24 hours
Black background	1,605±146	1,659±69	403±28	469±48
White background	1,579±163	1,672±62	543±99	462±46
Darkness	1,149±55	1,705±59	321±19	498±21

Table 5. Effects of extracts of fractionated hypothalamic tissue on synthesis of RNA and protein in cultured intermediate lobes (mean value of 3-4 specimens ± S.E.M.).

Treatment	^{3}H-uridine into RNA (dpm/piece)			^{14}C-proline into protein (dpm/piece)		
	0 hours	24 hours	48 hours	0 hours	24 hours	48 hours
None	1,357±178	1,360± 51	1,381±152	364±26	384±31	367±24
Fraction 3		366± 46	460± 37		217±28	207±38
Fraction 4		741± 31	521± 48		414±19	574±55
Brain extracts		1,369±186	1,440± 98		378±42	380±43
Actinomycin D		354± 52	247± 61		321±29	178±32

occur as a result of the inhibition of MSH synthesis. However, it is also probable that some kinds of protein must be synthesized for the release of MSH.

C.   *RNA and protein synthesis in the intermediate lobes*
In the intermediate lobes from the frogs adapted to various conditions, RNA and protein synthesis were compared (Table 4). The intermediate lobes from the frogs kept in darkness exhibited a lower rate of RNA and protein synthesis than the intermedia from frogs adapted on an illuminated black or white background. However, after they were cultured for 24 hours, the differences were no longer observed. These results coincide with those of electron microscopic observations, in which the endoplasmic reticulum was found to be very poor in the intermediate lobe cells of the frogs adapted to dim light or darkness.[6]

D.   *Effect of hypothalamic extracts on RNA and protein synthesis in the intermediate lobe*
As the results of the previous experiment suggested that RNA synthesis is closely related to the release and synthesis of MSH, experiments were carried out to see whether the extracts of the fractions of hypothalamus, which were effective in suppressing the release and/or synthesis of MSH, influence the synthesis of RNA and protein in the intermediate lobes *in vitro*. The results indicate that the extracts of fraction 3 were effective in suppressing both RNA and protein synthesis and that fraction 4 gradually inhibited RNA synthesis (Table 5). It is probable that this RNA synthesis is associated with the release and synthesis of MSH in the intermediate lobe.

IV.   SUMMARY

Hypothalamic extracts from bullfrogs were found to inhibit MSH release from the cultured intermediate lobes of frog pituitary glands and to decrease MSH content of the intermedia. By differential centrifugation of the hypothalamic tissue, two fractions, one that strongly inhibits MSH release, and one that decreases MSH content, were obtained.

Experiments designed to study the effects of actinomycin D on the release and synthesis of MSH in the cultured intermediate lobes revealed that the release and synthesis of MSH are closely related to RNA synthesis in the intermediate lobes.

In the intermediate lobes of frogs kept in darkness, RNA synthesis was less active than that taking place in those of frogs adapted on an illuminated white or black background.

Extracts of hypothalamic fractions suppressed RNA synthesis in the intermediate lobes suggesting that hypothalamic factors control the release and synthesis of MSH through regulating RNA synthesis in the intermedia.

REFERENCES

1. Etkin, W., Regulation of the pars intermedia to the hypothalamus. In *Neuroendocrinology* (L. Martini and W. F. Ganong, eds.), **2,** 261, Academic Press, New York (1967).
2. Kastin, A. J. and Schally, A. V., MSH activity in pituitaries of rats treated with hypothalamic extracts from various animals. *Gen. Comp. Endocr.*, **8,** 344 (1967).
3. Bercu, B. B. and Brinkley, H. J., Hypothalamic and cerebral cortical inhibitor of melanocyte-stimulating hormone secretion in the frog, *Rana pipiens*. *Endocrinology*, **80,** 399 (1967).
4. Tata, J. R., Requirements for RNA and protein synthesis for induced regression of the tadpole tail in organ culture. *Develop. Biol.*, **13,** 77 (1966).
5. Hogben, L. and Slome, D., The pigmentary effector system. VI. The dual character of endocrine coordination in amphibian color. *Proc. Roy. Soc.* Ser. B, **108,** 10 (1931).
6. Cohen, A. G., Observations on the pars intermedia of *Xenopus laevis*. *Nature*, **215,** 55 (1967).

Interrelationships of Vertebrate Chromatophores*

Joseph T. Bagnara and Wayne Ferris

Department of Biological Sciences, University of Arizona, Tucson, Arizona, U.S.A.

I. INTRODUCTION

Students of the biology of pigmentation have long been struck by the diversity of colors displayed by vertebrates. At one extreme exist the iridescent blues and greens of poikilotherms, and on the other are the browns, reds, and blacks of homeotherms. Despite these obvious differences, there are certain features that suggest that chromatophores of both groups are related. One of these is the fact that all known chromatophores are of neural crest orign; another is that melanin is a pigment prevalent in all groups. In view of the fact that during the last decade significant advances have been made in the cellular biology of pigment cells, it is appropriate to make a comparative investigation of these new findings with the aim of increasing our general understanding of the interrelationships of chromatophores.

Three levels of organization immediately suggest themselves. These are chemical, cellular, and organellar. The majority of this discussion will pertain to the cellular and organellar associations because the chemical affinities of pigment cells primarily relate to the bright colored pigment cells and this subject has been reviewed recently and extensively.[1-6] At the

---

* Supported by grants GB-8347, GB-4923 and GB-3330 from the National Science Foundation.

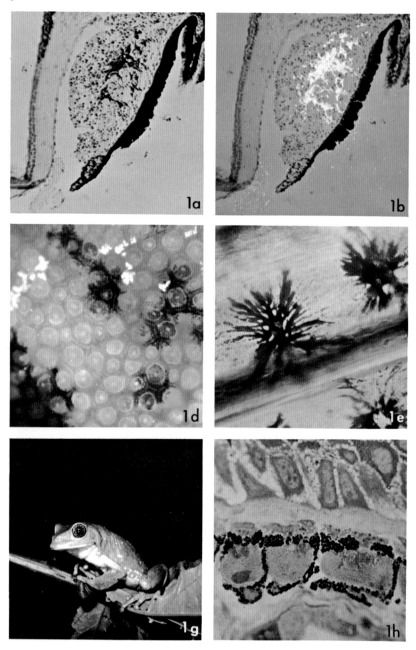

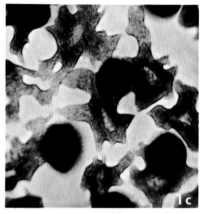

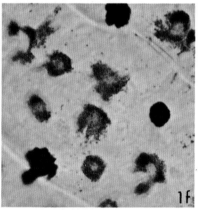

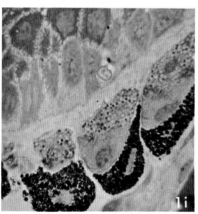

Fig. 1. a. Gross microscopic structure of the dove iris viewed with transmitted light. Brown mass in central region is a group of iridophores. ×200.

b. Same section as in 1a, viewed with reflected light under epi-illumination. The iridophores give the mass a silvery appearance. ×200.

c. Iridophores in a whole skin mount of adult *Rana pipiens* maintained in Ringer's solution. In the absence of MSH, melanophores are punctate while iridophores are dispersed. The blues, greens, and reds of the iridophores are structural colors. ×800.

d. Region on the dorsal surface of a live specimen of *Plethodon cinereus*. Silvery cells are iridophores. Melanophores and erythrophores around and between skin glands form solid masses of color in which it is difficult to discern discrete cells. ×20.

e. Erythrophores on scale of rosy barb, *Barbus conchonius*. ×400.

f. Erythrophores (xantho-erythrophores) on scale of the red swordtail, *Xiphophorus helleri*. Yellow component in center is carotenoid; peripheral red granules are pterinosomes. ×200.

g. Adult Mexican tree frog, *Agalychnis dachnicolor*. ×5.

h. Thick section (1μ) from MSH-treated adult *A. dachnicolor* showing large, red-colored melanosomes filling distal processes above iridophores and beneath xanthophores. Note scarcity of melanosomes in perinuclear area. ×800.

i. Same as 1h except adapted to white background (no hormone treatment). Melanosomes aggregated in perinuclear area leaving most distal processes free of obscuring pigment. ×800.

outset, it is necessary to establish a common ground of terminology as well as to review briefly the cytology of the various types of chromatophores.

II. SURVEY OF VERTEBRATE CHROMATOPHORES

There are three basic types of vertebrate chromatophores: melanophores, which are black, brown, or red in color; iridophores, which are primarily reflective; and xanthophores or erythrophores, which are yellow, orange, or red in appearance (Figs. 1 a–i). The basis for this terminology is presented elsewhere.[3] Basically, there are two types of melanophores. The dermal melanophore (Fig. 2), which has been studied most from the physiological standpoint, is typically stellate and represents the classic melanophore. In contrast, epidermal melanophores are long, slender, and generally spindle-shaped cells found in the epidermis of most vertebrates (Fig. 3). The term epidermal melanophore is used in this discussion; it may

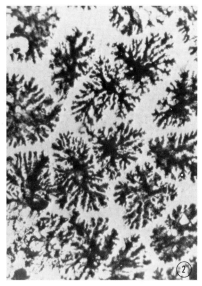

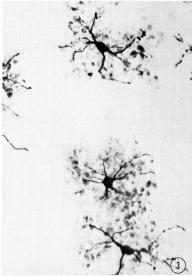

Fig. 2.   Dermal melanophores from dorsal surface of *Xenopus laevis* larva. ×150.

Fig. 3.   Epidermal melanophores from adult *Rana pipiens*. Dark blotches around each melanophore represent epidermal cells containing melanosomes. ×240.

be considered the equivalent of the "epidermal melanocyte." Because of the fact that dermal chromatophores are not typical of warm-blooded animals, and in view of our knowledge that these dermal cells are involved primarily in physiological color change (relatively rapid changes involving an intracellular mobilization of pigment), they will be considered separately from epidermal melanophores. The latter is an element concerned primarily with morphological color change (relatively slow changes involving the synthesis of pigment).

A. *Dermal chromatophores*
1. Melanophores

In keeping with their role in physiological color change, dermal melanophores contain large numbers of melanosomes and are quite large. In some forms they are broad and flat (Fig. 2) and in others they appear quite dendritic, so that in the dispersed state the body of the melanophore is deprived of melanosomes which move out to the distal processes (Fig. 1). The role of the dermal melanophore is primarily that of enabling the animal to adapt its color rapidly. While it does not deposit cytocrine melanin, the dermal melanophore is capable of contributing to morphol-

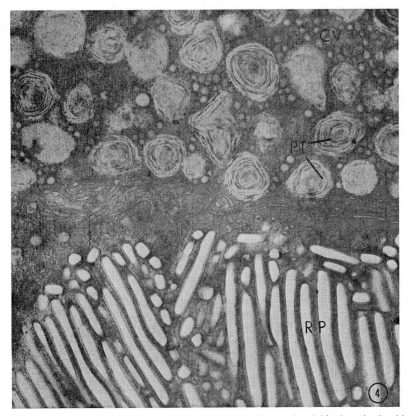

Fig. 4.   Transverse section of area between a xanthophore and an iridophore in the skin of the tree frog, *Hyla cinerea*. Note pterinosomes (PT) and carotenoid vesicles (CV) of the xanthophore and stacks of reflecting platelets (RP) in the iridophore. ×15,750.

ogical color change through the synthesis of more melanin. Accordingly, under prolonged hormonal stimulation, individual melanophores contain more pigment, new pigment cells appear from latent melanoblasts, and mitoses of existing melanized melanophores contribute heavily to increase the total melanophore population.

2. Iridophores

Pigment cells that function primarily through the reflection of light from the surface of orderly distributed organelles are called iridophores and their pigment containing organelles are known as reflecting platelets (Fig. 4).[8] The purines guanine, hypoxanthine, and adenine are deposited in crystalline form in the reflecting platelets which are usually arranged in oriented stacks (Fig. 4). The orientation of these stacks of platelets determines the nature of the pigmentary function of the iridophore. Normally, when viewed with reflected light, iridophores appear to contribute a metallic gold or silver luster (Fig. 1d). When viewed with transmitted light, iridophores exhibit blues, greens, and reds (Fig. 1c). These are structural colors probably arising from a diffraction of light by the stacks of reflecting platelets. Because of light scatter (Tyndall scatter), some iridophores examined with reflected light appear to be blue,[9] others appear

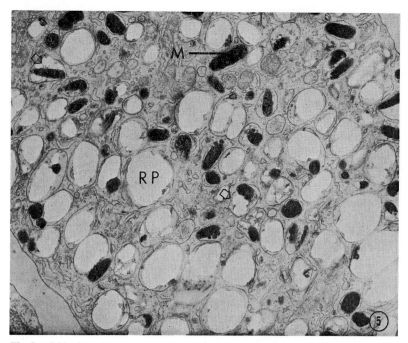

Fig. 5.   Iridophore from the dove iris containing typical reflecting platelets (RP), melanosomes (M), and other vesicular organelles which seem to be melanin-containing reflecting platelets (arrows). ×18,000.

to be khaki in color. These colors probably relate to the orientation and conformation of reflecting platelets.

Formerly, it was believed that iridophores are found only in invertebrates and lower vertebrates.[3] An exception to this generalization was found recently with the discovery that bright colors of the iris of two doves, the Inca dove (*Scardafella inca*) and the Mexican ground dove (*Columbigallina passerina*), are due to the presence of iridophores of the classic type.[10] Sections of the iris observed with the light microscope with transmitted light reveal the presence of a large semiopaque mass (Fig. 1a) which stands out as a silvery body when viewed with reflected light (Fig. 1b). Observation of this tissue with the electron microscope discloses the presence of iridophores containing large reflecting platelets (Fig. 5).

3. Xanthophores

The yellow, orange, or red chromatophores of lower vertebrates are usually called xanthophores or erythrophores (Figs. 1d, e, f). Their principal pigments are carotenoids or pteridines.[3] Carotenoids have long been attributed to xanthophores and are accumulated following their uptake from the diet. They are fat soluble and are stored in the xanthophore in either small or large vesicles that appear to be oil or fat droplets (Fig. 4).[9] Pteridines are synthesized by xanthophores[11] and are contained in organelles called pterinosomes (Fig. 4).[12] The chemical relationships between pteridines of xanthophores and purines of iridophores is a close one, and a possible interconversion of these substances through the mediation of MSH has been suggested.[1-3] The specific color of xanthophores or erythrophores depends on the pattern of pteridines or carotenoids it contains. Often, the cell contains only pteridines, such as is the case for the erythrophore shown in Fig. 1e. Drosopterins (red-colored pteridines) are the dominant members of the pteridine pattern of this erythrophore. Often, when pteridines and carotenoids are found in the same xanthophore or erythrophore, carotenoids are dispersed uniformly between the pterinosomes (Fig. 4). A prominent exception shown in Fig. 1f depicts the situation in the erythrophore-xanthophore of the swordtail, wherein the yellow carotenoid component is centrally located while the pteridine-containing region is peripheral.[12,13] The mobilization of the pigment-containing organelles of these cells during color change has not been studied.

III. CELLULAR ASSOCIATIONS

A. *The dermal chromatophore unit*

The primary function of dermal chromatophores is the regulation of rapid color changes among cold-blooded vertebrates. This is accomplished by the rapid mobilization of the pigment-containing organelles of the various chromatophores, leading to physiological color change. In the larvae of

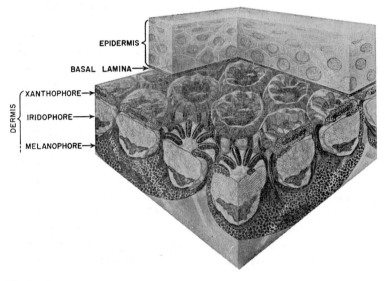

EPIDERMIS

BASAL LAMINA→

XANTHOPHORE→

DERMIS

IRIDOPHORE→

MELANOPHORE→

Fig. 6. Schematic interpretation of the dermal chromatophore unit from several different frogs. Adaptation to a dark background is represented.

most species, and in adults of many fishes, color change results from the individual response of certain pigment cells as they react to given stimuli such as hormone action or nervous stimulation. Among the adults of many amphibians and reptiles, notably frogs and lizards, color changes are brought about by coordinated responses of the three basic chromatophore types which are so situated that they comprise an integral, functional unit which has been designated "the dermal chromatophore unit" (Fig. 6).[9] Uppermost in the unit, just below the basement membrane, is a layer of xanthophores. Immediately beneath this layer of yellow pigment cells is found a layer of iridophores. In frogs, the iridophore layer that forms the reflecting component of the unit consists of a single layer of cells; however, in the lizard, *Anolis*, this region consists of several layers of iridophores.[14,15] The basal portion of the unit is composed of a layer of melanophores that have dendrites extending upward (Figs. 1h, i). In frogs, these dendrites terminate in fingerlike processes on the surface of the iridophore, just beneath the xanthophore layer. During adaptation to dark-colored backgrounds, melanosomes fill these processes (Fig. 1h), thus obscuring the reflecting surface of the iridophore and leading to consequent darkening of the animal (Figs. 7, 8). When the animal lightens, melanosomes move from the terminal processes and occupy a perinuclear position (Fig. 1i). As a result, their dermal melanophores are almost completely obscured by the overlying xanthophore and iridophore layers, and the animal appears

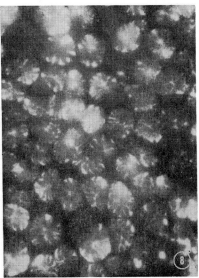

Fig. 7. Skin preparation from the dorsal surface of a partially darkened adult Mexican tree frog, *A. dachnicolor*, photographed with reflected light. Glaring white spots are iridophores reflecting light from their upper surfaces with only minimal interference from fingerlike distal melanophore processes that are only partially filled with melanosomes. ×700.

Fig. 8. Same as Fig. 7 except from fully darkened frog. Note—that iridophore surfaces are almost completely obscured as distal melanophore processes are fully laden with melanosomes. ×700.

light. The pigmentary role of the xanthophore layer relates to the establishment of the green color of many forms. In such animals, light waves leaving the iridophore layer appear blue because of Tyndall scattering[9] and, as the light waves pass through the overlying yellow pigment cells, the shortest wavelengths are absorbed so that finally the animal appears green (Fig. 1g). The importance of the xanthophore pigments in imparting green coloration is proved not only by the blue coloration of green frogs from which yellow pigments have been leached, but also by the existence of blue mutants of frogs or snakes. In such mutants, the pigment content of the overlying intact xanthophores or erythrophores is almost completely depleted.[16]

B. *The epidermal melanin unit*

Of all vertebrate chromatophores, only epidermal melanophores are truly ubiquitous. They are, of course, the primary, if not the only, type of melanophore found among homeotherms. Epidermal melanophores have been reported to occur in fishes,[17] but it is not generally realized that they occur commonly in most lower vertebrates. Hadley has studied them in frogs and has also obtained considerable data (unpublished) about their occurrence in other forms, including fishes, salamanders, turtles, and snakes.

The salient feature of epidermal melanophores is that they are associated with an adjacent pool of Malpighian cells to which they donate melanosomes by means of cytocrine activity (Fig. 3). This close integration of epidermal melanophores and Malpighian cells in mammals led to the concept of the "epidermal melanin unit," formulated by Fitzpatrick and Breathnach.[19] This concept was extended to vertebrates in later general studies by Hadley and Quevedo.[20] Probably this concept should include the cytocrine deposition of melanosomes from epidermal melanophores of the bird bill,[21] the feather shaft,[22] and the hair follicle.[23] In keeping with their cytocrine function, epidermal melanophores have a distinctive and consistent appearance among vertebrates, marked by the presence of slender dendritic processes that extend outward from the cell, well suited for making contact with adjacent epidermal cells. So similar are the epidermal melanophores from such widely separated forms as the frog and the mouse that it is difficult to distinguish between the two. From this vantage point, it seems reasonable that epidermal melanophores of lower vertebrates and those of homeotherms are truly homologous. At the same time, it does not appear that classical dermal melanophores are represented in homeotherms. Surely, melanophores are found in the mammalian dermis, but as Reams[24] points out these are apparently epidermal melanophores in transit to or from the epidermis.

Because of their cytocrine activities, epidermal melanophores are instruments of morphological color change.[20] This change is a relatively slow event involving the accumulation of considerable amounts of melanin in regions of the epidermis populated by large numbers of epidermal melanophores. In contrast, dermal chromatophores are elements of physiological color change involving the rapid mobilization of pigmentary organelles within the cell. In accordance with the nature of these two different systems of color change, a rather precise correlation exists between the localization of the two systems. From several unpublished investigations with Dr. Mac E. Hadley, we have concluded that when a well-developed system of dermal chromatophore units is present, epidermal melanophores and epidermal melanin units are lacking. A prime example is represented by tree frogs which have epidermal melanophores as larvae, but not as adults when they have acquired the ability to change color rapidly. The loss of epidermal melanophores as metamorphosis approaches is paralleled by the differentiation of dermal chromatophore units. In the frog, *Rana pipiens*, large spots and microspots on the dorsal surface are regions containing large numbers of epidermal melanin units. In the dermis immediately below these spots and in exact correspondence, less well-developed dermal chromatophore units are found. Conversely, in the interspot area which is involved in color adaptation, the dermis contains well-developed dermal chromatophore units, while epidermal melanin units are scarce. Apparently, some scheme of "communication" exists between the dermis and the epidermis and allows for the integration of pigment cell expression in these regions. It appears that the basis for this integration relates to the fact that in order for rapid color change to occur, the epidermis must be free of overlying melanin which would obscure the function of the dermal chromatophore unit. In the spot areas, the presence of only a few dermal chromatophore units probably relates to the presence of so many epidermal melanin units in the epidermis.

IV. SUBCELLULAR ASSOCIATIONS

A. *Origins of melanosomes, pterinosomes, and reflecting platelets*
Recently, it has been suggested that the pigment-containing organelles of dermal chromatophores are closely related. This is a conclusion arrived at and presented independently (unpublished) at the recent Seventh International Pigment Cell Conference at Seattle, Washington (September 1969), by Nancy Alexander and Joseph T. Bagnara. From her studies on the development of the skin of a chameleon, *Chameleo hohnelli*, Alexander concluded that all three chromatophore types of the dermis, melanophores, iridophores, and xanthophores are derived from a stem cell. Her conclusions were based largely upon the observation that in early development of the xanthophore, the formative stages of pterinosomes appeared to be identical to the premelanosomes of the melanophore layer. She also observed that young iridophores also contain these premelanosome structures. Bagnara submitted that the three pigmentary organelles, melanosomes, pterinosomes, and reflecting platelets, may be derived from a common equipotential primordial organelle. His conclusion was prompted by the frequent allusions to chromatophore metaplasia as suggested by Niu[25] and implied in the study of Loud and Mishima.[26] The strongest evidence in support of Bagnara's suggestion is derived from the observation of the existence of chromatophores of one type containing pigmentary organelles characteristic of another chromatophore type. Some examples follow.

1. Pterinosomes and melanosomes in erythrophores
Skin from the dorsal surface of the red form of the red-backed salamander, *Plethodon cinereus*, contains large numbers of erythrophores (Fig. 1d) that are laden with pterinosomes.[27] Like the pterinosomes of the other salamanders[28] these organelles lack the typical array of concentric internal lamellae. A striking feature of such erythrophores is that, in addition to pterinosomes, they frequently contain melanosomes and other electron-

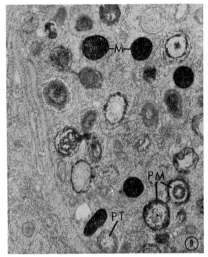 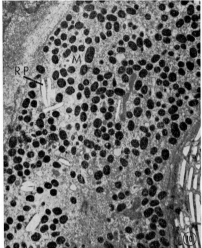

Fig. 9. Erythrophore from the dermis of the dorsal surface of an adult red-backed salamander, *P. cinereus*. M: melanosome; PM: unknown organelles thought to be partially melanized pterinosomes. ×12,000.

Fig. 10. Dermal melanophore from dorsal surface of the canyon tree frog, *Hyla arenicolor*, containing reflecting platelets (RP) interspersed between the melanosomes (M). Reflecting platelets in a neighboring iridophore can be used for comparison. ×5,600.

dense organelles that appear to be intergrades between the two (Fig. 9). It is possible that these three types of organelles have a common origin.

2. Melanosomes and reflecting platelets in iridophores

An examination with the electron microscope of the iridophores in the iris of the Inca dove and the Mexican ground dove reveals that, in addition to their content of reflecting platelets, these cells also contain melanosomes (Fig. 5). These melanosomes are interspersed between the reflecting platelets and closely resemble those found in nearby tapetal cells. A surprising discovery is that some of the reflecting platelets seem to contain melanin or at least an electron-dense material that appears to be melanin. It is interesting that only iridophores nearest to the tapetum contain melanosomes or melanin-containing iridophores. Taylor[29] has observed a similar occurrence of both reflecting platelets and melanosomes in the dermal melanophores of the tree frog, *Hyla arenicolor* (Fig. 10).

3. Pterinosomes and reflecting platelets in erythrophores

Recently, we have studied the dorsal red stripe of an adult ribbon snake, *Thamnophis proximus*, and have observed that it is formed from layers of dermal chromatophores, including erythrophores, iridophores, and melanophores.[30] The lowest of the erythrophore layers is contiguous with the iridophore layer beneath it, and it is unusual in that it contains reflecting platelets as well as pterinosomes. The pterinosomes appear to be normal, and the reflecting platelets are like those of adjacent iridophores. It is of particular interest that the erythrophores farthest from the iridophore layer lack reflecting platelets.

While all the evidence so far presented suggests strongly that organelles of all three dermal chromatophores are somehow closely related, not enough data are available about the immediate origin of each of these organelles. It is suspected that melanosomes of dermal melanophores are of golgi origin just as are those of epidermal melanosomes. Even less is known about the early development of pterinosomes and reflecting platelets. However, at our current level of knowledge, it is not unreasonable to consider that the existence of more than one type of pigmentary organelle might relate to a derivation from a common primordial organelle. This is perhaps more attractive than the possibility of an organellar metaplasia leading to the conversion of one pigment cell to another. Such has been suggested for the melanization of xanthic goldfish[26] wherein organelles of goldfish xanthophores, now known to be pterinosomes,[31] were thought to become melanosomes. This interpretation is still in doubt because Chavin[32]

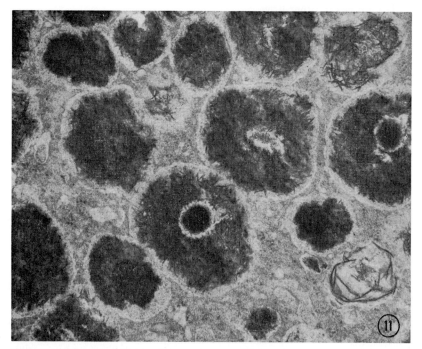

Fig. 11. Portion of a dermal melanophore of an adult *A. dachnicolor* showing the unusual giant melanosomes with their electron-dense kernals and dense outer region of fibrils. ×28,900.

explained melanization in terms of the appearance of a new population of melanophores. His observations are convincing even in the face of the discovery of tyrosinase activity in goldfish pterinosomes.[31] At present, the presence of tyrosinase in pterinosomes is possibly more pertinent to the interpretation that melanosomes and pterinosomes have a common intracellular origin.

B. *An example of melanosome polymorphism*
Of all vertebrate pigment-containing organelles, melanosomes are perhaps the most constant in their appearance. Generally, they are spherical or ellipsoidal in shape and have a diameter of about $0.5\,\mu$. Recently, a strikingly different melanosome was discovered in the dermis of adult Mexican tree frogs, *Agalychnis dachnicolor* (Figs. 1g,11). These melanosomes are about twice the size of most other melanosomes and are composed of an internal dense kernel and an outer fibrous mass.[33] Even at the level of the light microscope, the unusual size of these melanosomes is evident as is their peculiar red color (Figs. 1h,i). Whether this pigment is pheomelanin has not yet been determined. The physiology of melanophores containing these peculiar large melanosomes appears to be like that of dermal melanophores of other frogs.

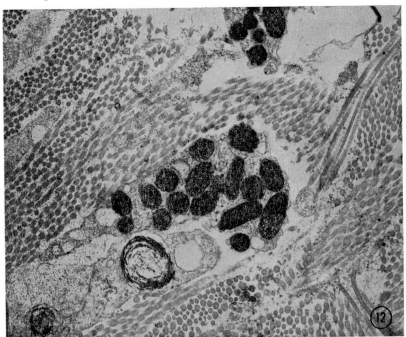

Fig. 12. Two adjacent melanophores in the dermis of an *A. dachnicolor* larva showing a comparison of the two types of melanosomes. $\times 20,400$.

Recently, we have studied the larvae of this species and, to our great surprise, we have observed the presence of two types of melanosomes. In addition to the large melanosome mentioned above, smaller melanosomes of the usual type are also present in the dermis (Fig. 12). Thus far, our preliminary studies have revealed that the two types are found in separate melanophores. It also appears that the large melanosomes are in early stages of development since none that we have found are as dense as those of the adult. It is noteworthy that the electron-dense regions of these melanosomes appear in the form of concentrically deposited fibers. Except for their large size, these organelles resemble pterinosomes. Because of the presence of what appear to be only early phases in the development of these large melanosomes, the impression is given that, possibly in younger stages, melanophores containing the usual melanosomes prevail, but in later stages, as metamorphosis approaches, melanophores with the large melanosomes become more prevalent until finally, in the adult, only the large form can be found. This contention is supported by observations of living tadpoles which appear, under the dissecting microscope, to contain melanophores of a wine-red color. Such larvae undergoing metamorphosis have a definite red cast. The fascinating melanosomes of *Agalychnis* are under further study.

V. CONCLUSIONS

Although the final appearance and function of the various kinds of pigment cells found in vertebrates differ widely, they are connected by several rather close relationships. The first, of course, is that they all have a common origin from the neural crest. In addition, they share remarkable abilities to migrate. Finally, they probably have, at some time during their life, the capacity for concentrating or dispersing their pigmentary organelles. Features that distinguish the various chromatophore types are based upon chemical differences of their pigments and on the form of their pigment-containing organelle. While these chemical variations result in gross differences in the final appearance of the chromatophore, it seems likely that all dermal chromatophores are closely related. Possibly, they all originate from a common stem cell or perhaps even their definitive organelles are derived from a pluripotential primordial organelle.

Topographical interrelationships of pigment cells are also evident in the association in the dermis of melanophores, iridophores, and xanthophores to form the dermal chromatophore unit, an element of physiological color change. A similar interrelationship between cells involves epidermal melanophores and adjacent Malpighian cells which together form the epidermal melanin unit, an element of morphological color change. An even more gross relationship concerns the disposition of dermal chromatophore units and epidermal melanin units; when one is present, the other is not.

In the course of this presentation, the biology of vertebrate chromatophores has been considered in broad terms. It seems appropriate as a final conclusion to emphasize the close connection that exists between the extremes represented by the pigmentation of the cold-blooded vertebrates and that of warm-blooded vertebrates. This connection is the epidermal melanophore, often referred to by mammalian investigators as the epidermal melanocyte. It is fruitless at this point to defend the use of one term over the other; the important thing is to understand that we are in reality dealing with the same cell. As we have pointed out in the body of this paper, the epidermal melanophores of birds, mammals, reptiles, amphibians, and fishes are truly homologous and, while dermal chromatophores are not generally represented in homeotherms, the existence of the ubiquitous epidermal melanophore provides a point of unity for all pigment cell biologists.

REFERENCES

1. Bagnara, J. T., Chromatotrophic hormone, pteridines and amphibian pigmentation. *Gen. Comp. Endocr.*, **1**, 124 (1961).
2. Stackhouse, H. C., Some aspects of pteridine biosynthesis in amphibians. *Comp. Biochem. Physiol.*, **17**, 219 (1966).
3. Bagnara, J. T., Cytology and cytophysiology of non-melanophore pigment cells. *Internat. Rev. Cytol.*, **20**, 173 (1966).
4. Bagnara, J. T., Hadley, M. E., and Taylor, J. D., Regulation of bright-colored pigments in amphibians. *Gen. Comp. Endocr.* suppl. **2** (1969).
5. Bagnara, J. T. and Hadley, M. E., Control of bright-colored pigments in fishes and amphibians. *Amer. Zool.*, **9**, 465 (1969).
6. Bagnara, J. T., Responses of pigment cells to intermedin. In *Colloq. Internat. du C.N.R.S.: La specificité zoologique des hormones hypophysaires et de leurs activités*, **177**, 153 (1969).
7. Pehlemann, F. W., Der morphologische farbewechsel von *Xenopus laevis* larven. *Z. Zellforsch.*, **78**, 484 (1967).
8. Taylor, J. D., The effects of intermedin on the ultrastructure of amphibian iridophores. *Gen. Comp. Endocr.*, **12**, 405 (1969).
9. Bagnara, J. T., Taylor, J. D., and Hadley, M. E., The dermal chromatophore unit. *J. Cell Biol.*, **38**, 67 (1968).
10. Ferris, W. and Bagnara, J. T., Reflecting pigment cells in the dove iris (in press) (1969).
11. Obika, M., Association of pteridines with amphibian larval pigmentation and their biosynthesis in developing chromatophores. *Develop. Biol.*, **6**, 99 (1963).
12. Matsumoto, J., Studies on the fine structure and cytochemical properties of erythrophores in the swordtail, *Xiphophorus helleri*, with special reference to their pigment. *J. Cell Biol.*, **27**, 493 (1965).
13. Goodrich, H. B., The development of hereditary color patterns in fish. *Amer. Nat.*, **69**, 267 (1935).

14. Alexander, N. J. and Fahrenbach, W. H., The dermal chromatophores of *Anolis carolinensis. Amer. J. Anat.,* **126,** 41 (1969).

15. Taylor, J. D. and Hadley, M. E., Z. *Zellforsch* (in press) (1970).

16. Matsumoto, J., Bagnara, J. T., and Hadley, M. E., unpublished observations.

17. Parker, G. H., *Animal Color Changes,* Cambridge University Press, Cambridge (1948).

18. Hadley, M. E., *Ph. D. diss.,* Brown University (1966).

19. Fitzpatrick, T. B. and Breathnach, A., Das epidermale melanin-einheit-system, **147,** 481 (1963).

20. Hadley, M. E. and Quevedo, W., Role of epidermal melanocytes in adaptive color changes in amphibians. *Adv. Biol. Skin,* **8,** 337 (1967).

21. Witschi, E. and Woods, R. P., Nuptial coloration of the bills of birds and their control by sex hormones. *Anat. Rec.,* **64,** 85 (1934).

22. Trinkhaus, J. P., Factors concerned in the response of melanoblasts to estrogen in the brown leghorn fowl. *J. Exp. Zool.,* **109,** 135 (1948).

23. Rawles, M. E., The origin of melanophores and their role in the development of color patterns in vertebrates (1948).

24. Reams, W. M., Jr., Pigment cell population pressure within the skin and its role in the pigment cell invasion of extraepidermal tissues. In *Advances in Biology of Skin,* vol. 8, *The Pigmentary System* (W. Montagna and F. Hu. eds.), Pergamon Press, Oxford (1937).

25. Niu, M. C., Further studies on the origin of amphibian pigment cells. *J. Exp. Zool.,* **125,** 199 (1954).

26. Loud, A. V. and Mishima, Y., The induction of melanization in goldfish scales with ACTH *in vitro. J. Cell Biol.,* **18,** 181 (1963).

27. Bagnara, J. T. and Taylor, J. D., Z. *Zellforsch.,* **106,** 412 (1970).

28. Potter, S., unpublished observations.

29. Taylor, J. D., unpublished observations.

30. Matsumoto, J., Bagnara, J. T., and Hadley, M. E., unpublished observations.

31. Matsumoto, J. and Obika, M., Morphological and biochemical characterization of goldfish erythrophores and their pterinosomes. *J. Cell Biol.,* **39,** 233 (1967).

32. Chavin, W., Pituitary-adrenal control of melanization in xanthic goldfish, *Carassius auratus* L. *J. Exp. Zool.,* **133,** 1 (1956).

33. Taylor, J. D. and Bagnara, J. T., *J. Ultrastruct. Res.,* **29,** 323 (1970).

DISCUSSION

DR. FITZPATRICK: What is the evidence that the iridophores and xanthophores come from the neural crest?

DR. BAGNARA: Very clearly shown. As a matter of fact, the cells used originally by Dushane were dermal pigment cells, and the epidermal pigment cell that we all know comes from the neural crest was never proven so in these early days by Dushane. But they certainly do arise from the neural crest. We know that is true for the mammal and for the bird. The evidence is very simple. You can do all kinds of tissue culture experiments, such as isolating the neural crest in culture, where you can get expressions of iridophores, xanthophores, etc.

DR. FITZPATRICK: This might be an example of pheomelanin in a frog. In other words, it might be a cysteinyldopa of the type that Prota and Nicolaus reported, because you know in the pheomelanin they do not occur in the epidermis either, but are limited to the dermis or the hair. The melanocytes of the epidermis do not contain pheomelanin. It is only the ones in the hair bulb. You really should look for cysteinyldopa in this preparation because it would be a striking finding.

DR. LERNER: Dr. Bagnara, to bring terminology up to date, what is the evolution of guanophore to iridophore?

DR. BAGNARA: I think the best way to designate cells is by their appearance. For instance, these red and yellow pigment cells that you have seen are fantastically complicated. For example, some of these cells have carotenoids in them and have been called lipophores. Some of them contain pteridines; some of them have both pteridines and carotenoids. Some of them change during ontogeny. It is hard to name a pigment cell in terms of what it has in it. It is easiest and best to say, if the pigment cell is black colored, let's call it a melanophore, or melanocyte if you like; if it is red, call it a red pigment cell, an erythrophore; if it is yellow, call it a yellow pigment cell; but don't try to indicate by its name what that pigment cell has in it. In other words, we call a guanophore an iridophore because it also contains purines other than guanine. Its most constant feature is its iridescence; hence our use of the term iridophore.

DR. MISHIMA: I am happy to hear about your beautiful results confirming our previous findings that xanthophores or xanthocytes can be converted to melanophores or melanocytes. In 1961, at the time the Fifth International Pigment Cell Conference (*Ann. N. Y. Acad. Sci.*, **100**, 607, 1963), we reported on evidence for the induction of melanization in the "large bodies" (pterinosomes) of yellow pigment cells treated with ACTH *in vitro*. What do you think of the conversion of the immature stage of xanthophores to melanophages?

DR. BAGNARA: In Japan, Dr. Obika and Dr. Matsumoto were working on this problem. Both of them have demonstrated the fact that isolated pterinosomes have tyrosinase activity. It is very attractive to extrapolate from this and to say that this is a good example of the possibility of melanization of a pterinosome.

DR. FITZPATRICK: Just to carry this unified hypothesis a little further, what happens in an albino frog?

DR. BAGNARA: Only the melanosome is affected. In fact, you can see melanophores in an albino frog.

DR. FITZPATRICK: But the other pigments are not affected?

DR. BAGNARA: No. Well, there is an effect, but I think it is an indirect effect. For instance, there is much more yellow pigment in an albino frog than there is in a normal frog, and I think this is a compensatory response, probably to protect it from radiation.

DR. LERNER: But you never assayed the pituitaries of those frogs?

DR. BAGNARA: No.

DR. SEIJI: In one of your electron microscopic pictures, there were a few reflecting crystals in a melanophore. Is there any possibility of transferring reflecting crystals from iridophores to melanophores?

DR. BAGNARA: This is one of the things that I originally thought, but a rather interesting thing is that the reflecting plates found in melanophores respond to MSH just as reflecting plates do normally. I doubt that this is an artifact. I thought at first that this is what was occurring in the bird eyes, but you saw the mixed organelles, the organelles containing both melanin and reflecting material; that cannot be a case of phagocytosis.

DR. CHAVIN: Can one be certain that "phagocytosis," or exchange of specific pigmentary organelles between different pigment cell types, does not occur? This appears to receive some support from an apparent "gradient"-type effect, i.e., cell types closest to one another show organelles of each type, while cells farthest apart do not show such organelle "swaps."

DR. BAGNARA: There are a number of explanations. Probably the most cogent argument is the last one I told you about, the mixed organelle. I cannot see how you are going to get a piece of one organelle being placed exactly into another organells as an artifact.

DR. CHAVIN: Well, not an artifact, just a routine exchange; exchange in life in the skin.

DR. BAGNARA: How is it going to get way into the middle of the pigment cell? Some of those cells are actually separated from one another. I think that the argument that this might possibly be some sort of phagocytosis is relatively weak. Of course, this occurred to me as one of the logical things that might happen.

DR. CHAVIN: It would be interesting if you could tag, or radiolabel, one of the subcellular particles and see if it proceeds through the plasma membrane.

Approaches to the Study of Melanocyte and Melanophore*

Walter Chavin

Department of Biology, Wayne State University, Detroit, Michigan, U.S.A.

The study of integumental pigment cell physiology is both stimulating and intriguing as these melanin-synthesizing cells have proven to be rather difficult subjects for study. For example, our early question dealing with the specific mode of action of hormones upon differentiation of melanin-synthesizing integumental pigment cells has not been resolved. Some of the problems encountered have resulted from the dearth of fundamental knowledge about the basic biochemical and physiological processes occurring in these cells. Thus, a number of diverse approaches were employed to determine the basic characteristics of melanocytes and melanophores. The propigment and melanin-synthesizing pigment cells of the goldfish (*Carassius auratus* L.) were used as a model system.

The early studies of endocrine action upon melanocyte differentiation in xanthic goldfish indicated the involvement of hypophyseal hormone.[1] Due to the diffuse nature of many of the endocrine tissues in this species, the *in vivo* experimental approach was not feasible. Therefore, subsequent work utilized organ culture of goldfish skin. It was shown in a series of collaborative studies with Dr. Funan Hu, Dr. T. T. Tchen, and other co-investigators[2-6] that ACTH and MSH stimulated melanocyte differentiation *in vitro*. However, the melanocyte precursor cell, the melanoblast,

---

* Contribution number 245, Department of Biology, Wayne State University.

was not identifiable in goldfish skin, despite utilization of histochemical methodology specific for propigment cells.[7] Recently, it has been possible to stimulate *in vitro* differentiation of melanocytes in xanthic goldfish skin in the presence of actinomycin D or puromycin or both. Scale explants in organ culture dishes were maintained at the environmental interface (95% O_2, 5% CO_2) in 1 ml of medium (199+5% fetal calf serum) containing colchicine (1 μg/ml) for 48 hours at 37°C. The explants were then rapidly washed several times with medium and subjected to 0.2 milli units–20 units ACTH+actinomycin D (1 μg/ml) and/or puromycin (0.1 μg/ml). The skin was maintained for 72 hours post-hormone treatment. The results showed that hormonal stimulation of differentiation occurred in the absence of protein synthesis. The extremely small melanocyte which was detectable by its melanin granule content (Fig. 1), therefore, may be considered the equivalent of the melanoblast as protein

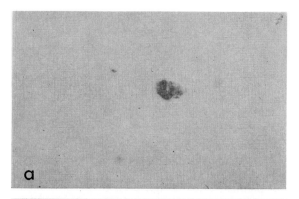

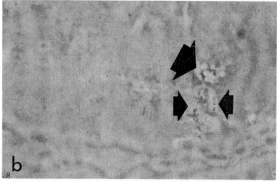

Fig. 1. Melanocytes in ACTH-treated xanthic goldfish skin (scale) in the presence of actinomycin D (1 μg/ml)+puromycin (0.1 μg/ml), grown in organ culture. These extremely small melanocytes reflect the shapes and sizes of the melanoblast, as protein synthesis is inhibited in the presence of these antibiotics. a: bright field; b: phase contrast. Bouin fixation, unstained. ×800.

synthesis has not occurred. Foote and Tchen,[8] using ACTH prior to antibiotic treatment, obtained a larger harvest of such cells in fin skin. Thus, in addition to the identification of this propigment cell equivalent, the findings also demonstrate that the synthesis of melanin does not require additional protein during this initial phase of cell development. Furthermore it is clear that under these circumstances both the proteinaceous premelanosomal lattice and the enzyme, tyrosinase, are present in the melanoblasts. This suggestion receives support from tyrosinase studies[6,9,10] of skins of white and xanthic goldfish, in which a considerable degree of enzymic activity is present despite the absence of melanin.

Continuing the question of the role of hormones in pigment cell physiology, the post-hypophysectomy death of most melanocytes and melanophores in black goldfish was of interest. To probe further the nature of such integumental melanocyte and melanophore loss in goldfish, an agent destroying the pigment cells, hydroquinone (1,4-dihydroxybenzene), was employed using black fish.[11] The cytological changes observed after these two treatments were different. Hypophysectomy produced perinuclear aggregation of the melanin granules in melanophores within 2–3 hours post-operatively. After several days these punctate cells began to lyse. The gradual decrease in pigment cell number proceeded for several weeks but a small pigment cell population remained. In contrast, hydroquinone produced rather rapid and dramatic cytotoxic effects detectable even within the earliest time interval studied, 1 hour post-treatment. Hu,[12] using B16 melanoma cells *in vitro*, found the action of hydroquinone abrupt and like that of a fixative. The sequential changes included melanin granule aggregation, vesiculation, pinching off of dendrites, and cell lysis. Similar cellular reactions occurred in human skin melanocytes.[13] It is interesting that during the course of hydroquinone treatment new melanocytes differentiated, but with continued treatment these cells also were destroyed. This will be discussed below. Extension of this work revealed a number of new and diverse chemical classes of depigmentary agents.[14-16] These compounds appeared to react with the melanocytes and melanophores to produce a pattern of cytological changes resulting in cell lysis similar to that described for hydroquinone. The findings suggest that the effective depigmentary agents in the black goldfish have a common mode of action. Using compounds from several of the chemical classes of these depigmentary agents, Fitzpatrick, Pathak, and co-workers[17,18] extended the findings to a mammal, the black guinea pig. Here also, the agents were found specifically to destroy integumental melanocytes, but again, they did not impede differentiation of new melanocytes. However, the mechanism of action of depigmentary agents in all species utilized to date remains speculative.

In studies of the enzyme fundamental to melanogenesis, tyrosinase, the highly specific, sensitive, and precise ^{14}C radiometric assay based upon the work of Kim and Tchen[19] was developed.[10] It became possible to definitively determine tyrosinase activity even in crude tissue homogenates. In fact, utilizing this assay procedure, it has been possible to measure the tyrosinase activity in a single melanophore of the goldfish integument.[20] Although the assay was developed for studies dealing with goldfish tyrosinase, its use has been extended to the entire subphylum vertebrata, from hagfish to man, as well as to studies dealing with melanomas of fish, mouse, and man. The general findings and derived concepts have been summarized recently.[21] Tyrosinase in the 30 vertebrate species studied and in melanoma was always associated with the particulate fraction of the skin $(144,000 \times g, 40$ minutes, $0-4°C)$. In some species and in certain physiological states, including some melanomas, the enzyme also occurred in the soluble fraction. As might be expected, a number of biological factors were found to affect both tyrosinase activity and subcellular distribution of the enzyme. These factors include the degree of melanin pigmentation,[22] anatomic location of the skin,[23] growth of hair containing melanin,[24] physiological state of the animal,[9, 25] etc. In addition, a number of tyrosinase inhibitor types were suggested from the reported findings. These include a type which rapidly destroyed tyrosinase as in the goldfish,[10] a type which may be diluted beyond its effective inhibitory concentration as seen in Amphibia,[26, 27] an oxidizable type liberating tyrosinase activity[2] as occurring in the Atlantic hagfish *(Myxine glutinosa)*, garpike *(Lepisosteus osseus)* and Australian lungfish *(Neoceratodus forsteri)*, a thermolabile type (human melanoma),[20] and a nonprotein, nonsulfhydril type (human melanoma).[20] Other types of integumental tyrosinase inhibitors described by other workers further increase the number of tyrosinase control mechanisms utilized by various vertebrate species. In view of their great importance, both theoretical and practical, it is surprising that these inhibitors have not been critically defined. Returning to the goldfish, at least two inhibitor systems appear to be present. It has been suggested[5, 8] that tyrosinase is present in inhibited state in the skin of xanthic goldfish. Hormone treatment, in a yet undefined manner, resulted in activation of the inhibited tyrosinase, thus permitting melanogenesis to occur. The work using antibiotics in organ culture revealed that *de novo* tyrosinase synthesis did not occur but melanin formation was apparent. Further, the presence of tyrosinase in white and xanthic goldfish skin[22] supports this hypothesis. In addition, a second type of inhibitor causing loss of tyrosinase activity also may be present, as goldfish tyrosinase was unstable in homogenates.[10] These findings, coupled with at least the two types of inhibitors occurring in human melanoma, indicate that more than one tyrosinase control mechanism may be present in the skin of one species. Multiple types of control mechanisms in the integument as a whole would provide

Table 1. Effects of hydroquinone and derivatives on tyrosinase activity.

	Tyrosinase activity, percent of control*									
Compounds	Mushroom tyrosinase		Black goldfish skin tyrosinase		Hybrid swordtail melanoma tyrosinase		B16 mouse (C57BK) melanoma tyrosinase		Human melanoma tyrosinase	
	$9 \times 10^{-4}M$	$9 \times 10^{-3}M$	$9 \times 10^{-4}M$	$9 \times 10^{-3}M$	$9 \times 10^{-4}M$	$9 \times 10^{-3}M$	$9 \times 10^{-4}M$	$9 \times 10^{-3}M$	$9 \times 10^{-4}M$	$9 \times 10^{-3}M$
Hydroquinone	0.1	0.0	17.0	221.7	22.1	4.2	3.0	5.7	9.0	2.2
Quinhydrone	0.0	0.0	600.0	—	0.0	0.0	16.4	66.9	105.5	13.9
Quinone	0.0	0.0	550.0	—	0.0	0.0	12.8	59.2	—	—
1,2,4-trihydroxybenzene	0.0	0.0	0.2	0.1	0.0	0.0	0.0	0.0	17.4	157.0
Di-Na-hydroquinone monophosphate	67.1	18.6	48.7	23.4	47.4	11.6	101.0	70.8	76.2	60.0
Chlorohydroquinone	0.2	0.0	24.7	18.6	7.6	0.0	7.4	0.0	63.5	137.4
2-hydroxy-1, 4-naphthoquinone	84.1	25.6	80.5	46.3	86.6	41.5	93.7	51.6	73.1	40.0
Monohydroxyethyl ether of hydroquinone	94.9	38.6	56.5	15.2	91.5	23.6	92.0	79.2	74.0	25.8
Tetrahydroxyquinone	94.0	96.1	102.0	102.2	87.2	109.6	95.6	127.1	94.6	96.2
2,6-dinitrohydroquinone	12.4	0.9	30.5	11.1	45.0	10.5	45.9	8.8	24.3	16.2
Control	100.0	100.0	100.0	100.0	100.0	100.0	100.0	100.0	100.0	100.0

* Control tyrosinase activity: 3,266 ± 161 T.U.

Table 2. Effects of catechol and derivative, related compounds of hydroquinone or catechol on tyrosinase activity.

Compounds	Mushroom tyrosinase		Black goldfish skin tyrosinase		Hybrid swordtail melanoma tyrosinase		B16 mouse (C57BK) melanoma tyrosinase		Human melanoma tyrosinase	
	$9 \times 10^{-4}M$	$9 \times 10^{-3}M$	$9 \times 10^{-4}M$	$9 \times 10^{-3}M$	$9 \times 10^{-4}M$	$9 \times 10^{-3}M$	$9 \times 10^{-4}M$	$9 \times 10^{-3}M$	$9 \times 10^{-4}M$	$9 \times 10^{-3}M$
Catechol	20.0	1.7	89.4	40.9	48.3	5.30	69.0	7.4	91.6	77.9
4-isopropyl catechol	27.2	4.0	31.4	9.6	5.4	0.39	25.9	0.0	43.3	21.9
3-isopropyl catechol	50.3	4.1	25.4	14.7	7.9	0.00	8.4	0.0	7.0	19.9
4-methyl catechol	3.6	0.3	21.1	9.7	6.2	0.00	10.1	0.0	34.3	18.0
3-methyl catechol	23.3	2.2	80.8	26.4	53.1	6.60	44.3	0.0	66.5	56.4
3,4-dihydroxyphenyl acetic acid	20.0	0.5	32.9	29.0	32.3	0.00	46.5	0.0	39.8	7.0
Resorcinol	0.6	0.0	30.5	4.1	15.7	0.80	13.5	0.0	30.6	4.9
Pyrogallol	0.5	0.1	9.3	4.4	0.3	0.00	0.0	0.0	—	—
Phloroglucinol	2.1	0.2	50.2	8.1	44.8	3.00	33.7	0.0	26.8	7.1
Control	100.0	100.0	100.0	100.0	100.00	100.00	100.0	100.0	100.0	100.0

Tyrosinase activity, percent of control*

* Control tyrosinase activity: 3,266 ± 161 T.U.

a more refined degree of enzymatic control under a variety of circumstances.

Attempting to gain insight into both the mechanism of action of depigmentary agents and tyrosinase control mechanisms, the *in vitro* effects of known *in vivo* depigmentary agents as well as some *in vitro* tyrosinase inhibitors upon tyrosinase of different origins were studied.[29] These tyrosinases included purified mushroom enzyme and homogenates of black goldfish skin, hybrid swordtail pigmented melanoma, B16 pigmented melanoma, and human pigmented malignant melanoma. An apparent shortcoming of any approach using homogenates is the unknown factor or factors present in the tissue which may affect drug action. Nevertheless, it is clear that the position of the hydroxyl groups in the hydroquinone drug series was important (Tables 1,2). The inhibitory effect decreased from hydroquinone (1,4-dihydroxybenzene) to resorcinol (1,3-dihydroxybenzene) to catechol (1,2-dihydroxybenzene). Addition of a third hydroxyl group (1,2,4-trihydroxybenzene) further increased the inhibition. However, tetrahydroxyquinone (1,2,4,5-tetrahydroxybenzene) appeared to produce little inhibition.

The studies dealing with various mercapto compounds have indicated the importance of free -SH and $-NH_2$ groups in tyrosinase inhibition (Table 3). Oxidation of the mercaptoethylamine -SH group to form the dimer, cystamine, produced a significant reduction of enzymic inhibitory activity. On the other hand, the effects of S-acylation or S-esterification increased the inhibitory effect, suggesting that such groups may aid in combination of the sulfhydryl sulfur with its target. In regard to the $-NH_2$ group, amino bound derivatives may affect the inhibitory potency. Lack of free $-NH_2$ reduced the inhibitory activity as in 2-(dimethylamino)-ethylthioacetate, compared to 2-aminoethylthioacetate, of which it may be considered a substitution product. Although continued exploration of depigmentary agents is necessary, these findings revealed a surprising number of similarities among the tyrosinases.[29] However, the differences between the purified and crude enzyme preparations in reaction to the drug used also revealed that a number of other factors in the tissue may be of considerable importance. The definition of such integumental or tissue factors affecting the enzyme will result in a better understanding of both natural and artificial tyrosinase control mechanisms in normal and abnormal pigment cells.

Attempting to clarify the reaction of the melanocytes and melanophores to depigmentary agents, the distribution and turnover of radiocarbon-labeled hydroquinone was studied in black goldfish.[30] The radiohydroquinone was administered subcutaneously at a depigmentary dose level (30 mg/kg) at a radionuclide level of 0.071 μCi/g (Table 4). Study of 15 tissues (Table 5) and the aquarium water at 12 intervals ranging from

Table 3. Effects of mercaptoethylamine (MEA) and derivatives on tyrosinase activity.

	Tyrosinase activity, percent of control									
	Mushroom tyrosinase		Black goldfish skin tyrosinase		Hybrid swordtail melanoma tyrosinase		B16 mouse (C57BK) melanoma tyrosinase		Human melanoma tyrosinase	
Compounds	$9 \times 10^{-4}M$	$9 \times 10^{-3}M$	$9 \times 10^{-4}M$	$9 \times 10^{-3}M$	$9 \times 10^{-4}M$	$9 \times 10^{-3}M$	$9 \times 10^{-4}M$	$9 \times 10^{-3}M$	$9 \times 10^{-4}M$	$9 \times 10^{-3}M$
Mercaptoethylamine HCl	0.4	0.1	12.9	5.6	65.6	14.8	28.1	13.5	14.0	3.3
Cystamine HCl	90.5	77.1	75.9	41.0	91.1	90.3	84.0	106.8	100.0	87.7
N-(2-mercaptoethyl)-dimethylamine HCl	10.6	0.2	73.8	9.1	62.7	7.6	82.1	3.9	43.3	17.7
2-(n-butylamino)-ethanethiol	73.6	17.7	7.0	2.1	91.5	8.3	76.1	1.4	3.3	0.0
2-(cyclohexylamino)-ethanethiol	0.4	0.2	7.0	1.5	11.6	0.5	21.9	0.4	63.0	21.3
2-(dimethylamino)-ethylthioacetate HCl	1.5	0.2	39.0	12.2	25.8	3.1	24.7	0.4	3.2	0.0
2-mercaptoisopropyl-amine HCl	0.0	0.1	24.5	12.5	49.9	9.1	28.0	4.8	29.6	9.7
3-mercaptopropylamine HCl	0.4	0.0	5.4	4.7	1.3	0.3	1.5	0.2	2.3	1.9

Compound									
N-(2-mercaptoethyl)-N-(dodecyl)-morpholinium bromide	77.9	19.1	5.6	2.9	87.1	12.4	86.9	58.4	17.0	0.0
N-(2-mercaptoethyl)-diethylamine HCl	1.3	—	41.8	7.9	29.9	3.5	50.2	1.6	36.7	8.1
2-(3-diethylamino-propylamino)-ethanethiol	52.3	0.2	40.8	6.7	77.7	11.7	61.8	2.4	30.6	0.0
2-(p-fluoroanilino)-ethanethiol	0.0	0.1	3.5	4.4	0.4	0.5	0.3	0.1	0.5	1.9
2-(d-α-methylphenyl-ethylamino)-ethanethiol	0.7	0.1	7.9	3.8	1.0	0.1	7.2	0.6	2.3	2.1
2-(2-picolylamino)-ethanethiol	2.4	0.0	16.1	2.6	32.0	0.8	49.5	0.8	12.0	0.0
S-MEA ester of L-proline di-HCl	0.1	0.0	8.7	3.0	1.0	0.3	0.9	0.7	2.8	0.0
S-MEA ester of DL-alanine di-HCl	0.0	0.0	4.1	3.9	1.0	0.3	0.2	0.3	0.9	0.0
S-MEA ester of DL-phenylalanine di-HCl	0.0	0.0	0.6	0.6	0.4	0.1	0.7	0.1	0.0	0.0
2-aminoethylthio-acetate HCl	—	—	—	11.7	1.0	0.0	0.7	0.1	—	—
Cysteine	1.3	0.0	16.8	0.6	39.3	0.8	75.2	2.3	—	—
Control	100.0	100.0	100.0	100.0	100.0	100.0	100.0	100.0	100.0	100.0

Table 4. Turnover of hydroquinone-2-3-5-6-^{14}C.

1. Acclimate black goldfish, 4 weeks.
2. Weigh animals.
3. Inject subcutaneously 0.071 μCi/g radiohydroquinone (S.A.: 0.261 mCi/mM). Hydroquinone dose equivalent to 30 mg/kg.
4. Sacrifice (freeze) at given post-injection interval (30 minutes to 96 hours); 4–5 fish per group.
5. Autopsy and weigh 15 organ and tissue samples collected per animal (wet weights).
6. Dry samples and weigh (dry weights).
7. Dissociate sample in 0.6N quaternary ammonium hydroxide (NCS solubilizer), bleach with H_2O_2, and add scintillator.
8. Count at 0°C in liquid scintillation counter to 1×10^6 counts or 10 minutes. Duplicate counts.
9. Correct for quenching, background, etc.

30 minutes to 96 hours post-injection revealed that the hydroquinone uptake and turnover were rapid (Table 6). The peak uptake level was achieved at the earliest time interval of the present study in all the tissues studied except the liver and gall bladder, with or without melanin-synthesizing abilities. The hydroquinone was essentially eliminated at 48 hours. A major route of hydroquinone excretion is via the bile. The rapid tissue hydroquinone turnover appeared to be a function of circulation time rather than a specificity of hydroquinone uptake, except for the liver and gall bladder. The intestine followed the gall bladder turnover pattern. Therefore, compared to most tissues, the melanocytes and melanophores showed no special affinity for hydroquinone when the radiolabeled compound was administered at a depigmentary dose level (Figs. 2,3,4). However, these pigment cells show rapid cytological responses to hydroquinone.[11] It appears, therefore, that a unique biochemical or biophysical property of melanocytes and melanophores may be involved in the drug-cell interaction. Under these circumstances, tyrosinase would appear to be a reasonable target of drug action, as has been

Table 5. Organ and tissue samples used in radiohydroquinone turnover study.

Samples containing melanin	Samples not containing melanin
Skin, fin	Liver
Skin, body	Gall bladder
Eye	Intestine
Peritoneum	Gonad
	Spleen
	Heart (with blood)
	Kidney
	Lymphoidal pronephric derivative
	Gill
	Brain
	Skeletal muscle
Aquarium water from each group	

Table 6. Calculated quadratic (parabolic) regression curves by the method of least squares of the ^{14}C-hydroquinone uptake in goldfish tissue as a function of time. Independent variable: time (X); dependent variable: uptake (Y)

Tissue	Quadratic regression	Standard error
Integument[a]	$Y = 20.02X^{-0.4186}$	0.2169
Eye	$Y = 1.1553X^{-0.5075}$	0.2184
Peritoneum	$Y = 0.5681X^{-0.3647}$	0.3760
Liver[b]	$Y = 2.1196X^{-0.4538}$	0.3957
Gall bladder[c]	$Y = 5016X^{-1.2274}$	0.6294
Intestine[d]	—	—
Gonad[b]	$Y = 1.1187X^{-0.6454}$	0.4225
Spleen[b]	$Y = 0.9559X^{-0.4257}$	0.3640
Heart	$Y = 1.3715X^{-0.6429}$	0.3469
Kidney	$Y = 0.9247X^{-0.5081}$	0.3081
Pronephric area	$Y = 1.0719X^{-0.5697}$	0.3334
Gill	$Y = 0.3030X^{-0.4687}$	0.3213
Brain	$Y = 0.7698X^{-0.5815}$	0.2770
Muscle, skeletal	$Y = 0.3277X^{-0.5898}$	0.3805

[a] Combined body skin and fin skin data.
[b] Data from 1 hour (peak uptake).
[c] Data from 3 hours (peak uptake).
[d] Calculation omitted due to retention plateau. See text.

suggested previously.[31-33] The effect of hydroquinone upon black goldfish tyrosinase or even other tyrosinases demonstrates that the depigmentary action of hydroquinone is not mediated via tyrosinase.[11, 16, 20] However, the melanosome with its membrane and melanin polymer appears to be an alternate specific target for drug action. Nevertheless, as the depig-

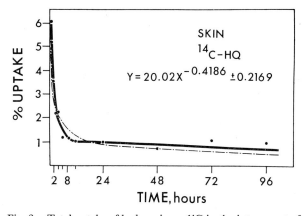

Fig. 2. Total uptake of hydroquinone-^{14}C in the integument of black goldfish normalized to 5 g. Percent injected dose as a function of time. Black circles: group means; solid line: fitted curve; broken line: regression curve.

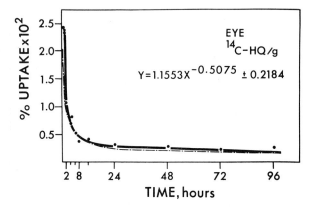

Fig. 3. Uptake of hydroquinone-^{14}C in the eye of the black goldfish normalized to 5 g fish. Percent injected dose per g eye. Black circles: group means; solid line: fitted curve; broken line: regression curve.

mentary effects of hydroquinone upon the melanosomes have not been reported, the site and mechanism of hydroquinone action remain speculative.

The lack of hydroquinone affinity for the melanin-synthesizing cells coupled with the rapid turnover of hydroquinone in the skin indicates that the effective time of hydroquinone action upon pigment cells is very short. In black goldfish, cytologically detectable changes in the living melanocytes and melanophores were present within 1 hour after hydro-

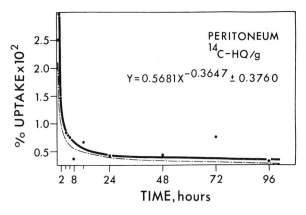

Fig. 4. Uptake of hydroquinone-^{14}C in the peritoneum of black goldfish normalized to 5 g fish. Percent injected dose per g peritoneal tissue. Black circles: group means; solid line: fitted curve; broken line: regression curve.

quinone administration.[11] At a hydroquinone dose level (20–40 mg/l) comparable to that used in the present study (30 mg/kg), Hu[12] found that melanin granule movement in cultured B16 mouse melanoma cells was abruptly halted in the same manner as a fixative affects the cells. Further, the similarity of these fish and mammalian pigment cells is also indicated by the hydroquinone titer necessary to produce the observed cytoxicity. As the peak uptake of radiohydroquinone in goldfish skin is 6.1% of the injected dose at 30 minutes post-injection (Fig. 2), the hydroquinone concentration in the skin at this time is 1.83 μg/g skin (wet). The pigment cells are lysed at this *in vivo* hydroquinone dose level. Hu[12] has shown that B16 melanoma cells degenerate *in vitro* when subjected to hydroquinone in the range of 1.25–2.50 μg/ml culture medium. The rapidity of the hydroquinone action both *in vivo* and *in vitro* upon fish and mammalian cells, respectively, and the similarity in cytological responses of these cultured mammalian tumor cells and the normal fish pigment cells *in vivo* at comparable hydroquinone levels are indicative of a common sensitive intracellular target with a given hydroquinone sensitivity. The evolutionary disparity and the obvious differences in the physiological states of the cells of these species make the findings even more remarkable, thus reinforcing the hypothesis of a common and specific target.

The appearance of newly developing melanocytes in the integument of black goldfish during the course of hydroquinone depigmentary treatment has been an intriguing if puzzling phenomenon.[11] The short biological half-life of hydroquinone in the skin coupled with the rapidity of hydroquinone action upon the pigment cells indicates that only the pigment cells present *at the time of hydroquinone treatment* are destroyed. As hydroquinone has been injected three times weekly and as the hydroquinone level in the skin decreased rapidly, the development of new melanocytes was not impeded. Thus, the appearance of new melanocytes among the debris of lysed melanocytes occurring in skin biopsies of treated animals is now explicable. Continued injection of hydroquinone would destroy such newly developed cells but would not impede further differentiation of additional melanocytes. With cessation of hydroquinone treatment, the newly developing melanocytes would regenerate the pigment cell population of the skin. Thus, to achieve skin devoid of pigment cells long-term hydroquinone treatment has been found necessary, but regeneration of melanocytes and melanophores to the normal level occurred within 3 weeks post-treatment.[11] The present findings suggest that rapid and complete destruction of the integumental melanin-synthesizing pigment cells in the goldfish may possibly be achieved by continuous infusion of hydroquinone at a level sufficient to maintain the drug titer at approximately 1.8 μg/g skin. Such complete depigmentation would be maintained only for a limited period after cessation of hydroquinone treatment.

Table 7. *In vivo* labeling of melanin.

1. Acclimate black goldfish, 4 weeks.
2. Weigh animals.
3. Depigment by IM injection of hydroquinone (40 mg/kg); 3 times weekly; total, 10 injections.
4. Two-day recuperation.
5. Inject (IM) 1 μCi/g uniformly labeled L-tyrosine-^{14}C (S.A.: 360 mCi/mM).
6. Sacrifice at given post-injection interval (3 hours to 900 days).
7. Extract integumental and retinal melanin.

In addition to yielding information as to possible modes of depigmentary action and possible approaches to the control of tyrosinase and pigment cell activities, depigmentary agents are useful in the partial or complete elimination of an existing pigment cell population. As many agents do not produce permanent effects, a new population of melanocytes will differentiate after treatment is terminated so that a pigment cell population of fairly uniform age is available for study of certain fundamental characteristics. Using hydroquinone to depigment black goldfish, the melanin-synthesizing activity and life span of the cells were studied with the use of uniformly labeled L-tyrosine-^{14}C. The animals were subjected to 10 injections of hydroquinone on alternate days and injected with a tracer dose of L-tyrosine-^{14}C (1 μCi/g) 48 hours post-hydroquinone

Table 8. Melanin extraction procedure.

1. Weigh animals and sacrifice.
2. Remove and weigh skin and eyes (wet weight).
3. Dehydrate and defat in chloroform-methanol (2:1, v/v), 8 changes, 15 ml/change, 25°C.
4. Dry tissue at 80°C, 18 hours. Weigh (dry weight).
5. Hydrate tissue in 4.5 ml distilled water at 95°C, 1 hour.
6. Add 0.5 ml 5N HCl and incubate at 37°C, 18 hours, with constant stirring. (Removes bone in scales and fin rays.)
7. Adjust to pH 7.8 with NaOH.
8. Add pronase to 0.5% and neomycin sulfate to 5 μg/ml. Incubate 48 hours at 37°C with constant stirring.
9. Add sodium EDTA (pH 7.8) to 5% for removal of calcium precipitate.
10. Add 5 ml 0.5% triton X-100.
11. Filter through two 100 mμ pore membrane filters (preweighed). Lower filter is control.
12. Wash 15 times with 10 ml distilled water/wash.
13. Air dry filters: a. 48 hours at 25°C.
    b. 20 minutes at 90°C.
14. Weigh and correct for filter weight change (0.0–0.65 mg). Residue weight.
15. Count residue to 10,000 counts or 30 minutes. Triplicate counts. Thin window gas flow counter.
16. Correct residue weight for non-melanin material.
17. Correct melanin CPM for self absorption.

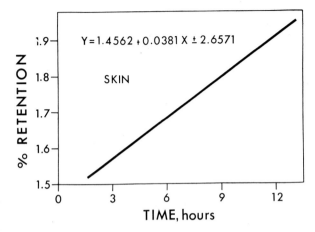

Fig. 5. Uptake of L-tyrosine-^{14}C per mg integumental melanin in hydroquinone-pretreated black goldfish (normalized to 5 g) as a function of time, to 12 hours peak uptake.

treatment (Table 7). The animals were sacrificed at 22 intervals from 3 hours to 900 days. Normal untreated goldfish (10 xanthic, 15 black) were similarly injected and sacrificed at 12, 24, and 72 hours (black fish only). The melanin was extracted as indicated in Table 8.

The developing population of melanocytes and melanophores after hydroquinone pretreatment converted radiotyrosine into melanin granules. These goldfish, compared to normal black animals, showed approximately a 50-fold increase in L-tyrosine-^{14}C incorporation at 12 hours. The developing new pigment cell population, therefore, synthesized melanin more actively than the established, predominantly mature pigment cell population of the untreated animals. The location of the radio-label of these melanin granules may be either in the melanin polymer or in the protein lattice of the melanosome. In either case, the granule itself is permanently labeled. The rapidity of melanin synthesis as indicated by radiotyrosine conversion (Fig. 5) may be greater than that occurring normally in mature pigment cells but provides an indication of the physiological potential of the developing pigment cells. As the melanin granules of black goldfish melanocytes and melanophores are normally retained until the death of the cell, retention of the label is indicative of cell viability. The rather slow loss of labeled melanin granules revealed the melanophore life span to be in excess of 1.75 years (639 days) (Fig. 6). The long life of these cells was not unexpected, as lysed melanophores are only very rarely found in the integument of normal black fish.

In contrast to the integumental melanin-synthesizing cells, the retinal pigment cells of hydroquinone-pretreated fish showed only a 5.3-fold

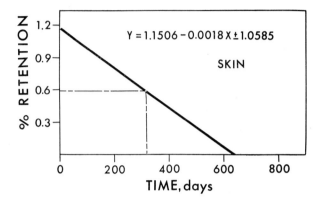

Fig. 6. Retention of L-tyrosine-^{14}C per mg integumental melanin in hydroquinone-pretreated black goldfish (normalized to 5 g) as a function of time post-peak uptake, 12 hours to 900 days. The biological half-life of the radiomelanin is 319 days. Complete loss of integumental radiomelanin occurs at 639 days.

increase in radiotyrosine conversion to melanin, compared to the normal fish. This increase is of importance as it revealed that hydroquinone affected the retinal pigment cells, but to a considerably lesser degree ($\sim10\%$) than the integumental pigment cells. The peak uptake of L-tyrosine-^{14}C, which was only 20% that of the integument, was achieved more slowly in the retina, 3 days, than in the integument (Fig. 7). It appears, therefore, that the retinal pigment cells were physiologically less active than their integumental counterparts in regard to both hydro-

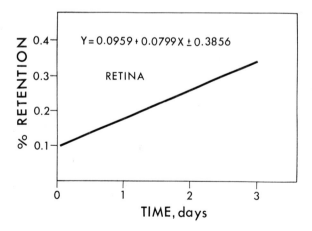

Fig. 7. Uptake of L-tyrosine-^{14}C per mg retinal melanin in hydroquinone-pretreated black goldfish (normalized to 5 g) as a function of time to 3-day peak uptake.

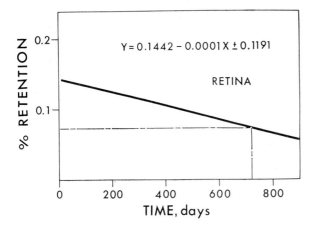

$$Y = 0.1442 - 0.0001X \pm 0.1191$$

RETINA

Fig. 8. Retention of L-tyrosine-^{14}C per mg retinal melanin in hydroquinone-pretreated black goldfish (normalized to 5 g) as a function of time post-peak uptake, 3–900 days. The biological half-life of the radiomelanin is 720 days. Complete loss of retinal radiomelanin is calculated to occur at 1,441 days.

quinone effects and pigment synthesis, despite similar radiohydroquinone turnover times. Little is known of melanin synthesis and life span of the pigment cells in the goldfish retina. Thus, the long retention of radio-melanin granules (1,441 days) may reflect the longevity of the cell or the very slow uptake and turnover (loss) of melanin in even longer-lived cells (Fig. 8). However, the relatively long life spans of the integumental (639 days) and the retinal (1,441 days) pigment cells, as derived from the present data, indicate that these cells of neural origin are normal despite the contained radiolabeled melanin granules.

At the present time, a number of advances have been made in regard to understanding the pigment cell in the model system we have used. The initial question remains unresolved, but in the course of approaching the question we have learned that the existing controls of the pigment cell may be many, even in a given species. The use of various pharmacological agents provides some guidance as to the possible chemical structural features which may affect pigment cells. This may be useful not only in development of potent depigmentary agents, but also in yielding insight into mechanism(s) of action and integumental factors affecting the pig-ment cells. Further, such agents furnish the opportunity to develop fairly uniform populations of pigment cells *in vivo* so that a variety of methods of study may be used. As the physiological regulation of melanocytes and melanophores is a complex of interrelated factors, the mode of action of hormones upon differentiation of these cells cannot be relegated to a relatively simple action upon propigment cells but must include also the

93

internal and external environments of the cell. In addition, as hormones in a living vertebrate rarely act independently of one another, it is clear that meaningful studies must consider the effects of concerted hormonal action in both chemoregulatory balance and imbalance. Such complexities offer new and exciting but rather sophisticated challenges to occupy the pigment cell workers of the future.

SUMMARY

1. *In vitro* studies of xanthic goldfish skin reveal that ACTH elicits melanin synthesis despite inhibition of protein synthesis. The propigment cell must contain the proteinaceous melanosomal lattice and the enzyme, tyrosinase, in an inhibited state. The hormone appears to act directly or indirectly upon release of enzymatic activity.

2. The various depigmentary agents studied act in a cytologically similar manner which is different from the effects of post-hypophysectomy hormonal deficiency.

3. Phylogenetic evaluation of tyrosinase activity in the vertebrate integument and melanomas reveals a number of different enzymatic control systems. Further, more than one such system may be present in a species.

4. The *in vitro* effects of depigmentary agents upon tyrosinase preparations reveal a number of similarities as well as differences. Some of the differences in effects upon the purified and crude tyrosinase preparation appear to be the result of tissue factors, thus, revealing also their importance in control of pigmentation.

5. Using radiolabeled hydroquinone as a depigmentary agent, it is essentially eliminated from the fish 48 hours post-injection. A major route of excretion is the liver via the bile. Melanin-containing tissue shows no preferential uptake of hydroquinone. The time of effective action is extremely short, so that after the rapid decrease in hydroquinone titer, melanogenesis may occur.

6. The differentiating new pigment cell population (melanocytes) after hydroquinone treatment rapidly converts administered L-tyrosine-^{14}C to melanin. The peak uptake occurs at 12 hours. The label is retained for 639 days, indicating this period to be the cell life span. Retinal pigment cells retain the label for 1,441 days.

REFERENCES

1. Chavin, W., Pituitary-adrenal control of melanization in xanthic goldfish, *Carassius, auratus* L. *J. Exp. Zool.*, **133,** 1 (1956).

2. Hu, F. and Chavin, W., Induction of melanogenesis *in vitro. Anat. Rec.*, **125,** 600 (1956).

3. Hu, F. and Chavin, W., Hormonal control of melanogenesis *in vitro*. *Anat. Rec.,* **131,** 568 (1958).

4. Hu, F. and Chavin, W., Hormonal stimulation of melanogenesis in tissue culture. *J. Invest. Derm.,* **34,** 377 (1960).

5. Tchen, T. T., Ammeraal, R. N., Kim, K., Wilson, C. M., Chavin, W., and Hu, F., Studies on the hormone-induced differentiation of melanoblasts into melanocytes in explants from xanthic goldfish tailfin. *Nat. Cancer Inst. Monogr.,* **13,** 67 (1964).

6. Kim, K., Tchen, T. T., Chavin, W., and Hu, F., Tyrosinase and melanogenesis in goldfish. *Fed. Proc.,* **23,** 223 (1961).

7. Chavin, W., Pituitary hormones in melanogenesis. In *Pigment Cell Biology* (M. Gordon, ed.), p. 63, Academic Press, New York.

8. Foote, C. D. and Tchen, T. T., Studies on the mechanism of hormonal induction of the melanoblast-melanocyte transformation in organ culture. *Exp. Cell Res.,* **47,** 596 (1967).

9. Chavin, W., Kim, K., and Tchen, T. T., Endocrine control of pigmentation. *Ann. N. Y. Acad. Sci.,* **100,** 678 (1963).

10. Chen, Y. M. and Chavin, W., Radiometric assay of tyrosinase and theoretical considerations of melanin formation. *Anal. Biochem.,* **13,** 234 (1965).

11. Chavin, W., Effects of hydroquinone and of hypophysectomy upon the pigment cells of black goldfish. *J. Pharmacol. Exp. Ther.,* **142,** 275 (1963).

12. Hu, F., The influence of certain hormones and chemicals on mammalian pigment cells. *J. Invest. Derm.,* **46,** 117 (1966).

13. Spencer, M. C., Topical use of hydroquinone for depigmentation. *J. Amer. Med. Ass.,* **194,** 962 (1965).

14. Chavin, W. and Schlesinger, W., A new series of depigmentational agents in the black goldfish. *Naturwissenschaften,* **53,** 163 (1966).

15. Chavin, W. and Schlesinger, W., Some potent depigmentary agents in the black goldfish. *Naturwissenschaften,* **53,** 413 (1966).

16. Chavin, W. and Schlesinger, W., Effect of melanin depigmentational agents upon normal pigment cells, melanoma and tyrosinase activity. In *Advances in Biology of Skin,* vol. 8, *The Pigmentary System* (W. Montagna and F. Hu, eds.), p. 421, Pergamon Press, Oxford (1967).

17. Frenk, E., Pathak, M. A., Szabó, G., and Fitzpatrick, T. B., Selective action of mercaptoethylamines on melanocytes in mammalian skin. *Arch. Derm.,* **97,** 465 (1968).

18. Bleehen, S. S., Pathak, M. A., Hori, Y., and Fitzpatrick, T. B., Depigmentation of skin with 4-isopropylcatechol, mercaptoamines, and other compounds. *J. Invest. Derm.,* **50,** 103 (1968).

19. Kim, K. and Tchen, T. T., Tyrosinase of the goldfish *Carassius auratus* L. I. Radioassay and properties of the enzyme. *Biochim. Biophys. Acta,* **59,** 569 (1962).

20. Chen, Y. M. and Chavin, W., unpublished observations.

21. Chavin, W., Fundamental aspects of morphological melanin color changes in vertebrate skin. *Amer. Zool.,* **9,** 505 (1969).

22. Chen, Y. M. and Chavin, W., Tyrosinase activity in goldfish skin. *Proc. Soc. Exp. Biol. Med.,* **121,** 497 (1966).

23. Chen, Y. M. and Chavin, W., Comparative biochemical aspects of integumental and tumor tyrosinase activity in vertebrate melanogenesis. In *Advances in Biology of Skin,* vol. 8, *The Pigmentary System* (W. Montagna and F. Hu, eds.), p. 253, Pergamon Press, Oxford (1967).

24. Chen, Y. M., Chavin, W., Ontogenetic alterations in tyrosinase activity. *J. Invest. Derm.,* **50,** 289 (1968).

25. Kim, K., Tchen, T. T., and Chavin, W., Tyrosinase of the goldfish, *Carassius auratus* L. II. Correlation of tyrosinase activity with pigmentation. *Biochim. Biophys. Acta,* **59,** 577 (1962).

26. Chen, Y. M. and Chavin, W., Melanogenesis in frog skin. *Experientia,* **24,** 332 (1968).

27. Chen, Y. M. and Chavin, W., Integumental tyrosinase activity in amphibians. *Experientia,* **24,** 31 (1968).

28. Chen, Y. M. and Chavin, W., Activation of skin tyrosinase. *Experientia,* **24,** 550 (1968).

29. Chen, Y. M. and Chavin, W., Effects of depigmentary agents and related compounds upon *in vitro* tyrosinase activity. Proceedings of the Seventh International Pigment Cell Conference, Seattle, Washington (in press) (1969).

30. Chavin, W., Utility of a depigmentary agent in the study of some basic characteristics of melanin synthesizing cells in the integument and the retina. Proceedings of the Seventh International Pigment Cell Conference, Seattle, Washington (in press) (1969).

31. Denton, C. R., Lerner, A. B., and Fitzpatrick, T. B., Inhibition of melanin formation by chemical agents. *J. Invest. Derm.,* **18,** 119 (1952).

32. Ito, M. and Nakajo, A., Relationship between melanin production and dioxybenzenes. *Tohoku J. Exp. Med.,* **55,** 88 (1952).

33. Iijima, S., and Watanabe, K., Studies on DOPA reaction. II. Effect of chemicals on the reaction. *J. Invest. Derm.,* **28,** 1 (1957).

DISCUSSION

DR. LERNER: I would like to make two comments. In the work that you are doing on the effects of ACTH and MSH on goldfish, have you ever tried caffein in your system? The action of caffein in increasing cyclic AMP is different from that of ACTH or MSH, and all darken frog skin. The second comment concerns your tyrosinase assays. In the research of Dr. Joseph S. McGuire of our department on tyrosinase in frog skin before and after exposure to MSH, it was necessary that trypsin be present. Without trypsin the assays for tyrosinase are not consistent.

DR. CHAVIN: The answer to the first question is no. We have not used caffein. In regard to the preparation of the homogenate for tyrosinase assay, we are very fussy about such enzymes in the homogenate preparations. We routinely homogenize in phosphate buffer using a micro-Waring blender followed by a Potter-Elvehjem homogenizer. We use nothing else.

DR. LERNER: Maybe you should try and see what happens.

DR. CHAVIN: Well, this is an excellent suggestion, but with additional treatment we could not be sure of what we were doing to the enzyme itself. Perhaps in Dr. McGuire's case the enzyme was liberated, but where the enzyme is free (soluble fraction) after trypsinization it is possible that the enzyme may be inactivated. In addition, the effects of trypsin upon the assay system would also need to be determined as trypsin could not readily be removed. Further similar evaluation of trypsin inhibitors would be required. Therefore, we take the more conservative approach and simply use physical treatment in the preparation of the enzyme rather than chemical treatment in our tyrosinase studies.

DR. QUEVEDO: In view of the demonstrated autophagocytosis of melanosomes in melanoma cells, have you any evidence that would suggest that the loss of labeled tyrosine from goldfish skin is not due to cell death but rather the result of melanin degradation within lysosomes?

DR. CHAVIN: This is a good point, but if one looks at a normal, untreated goldfish skin biopsy, one rarely, if ever, sees free melanin granules. On rare occasions melanin granules may occur in macrophages.

DR. QUEVEDO: It is known now that melanocytes or melanoma cells can put their melanosomes into lysosomes within the cell and presumably destroy them via this mechanism. I was wondering if there is any evidence for this in the retina, or, for that matter, in the skin melanophores.

DR. CHAVIN: We have not examined the retinal or dermal pigment cells in the goldfish ultrastructurally nor are descriptions of lysosomes present in the literature dealing with goldfish melanophores. The retinal pigment cells of the goldfish remain to be described.

DR. BLOIS: Would you comment on the specificity of your assay system as it may distinguish between tyrosinase itself and other nonspecific hydroxylases? And are you aware of the effect of the quinones and sulfhydryl compounds you have used upon enzymes other than tyrosinase?

DR. CHAVIN: I will answer your second question first. Since we are not using pure enzyme preparations, we do not know the content of any of the other factors in these tissues which may affect the enzyme itself or which may affect the drug.

DR. BLOIS: So you do not know whether these drugs are acting on the enzyme or acting on an inhibitor or something else in the system?

DR. CHAVIN: That is correct. So I really cannot answer the second question except to agree that it is time to do some more work with these compounds and systems. In regard to your first question dealing with the assay, the assay has been shown to be specific. The tyrosinase measured in

97

the assay will not utilize aliphatic amino acids but may utilize some of the aromatic amino acids (histidine, phenylalanine, and tryptophane) but to a much smaller degree than tyrosine. After incubation with radio-tyrosine, hydrolysis of the end product with 6 N HCl does not result in the loss of radioactivity. Inhibitors of protein synthesis do not inhibit the reaction. The reaction is both dopa dependent and oxygen dependent. Also copper binding and reducing agents inhibit the reaction.

DR. BLOIS: The assay consists of converting tyrosine to an insoluble product that you then count?

DR. CHAVIN: That is correct. The material that is counted can be considered to be melanin only by the indicated criteria, as no analytical procedure for identifying melanin itself is available. Nevertheless, the material counted meets all the known criteria required for melanin synthesis.

Genetic Regulation of Pigmentation in Mammals*

Walter C. Quevedo, Jr.

Division of Biological and Medical Sciences, Brown University, Providence, Rhode Island, U.S.A.

I. INTRODUCTION

Central to a consideration of the genetic regulation of melanin pigmentation in present-day mammals is the recognition that they constitute a highly diversified group of organisms. Mammals originated from reptilian ancestors early in the Mesozoic era of geological time and most of the various orders of placental mammals arose by evolutionary radiations during the late Cretaceous of the Mesozoic through the Paleocene of the succeeding Cenozoic era.[1,2] It is clear that the great majority of the existing orders of placental mammals have been taxonomically distinct for at least the past 54 million years. During this period of time ample opportunity has existed for gene mutation and natural selection to produce variations in the manner in which pigmentation of the skin and hair is accomplished among different mammalian species. Pigment cell workers in the field of comparative genetics have as one of their major goals the determination of the extent to which specific pigment genes are broadly or narrowly distributed among different groups of mammals.[3] Since it is not possible to make direct genetic tests by crossbreeding remotely related

* In memory of Dr. Margaret M. Dickie, who died on July 4, 1969. Various phases of the author's work reported herein were supported by U.S. Public Health Research Grant No. CA 06097 from the National Cancer Institute, Training Grant GM-00582 from the Division of General Medical Sciences, United States Public Health Service, and American Cancer Society Grant PS-42.

mammals, arguments for gene homology, i.e., genic identity through common evolutionary derivation, must largely be based on similarities in the developmental pathways by which they are expressed.

Detailed inquiries into the developmental genetics (phenogenetics) of melanin pigmentation have been largely restricted to rodents, most particularly to the house mouse *(Mus musculus)*, where more than 70 genes influencing melanin pigmentation are known to exist at approximately 40 loci.[4,5] Creative utilization of the diverse pigmented stocks of mice in experimental research has led to an increasingly clearer understanding of how genes act in the development and maintenance of ocular and integumentary pigmentation. Although considerably less is known about the genetic regulation of melanocyte performance in man, the total knowledge of melanin pigmentation in man is the clearest for mammals other than rodents. Certain hereditary defects of human pigmentation clearly parallel those found in rodents and may constitute evidence for gene homology.

The present paper outlines some recent advances in the study of the phenogenetics of melanin pigmentation in mice, particularly as they help to elucidate the form and function as well as evolution of pigmentary systems in man and other mammals. Recently, two more comprehensive reviews of the control of color in mammals have been published elsewhere.[6,7]

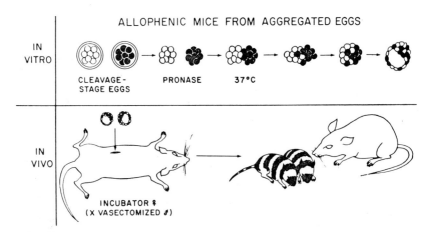

Fig. 1. Diagram of procedures followed by Mintz in constructing mosaic mice by combining early embryos of differing genotypes. The combination of embryos of contrasting color types produces adults with a striped hair coat. It appears that all of the melanocytes in the adult are derived from 34 stem melanoblasts which originate in the neural crest of the embryo. Reproduced with permission.[9]

II. ONTOGENY OF THE MELANOCYTE SYSTEM

In mammals, the melanocytes of the skin and hair are derived from melanoblasts which migrate from the embryonic neural crest early in development.[8] Those of the eye have a dual origin involving the optic cup (retinal melanocytes) and the neural crest (uveal melanocytes). Evidence indicates that a small number of primary melanoblasts from the neural crest undergo proliferation and differentiation to pigment local regions of the skin (Fig. 1).[9] The differentiation of melanoblasts into melanocytes is signaled by the synthesis of specialized melanin-containing organelles (melanosomes). This cellular transformation is influenced first by the genotype of the melanoblast which sets the minimal environmental requirements necessary for differentiation and, second, by the adequacy of the local tissue environment which may be subject to change during development. Various types of heritable "white-spotting" of the pelage in mice trace their developmental origin to either a neural crest defect or hostility of the local tissue environment to melanoblast differentiation and/or survival.[10, 11]

III. BIOSYNTHESIS OF MELANOSOMES

A. *General considerations*

The ultrastructural events related to melanosome synthesis within mouse melanocytes have been outlined by Moyer[12, 13] and Rittenhouse.[14, 15] The melanosome is considered to be a membrane-limited, round to elliptical cytoplasmic organelle in which a matrix consisting of either a single spiralized sheet or several concentrically arranged sheets of structural and enzymic (tyrosinase) protein serves as a scaffold upon which melanin is deposited.[7] The union of matrix proteins and the melanin polymer formed from tyrosine in the presence of tyrosinase gives rise to the melanoprotein identified on chemical analysis of melanosomes.[16, 17] The precise distribution of tyrosinase within melanosomes as well as the molecular composition of the matrix remains uncertain. There is some debate as to whether tyrosinase is structurally integrated into the protein matrix or is limited to the outer membrane of the melanosome.[18]

Genes at a number of loci in the mouse are known to influence specific ultrastructural features of melanosomes (Fig. 2).[12-15] To account for the diversity of heritable differences in melanosomes, it has been proposed that tyrosinase and an array of structural proteins become associated to form the melanosome matrix.[13, 19] Analytical methods have yet to be devised which will reveal the extent to which melanosomal proteins are unequivocally divisible into structural and enzymic varieties.

Moyer[12] and Rittenhouse[14] have suggested an involvement of free

101

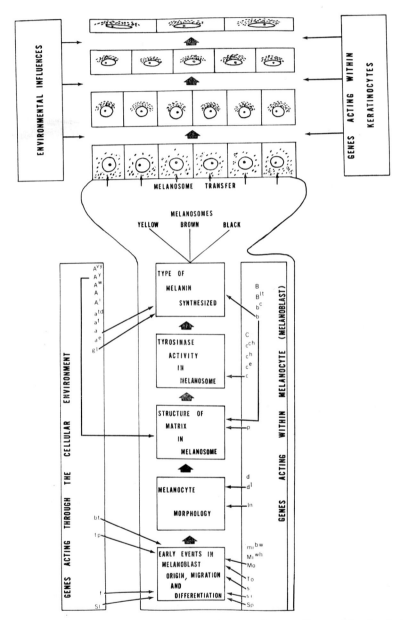

Fig. 2. Probable times and sites of gene action on melanocyte form and function in the house mouse. Some genes act directly through the genome of the melanocyte, others by way of the tissue environment. As illustrated, an epidermal melanocyte together with an associated pool of keratinocytes forms an *epidermal melanin unit,* a fundamental multicellular system programed for the synthesis and transport of melanosomes.[6]

102

ribosomes in the synthesis of melanosomal proteins within mouse melano-cytes. The bulk of evidence from other organisms, however, indicates that melanosomal proteins are synthesized during translation of the gene-specified sequences of nucleotides in messenger RNA (mRNA) on the ribosomes of the rough endoplasmic reticulum (Fig. 3). From there they are transferred to the region of the Golgi apparatus where melanosomes are formed by the sequestering of appropriate proteins within membrane-limited vesicles.[20] The "unit fibers" of melanosomal protein aggregate into "compound fibers." In turn, compound fibers are thought to become aligned in parallel and to develop cross-linkages at regular intervals with the result that a lattice-like matrix is formed within the melanosome. In the presence of active tyrosinase, melanin is deposited on and between the layers of matrix. The details of melanosome formation are no doubt more complex than outlined, for new evidence suggests that early stages of melanin synthesis may be initiated before formation of the matrix and perhaps even prior to the formation of the melanosome.[18, 21] Electron microscopic studies on mouse and human melanoma cells also indicate that all melanosomes do not undergo their major development as inde-pendent vesicles (Fig. 3). On the contrary, they reveal that sequestration of melanosomal proteins may occur within dilatations of the cisternae of smooth endoplasmic reticulum adjacent to the Golgi apparatus.[22, 23] Such melanosomes appear to become detached from continuity with the Golgi-associated endoplasmic reticulum after the matrix is formed but before melanin deposition has been completed.[23] The genetic basis for possible variations in the sites of melanosome formation within normal melanocytes and melanoma cells remains to be established.

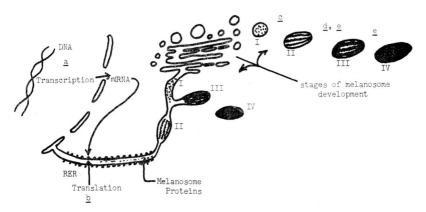

Fig. 3. Diagrammatic representation of possible alternative modes of melanosome for-mation within melanocytes and melanoma cells. See text for an explanation.

103

B. *Genetic control*

Comparative analyses of color mutants of the mouse have provided a tentative outline of the step by step action of genes in the origin and development of melanosomes.[3-6,12-15,19] Alleles at the *a, b, c, d,* and *p* loci appear to influence the affinities of the fibers which associate during the formation of the melanosome matrix. Specific departures from the normal pattern of aggregation of melanosomal fibers into higherorder structures parallels appropriate allelic substitution at these loci. To illustrate, in lethal yellow (A^y/a) mice, the unit fibers do not appear to aggregate properly into compound fibers within the melanosomes.[13] In the absence of a definitive matrix, yellow pigment (pheomelanin) is deposited in a tangled mat of very fine fibers. In black mice manifesting pink-eyed dilution (p/p), the compound fibers often fail to form parallel arrays and to cross-link properly; the result is an abnormal matrix which contains fibers oriented in several different planes.[13] Reduced amounts of eumelanin are found within the melanosomes. Although Sidman and Pearlstein[24] were able to increase melanin deposition within p/p melanosomes by providing tyrosine *in vitro*, they did not observe any correction in the irregularities of matrix structure.

Alleles at the *b* locus appear to influence the structure of melanosomes as well as the polymerization of melanin.[14] The melanosomes of black (B/B) mice may differ from the melanosomes of brown (b/b) mice not only in the type of eumelanin they contain (black vs. brown) but also, at least in some cases, in the organization of the matrix.

The recessive beige gene (bg) in mice leads to giant melanosome formation within melanocytes apparently by conditioning the fusion of melanosomes at early stages in their development.[25,26] The pigmentary findings on the beige mutant of the mouse are comparable to those observed in Chediak-Higashi syndrome which is inherited as an autosomal recessive trait in man.[27] The giant melanosomes found in melanocytes of humans afflicted with Chediak-Higashi syndrome appear to represent both abnormally large single melanosomes and aggregates of melanosomes.[28,29]

The structural gene for tyrosinase appears to be located at the albino (c) locus.[4,19,30] The quality of tyrosinase is influenced by allelic substitutions at this locus.[4,30] Maximum tyrosinase activity for a given genotype is found in full color (C/C) mice. In mice homozygous for the albino allele *c,* the lowest member of the allelic series, active tyrosinase is absent and the animals are totally devoid of melanin. Except for their lack of melanin, melanosomes of albino (c/c) mice appear to be normal in structure.[13] The aggregation of unit and compound fibers during matrix formation apparently proceeds normally, but tyrosinase either is not present or is not activated. Significantly, radiation-induced mutations of C to c have been reported to result in structural aberrations in melanosomes as well as in

the loss of tyrosinase activity.[31] It is possible that differences between expression of the normal c allele and those induced by radiations may indicate
the degree to which nucleotide sequences have been changed within the
DNA of the c locus; one consequence would be limited or extensive
alterations in the primary structure of the protein specified by the c locus.
The normal c allele may elicit only a minor change in the tyrosinase
molecule which, although sufficient to cause loss of enzymic activity, does
not compromise its structural contribution during melanosome formation. On the other hand, the radiation-induced "c-like" alleles may involve more major changes in nucleotide sequence at the c locus resulting
in tyrosinase molecules defective in both their enzymic and structural
functions.

It is highly significant that Witkop et al.[32] have now clearly established
that two types of recessive oculocutaneous albinism occur in man. Both
traits are inherited as autosomal recessives and are controlled by nonallelic genes. Tyrosinase-negative oculocutaneous albinism is characterized by the presence of nonmelanized melanosomes within melanocytes
and a lack of histochemically demonstrable tyrosinase activity within
plucked hair bulbs incubated in tyrosine-containing solutions. As in the
albino mouse, human tyrosinase-negative oculocutaneous albinism appears to involve a defect in tyrosinase. In contrast, tyrosinase-positive
oculocutaneous albinism in man is defined by the existence of partially
melanized melanosomes within melanocytes and evidence of tyrosinase
activity within hypomelanotic hair bulbs incubated in tyrosine-containing
solutions. Tyrosinase-positive oculocutaneous albinism may result from
limitations in the availability of tyrosine within melanosomes.[32] In this
sense, the expression of the recessive gene for tyrosinase-positive oculocutaneous albinism appears to parallel that for pink-eyed dilution (p) in
mice. Abnormal matrices have been reported to be present in some
melanosomes of human oculocutaneous albinos, but the "tyrosinase
status" of the subjects examined was not determined.[33]

It is possible that genes of the several loci known to influence melanosome morphology do so by regulating the structure of polypeptides which
become associated with tyrosinase either prior to or during melanosome
formation. Recent demonstrations of heritable differences in the multiple
forms of tyrosinase within the follicular melanocytes of mice lend some
support to this view.

It is now clear that tyrosinase extracted from mammalian melanocytes
is separable into several discrete forms by acrylamide-gel electrophoresis.[34-37] The location of tyrosinase within acrylamide-gels is evident on
incubation of the gels in a buffered solution of L-3,4-dihydroxyphenyla-
lanine (dopa). Dopa-melanin is deposited in the form of narrow dark
bands in regions of the gel where active tyrosinase is present (Fig. 4).

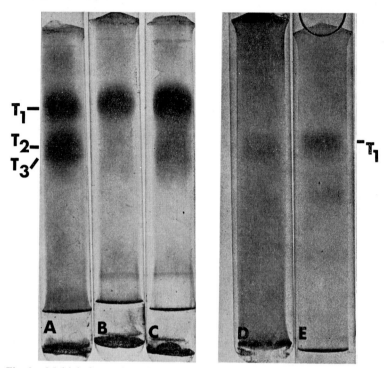

Fig. 4. Multiple forms of tyrosinase from hair-bulb melanocytes of mice as visualized in acrylamide-gels. A: black (B/B) 10 days after plucking; B: black (B/B) 15 days after plucking; C: pink-eyed dilution (p/p) 10 days after plucking; D: lethal yellow (A^y/a), 10 days after plucking; E: viable yellow (A^{vy}/a), "clear yellow" phenotype, 10 days after plucking.

Allelic substitutions at the a, b, c, d, bg, ru, and p loci in mice indicate that the multiple forms of tyrosinase demonstrable in extracts of pigmented hair bulbs are under genetic control.[34-36] For the comparative genetic studies in the mouse, all hair-bulb preparations have been obtained between the 8th and 10th day of the approximately 21-day-long hair growth cycle. Irrespective of genotype, a maximum of three electrophoretically separable forms of tyrosinase have been demonstrated in any given hair-bulb preparation. The fastest migrating tyrosinase band has been designated as T_1 $(R_x \cong 0.70)$, the intermediate one as T_2 $(R_x \cong 0.57)$, and the slowest one as T_3 $(R_x \cong 0.54)$. Consistent with other impressive evidence that the c locus contains the structural gene for tyrosinase, no dopa-reactive bands are found after electrophoresis of acrylamide-gels treated with extracts of hair bulbs from albino (c/c) mice.

All combinations of alleles at the b locus, with the exception of brown (b/b), do not perceptibly modify the typical tyrosinase pattern consisting of three uniformly darkened bands of dopa-melanin. In the case of the brown phenotype, although the T_1 band exhibits normal darkening, considerably less dopa-melanin is deposited at the T_2 and T_3 positions.

Alleles at the agouti (a) locus also have a profound influence on the patterns of tyrosinase resolvable by acrylamide-gel electrophoresis. Agouti (A/A) hair follicles yield a typical tribanded pattern when assayed on the 10th day of the hair growth cycle, a period of eumelanin synthesis within follicular melanocytes.[34,35] Genetically identical viable yellow ($A^{vy}/-$) mice vary from "clear yellow" to "agouti" in coat color.[38] Viable yellow mice with "mottled" coat colors intermediate between the clear yellow and agouti phenotypes are also commonly observed. Hair bulbs from agouti viable yellow (A^{vy}/a) mice reveal a typical tribanded pattern of tyrosinase activity when assayed on the 10th day of the hair growth cycle, a time, as in A mice, when eumelanin is synthesized within follicular melanocytes. Hair-bulb preparations from mottled (A^{vy}/a) viable yellow mice contain a mixture of follicles engaged either in pheomelanin or eumelanin synthesis. The T_1 band isolated from such hair-bulb preparations appears to be normal in dopa-reactivity, but T_2 and T_3 are definitely reduced. In clear yellow (A^{vy}/a) viable yellow mice where almost all of the hair bulbs are engaged in pheomelanin synthesis, T_1 is reduced in intensity and T_2 and T_3 are barely detectable (Fig. 4). In lethal yellow (A^y/a) mice, all hair follicles produce pheomelanin and the tyrosinase pattern is characterized by a weak T_1 band with no discernible dopa-melanin deposited at the T_2 and T_3 positions (Fig. 4). In general, it appears that the production of pheomelanin under the influence of A^{vy} and A^y alleles is associated with a reduction in dopa-reactivity of T_1 and a loss in reactivity of T_2 and T_3.[35] In black mice homozygous for p or d, the T_2 and T_3 bands are also greatly reduced (Fig. 4).

As in the case of the b, d, and p genes, A^y is associated with morphological defects in melanosomes.[13] Alleles at a, b, d, and p loci may influence the composition of one or more structural components which combine with tyrosinase either prior to or during the formation of the melanosome matrix with resultant influences on both tyrosinase patterns and melanosome structure. Possibly some of the genes influence melanosome structure by modifying the activity of other enzymes co-requisite for melanosome assembly.

It is noteworthy that multiple forms of tyrosinase are found in a variety of mammals, including man.[39] Like the mouse, the black rat possesses a T_1, T_2, and T_3 system within hair follicles. However, there is an additional band (T_0) which migrates slightly more rapidly than T_1. Black rabbits are characterized by a T_1 band and a single "T_2–T_3" band, both of which are at best lightly darkened on incubation in dopa. Two bands corresponding to T_1 and T_2–T_3 are found in hair-bulb preparations obtained from gerbils and hamsters at a time when follicular melanocytes are engaged in the production of eumelanin. For the gerbil, it is now clear that the T_2–T_3 band is greatly reduced during pheomelanin synthesis. The results suggest that the T_1 band has approximately the same mobility in all species examined, including man. The T_2–T_3 tyrosinase(s) are more variable in expression and mobility. The constancy of T_1 in a broad variety of animals suggests that this form of tyrosinase may stand very close to the primary action of the albino (c) locus; the c locus is known to be widely distributed among mammals.[3] T_1 would seem to represent a conservative form of tyrosinase that has remained relatively unmodified throughout the long evolutionary history of many mammals; T_0 and T_2–T_3 are possibly indicative of broader evolutionary experimentation. The multiple forms of tyrosinase present exciting possibilities for future studies on the evolutionary genetics of mammalian pigmentation.

The functional significance of the multiple forms of tyrosinase is still obscure. The tyrosinase pattern has been demonstrated to undergo change during the hair growth cycle in black (C57BL) mice.[36] There is no definite evidence, however, that the observed variations in the multiple forms of tyrosinase reflect functional adaptations of melanocytes to cyclic changes in the biochemistry and physiology of the skin. It would be of considerable interest to determine whether T_1, T_2, and T_3 are each tailored to function maximally under different conditions specified by the hair-follicle environment.

IV. REGULATION OF MELANOCYTE FUNCTION

A number of lines of indirect evidence lead toward the conclusion that melanogenically inactive cells of the melanocyte lineage (melanoblasts?) exist in the epidermis and quiescent hair follicles of mice.[40–42] Zelickson and Mottaz[43] have identified "indeterminate" cells in human epidermis which they speculate may be premelanocytes. Mishima and Widlan[44] and Schroeder[45] have demonstrated the presence of relatively inactive melanocytes in the epidermis and gingiva of man.

As illustrated in Fig. 3, melanogenesis within melanocytes, in theory, may be inhibited at several key steps; it might be inhibited at the level of (a) pigmentary genes (transcription), (b) protein synthesis (translation), (c) assembly of protein subunits into melanosomes, (d) deposition of melanin within melanosomes as a result of a tyrosinase defect/inhibition, or (e) deposition of melanin owing to the lack of available substrate. Melanocytes producing only a few melanosomes may reflect an incomplete inhibition of one or more of these processes. The failure of relatively inactive gingival melanocytes to synthesize melanin within melanosomes[45] may involve an inhibition of tyrosinase or a lack of sufficient amounts of tyrosine.

Whittaker[46] has proposed that the reduction in melanin synthesis within rapidly growing cultures of chick retinal melanocytes is not the result of the repression of the structural gene for tyrosinase, but rather the consequence of dilution of tyrosinase synthesis through the failure of messenger RNA (mRNA) coding for tyrosinase to compete successfully with a flood of "growth protein" mRNA for translation on the ribosomes, the seats of cellular protein synthesis. Evidence for the presence of a specific diffusible agent which represses the function of genes related to melanogenesis has been submitted by Davidson et al.[47, 48] They reported on the specific loss of melanin-forming capacity in hybrid cells formed by the union of Syrian hamster melanoma cells and nonmelanocytic mouse cells *in vitro*. The chemical composition and mode of action of the postu- lated repressor substance has yet to be determined. It is possible that in normal skin, diffusible repressor substances are produced by or penetrate melanocytes at certain times (e.g., telogen stage of the hair cycle) and act to suppress further melanogenesis. The shift from the synthesis of pheo- melanin to eumelanin within agouti hair follicles may also involve the action of repressor substances on specific genes.[49]

Mitotic activity of melanocytes and keratinocytes appears to be under "negative control".[50, 51] Bullough and Laurence[51] have reported the existence of a melanocyte chalone in normal pig skin and rodent melano- mas which suppresses mitotic activity of melanoma cells and, by inference, which may also regulate the turnover of normal epidermal melanocytes. The evidence suggests that melanocytes synthesize a specific chalone which is different from the chalone produced by keratinocytes.[50, 51]

The role of chalones in the interactions of keratinocytes and melano- cytes can only be speculated on at the moment. The melanogenic activity of melanocytes may be linked to increased mitotic activity of keratinocytes. The loss of melanogenic activity in the interfollicular epidermal melano- cytes of neonatal mice is associated with the thinning of the epidermis during the first three weeks of life.[40] Conversely, chemical and physical agents which stimulate melanogenesis within melanocytes also evoke hyperplasia of keratinocytes.[40, 52] Melanocytes normally initiate melano- genesis within hair follicles paralleling the onset of keratinocyte prolifera- tion.[53] The molecular events at the level of genes relevant to melanin synthesis remain unclarified. It may be of significance that follicular melanocytes undergo proliferation at the onset of the hair growth cycle in the normal skin of mice;[54, 55] similarly, melanocytes and keratinocytes proliferate in the wounded epidermis of primates.[56] It is not clear whether the proliferation of keratinocytes and melanocytes is causally related.

Frenk and Schellhorn[57] have found that the ratio of melanocytes to keratinocytes is a constant throughout the human epidermis, suggesting some interaction between the two populations which may involve their respective chalones.

During suntanning of the skin, there is an increase in the number of melanogenic melanocytes observed at the light microscope level.[58] They appear to be recruited from an extensive population of melanocytes which is normally relatively inactive melanogenically.[44,58] In addition, some limited division of melanocytes may also contribute to the increase observed.[44] The mechanism by which melanogenesis is induced by UV remains to be elucidated.

V. ORIGIN AND SIGNIFICANCE OF RACIAL DIFFERENCES IN HUMAN PIGMENTATION

Significant disagreement exists over the adaptive significance of racial differences in melanin pigmentation in man.[59-62] Much attention has been directed toward the possibility that the marked melanization of the epidermis in tropical Negroids and Australoids is an evolutionary adaptation providing protection against the deleterious effects of UV which may involve frank damage to the skin or elevation of vitamin D synthesis to harmful levels.[60,62] The reduced amounts of melanin in the skin of Caucasoids may reflect a decline in selection pressure for melanized skin during the early migrations of man into more northern latitudes, where UV posed less of a hazard for the skin. It is also conceivable that decreased melanization was a strict requirement for increased efficiency of vitamin D synthesis, owing to reduced solar UV in northern latitudes. The reversible tanning mechanism particularly evident in Caucasoids may indicate an adaptation to seasonal highs and lows in solar UV reaching the earth's surface in temperate zones. It is important to note that arguments for the significance of the role of vitamin D in the origin of racial differences in human pigmentation have not been unchallenged.[63]

In man the transport of melanosomes within keratinocytes varies with race and presumably reflects underlying genetic mechanisms of control. The melanosomes of Caucasoids and Mongoloids form aggregates within lysosome-like vesicles of keratinocytes.[64] There is the suggestion that at least some melanosomes may be broken down within such vesicles and melanin fragments released into the keratinocyte cytoplasm[7,65,66] (Fitzpatrick et al., personal communication). The melanosomes of Negroids and Australoids generally do not form aggregates within keratinocytes but rather are transported as single entities.[64,67] Melanosomes have been identified in the cornified epidermis of Negroids, Caucasoids, Mongoloids, and Australoids.[67-70] If it should be shown that melanosomes are broken down to a greater extent within the keratinocytes of Caucasoids and Mongoloids, the process may be of considerable physiological significance. A variety of interpretations might be entertained at the moment. One possibility is that melanin is resistant to the action of the lysosomal hydrolases but that the fibers of the matrix are susceptible at least at certain sites. One result of melanosome degradation could be the release of melanin fragments which might greatly extend the protection against UV afforded by a limited number of melanosomes. The melanin fragments might spread throughout the cornifying keratinocytes tending to maximize the surface area of "melanin cover." The Negroid and Australoid with a greater inherent capacity for melanosome production might not require melanosome degradation to achieve adequate UV protection.[68,69] An early genetic selection toward hypopigmentation of the skin in certain human populations, possibly as a requirement for increased efficiency of vitamin D synthesis, may have proceeded too far for some unknown reason, rendering the skin sensitive to UV. Melanosome degradation might then have been derived secondarily as a means of enhancing the protective value of a limited amount of melanin synthesized by melanocytes in accordance with their genome. Selection would have been in favor of those additional genes which programed limited melanosome degradation with resulting increased efficiency of the suntanning mechanism.

An alternative possibility is that natural selection has led to two hereditary mechanisms which reduce the amount of "physiologically effective" melanin in the epidermis of Caucasoids and Mongoloids. The first has acted by lowering the amount of melanin synthesized by melanocytes, the second by reducing the opacity of the epidermis through the aggregation and perhaps breakdown of melanosomes within keratinocytes. This may have been a requirement for increased efficiency in vitamin D synthesis paralleling a decline in available UV. Thus there is a suggestion that genes of at least two types may regulate the melanization of the epidermis: those that regulate the numbers of melanosomes synthesized and those that regulate melanosome aggregation and perhaps degradation in keratinocytes. Their general effects may be additive but involving two rather than one pathway of cytological control.

The alternative hypotheses relating to racial differences in human pigmentation are currently under investigation. Further comparative studies on the biology of melanin pigmentation in mammals should clarify, not only the mechanisms that currently govern melanocyte form and function in man, but also the nature of the historic processes by which they came to be established.

REFERENCES

1. Romer, A. S., *Vertebrate Paleontology* (3d ed.), University of Chicago Press, Chicago (1966).
2. McKenna, M. G., The origin and early differentiation of Therian mammals. *Ann. N. Y. Acad. Sci.*, **167,** 217 (1969).
3. Searle, A. G., *Comparative Genetics of Coat Colour in Mammals*. Academic Press, New York (1968).
4. Wolfe, H. G. and Coleman, D. L., Pigmentation. In *Biology of the Laboratory Mouse* (E. L. Green, ed.), p. 405 (2d ed.), McGraw-Hill, New York (1966).
5. Quevedo, W. C., Jr., Genetics of mammalian pigmentation. In *The Biologic Effects of Ultraviolet Radiation (With Emphasis on the Skin)*, (F. Urbach, ed.), p. 315, Pergamon Press, Oxford (1969).
6. Quevedo, W. C., Jr., The control of color in mammals. *Amer. Zool,* **9,** 531 (1969).
7. Breathnach, A. S., Normal and abnormal melanin pigmentation of the skin. In *Pigments in Pathology* (M. Wolman, ed.), p. 353, Academic Press, New York (1969).
8. Rawles, M. E., Origin of pigment cells from the neural crest in the mouse embryo. *Physiol. Zool.,* **20,** 248 (1947).
9. Mintz, B., Gene control of mammalian pigmentary differentiation. I. Clonal origin of melanocytes. *Proc. Nat. Acad. Sci. U.S.A.,* **58,** 344 (1967).
10. Mayer, T. C., Temporal skin factors influencing the development of melanoblasts in piebald mice. *J. Exp. Zool.,* **166,** 397 (1967).
11. Mayer, T. C. and Green, M. C., An experimental analysis of the pigment defect caused by mutations at the W and Sl loci in mice. *Develop. Biol.,* **18,** 62 (1968).
12. Moyer, F. H., Genetic effects on melanosome fine structure and ontogeny in normal and malignant cells. *Ann. N. Y. Acad. Sci.,* **100,** 584 (1963).
13. Moyer, F. H., Genetic variations in the fine structure and ontogeny of mouse melanin granules. *Amer. Zool.,* **6,** 43 (1966).
14. Rittenhouse, E., Genetic effects on fine structure and development of pigment granules in mouse hair bulb melanocytes. I. The b and d loci. *Develop. Biol.,* **17,** 351 (1968).
15. Rittenhouse, E., Genetic effects on fine structure and development of pigment granules in mouse hair bulb melanocytes. II. The c and p loci, and $ddpp$ interaction. *Develop. Biol.,* **17,** 366 (1968).
16. Zelickson, A. S., *Ultrastructure of Normal and Abnormal Skin*. Lea and Febiger, Philadelphia (1967).
17. Duchon, J., Fitzpatrick, T. B., and Seiji, M., Melanin 1968: Some definitions and problems. In *The Year Book of Dermatology*, 1967-1968 Year Book Series, no. 6 (A. W. Kopf, and R. Andrade, eds.), p. 6, Year Book Medical Publishers, Chicago.
18. Toda, K. and Fitzpatrick, T. B., Ultrastructural and biochemical studies of the formation of melanosomes in the embryonic chick retinal pigment epithelium (ECRPE). *J. Invest. Derm.,* **54,** 99 (1970) (abstract).
19. Foster, M., Mammalian pigment genetics. *Advances Genet.,* **13,** 311 (1965).
20. Fitzpatrick, T. B., Mammalian melanin biosynthesis. *Trans. St. John Hosp. Derm. Soc.,* **51,** 1 (1965).
21. Zelickson, A. S., Hirsch, H. M., and Hartmann, J. F., Melanogenesis: An autoradiographic study at the ultrastructural level. *J. Invest. Derm.,* **43,** 327 (1964).
22. Novikoff, A. B., Albala, A., and Biempica, L., Ultrastructural and cytochemical observations on B-16 and Harding-Passey mouse melanomas. *J. Histochem. Cytochem.,* **16,** 299 (1968).

23. Maul, G. G., Golgi-melanosome relationship in human melanoma *in vitro*. *J. Ultrastruct. Res.*, **26**, 163 (1969).

24. Sidman, R. L. and Pearlstein, R., Pink-eyed dilution (*p*) gene in rodents: Increased pigmentation in tissue culture. *Develop. Biol.*, **12**, 93 (1965).

25. Pierro, L. J., Pigment granule formation in slate, a coat color mutant in the mouse. *Anat. Rec.*, **146**, 365 (1963).

26. Lutzner, M., Ultrastructure of giant melanin granules in the beige mouse during ontogeny. *J. Invest. Derm.*, **54**, 91 (1970) (abstract).

27. Leader, R. W., The Chediak-Higashi anomaly—an evolutionary concept of disease. *Nat. Cancer Inst. Monogr.*, **32**, 337 (1969).

28. Windhorst, D. B., Zelickson, A. S., and Good, R. A., Chediak-Higashi syndrome: Hereditary gigantism of cytoplasmic organelles. *Science*, **151**, 81 (1966).

29. Zelickson, A. S., Windhorst, D. B., White, J. G., and Good, R. A., The Chediak-Higashi syndrome: Formation of giant melanosomes and the basis of hypopigmentation. *J. Invest. Derm.*, **49**, 575 (1967).

30. Coleman, D. L., Effect of genic substitution on the incorporation of tyrosine into the melanin of mouse skin. *Arch. Biochem.*, **96**, 562 (1962).

31. Rittenhouse, E., Effects of four radiation-induced lethal alleles at the albino locus on the fine structure of melanin granules in the mouse. *J. Invest. Derm.*, **54**, 96 (1970) (abstract).

32. Witkop, C. J., Albinism. In *Advances in Human Genetics*, vol. 2 (in press).

33. Breathnach, A. S., and Robins, J., Ultrastructure of melanocytes and melanosomes in human oculo-cutaneous albinism. *J. Anat.*, **103**, 387 (1968).

34. Holstein, T. J., Burnett, J. B., and Quevedo, W. C., Jr., Genetic regulation of multiple forms of tyrosinase in mice: Action of *a* and *b* loci. *Proc. Soc. Exp. Biol. Med.*, **126**, 415 (1967).

35. Holstein, T. J., Genetic control and developmental variations of the multiple forms of tyrosinase in the melanocytes of mice. Ph.D. diss., Brown University (1969).

36. Burnett, J. B., Holstein, T. J., and Quevedo, W. C., Jr., Electrophoretic variations of tyrosinase in follicular melanocytes during the hair growth cycle in mice. *J. Exp. Zool.*, **171**, 369 (1969).

37. Burnett, J. B. and Seiler, H., Multiple forms of tyrosinase from human melanoma. *J. Invest. Derm.*, **52**, 199 (1969).

38. Wolff, G. L., Body composition and coat color correlation in different phenotypes of "viable yellow" mice. *Science*, **147**, 1145 (1965).

39. Holstein, T. J., Burnett, J. B., and Quevedo, W. C., Jr., The developmental genetics of tyrosinase within the melanocytes of mice. *J. Invest. Derm.*, **54**, 88 (1970) (abstract).

40. Quevedo, W. C., Jr., Youle, M. C., Rovee, D. T., and Bienieki, T. C., The developmental fate of melanocytes in murine skin. In *Structure and Control of the Melanocyte* (G. Della Porta and O. Mühlbock, eds.), p. 228, Springer-Verlag, Berlin (1966).

41. Quevedo, W. C., Jr., and Smith, J., Electron microscope observations on the postnatal "loss" of interfollicular epidermal melanocytes in mice. *J. Cell Biol.*, **39**, 108a (1968).

42. Silver, A. F., Chase, H. B., and Potten, C. S., Melanocyte precursor cells in the hair follicle germ during the dormant state (telogen). *Experientia*, **25**, 299 (1969).

43. Zelickson, A. S. and Mottaz, J. H., Epidermal dendritic cells. *Arch. Derm.*, **98**, 652 (1968).

44. Mishima, Y. and Widlan, S., Enzymically active and inactive melanocyte populations and ultraviolet irradiation: Combined dopa-premelanin reaction and electron microscopy. *J. Invest. Derm.*, **49**, 273 (1967).

45. Schroeder, H. E., Melanin containing organelles in cells of the human gingiva. I. Epithelial melanocytes. *J. Periodont.*, **4,** 1 (1969).

46. Whittaker, J. R., Translational competition as a possible basis of modulation in retinal pigment cell cultures. *J. Exp. Zool.*, **169,** 143 (1968).

47. Davidson, R., Ephrussi, B., and Yamamoto, K., Regulation of melanin synthesis in mammalian cells, as studied by somatic hybridization. I. Evidence for negative control. *J. Cell. Physiol.*, **72,** 115 (1968).

48. Davidson, R. and Yamamoto, K., Regulation of melanin synthesis in mammalian cells, as studied by somatic hybridization. II. The level of regulation of 3,4-dihydroxyphenylalanine oxidase. *Proc. Nat. Acad. Sci. U.S.A.*, **60,** 894 (1968).

49. Takeuchi, T., Regulatory function of the agouti locus in the mouse melanocyte. *J. Invest. Derm.*, **54,** 98 (1970) (abstract).

50. Bullough, W. S. and Laurence, E. B., Control of mitosis in rabbit Vx2 epidermal tumours by means of the epidermal chalone. *Europ. J. Cancer,* **4,** 587 (1968).

51. Bullough, W. S. and Laurence, E. B., Control of mitosis in mouse and hamster melanomata by means of the melanocyte chalone. *Europ. J. Cancer,* **4,** 607 (1968).

52. Quevedo, W. C., Jr., The role of melanocytes in skin carcinogenesis. *Nat. Cancer Inst. Monogr.,* **10,** 561 (1963).

53. Chase, H. B., Rauch, H., and Smith, V. W., Critical stages of hair development and pigmentation in the mouse. *Physiol. Zool.,* **24,** 1 (1951).

54. Potten, C. S. and Howard, A., Radiation depigmentation of mouse hair: The influence of local tissue oxygen tension on radiosensitivity. *Radiat. Res.,* **38,** 65 (1969).

55. McGrath, E. P. and Quevedo, W. C., Jr., Genetic regulation of melanocyte function during hair growth in the mouse: Cellular events leading to "pigment clumping" within developing hairs. In *Biology of the Skin and Hair Growth* (A. G. Lyne, and B. F. Short, eds.), p. 727, Angus and Robertson, Sydney (1965).

56. Giacometti, L. and Montagna, W., The healing of skin wounds in primates. II. The proliferation of epidermal cell melanocytes. *J. Invest. Derm.,* **50,** 273 (1968).

57. Frenk, E. and Schellhorn, J. P., Zur Morphologie der epidermalen Melanineinheit. *Dermatologica,* **139,** 271 (1969).

58. Quevedo, W. C., Jr., Szabó, G., and Virks, J., Influence of age and UV on the populations of dopa-positive melanocytes in human skin. *J. Invest. Derm.,* **52,** 287 (1969).

59. Blum, H. F., Is sunlight a factor in the geographical distribution of human skin color? *Geog. Rev.,* **59,** 557 (1969).

60. Loomis, W. F., Skin pigment regulation of vitamin-D biosynthesis in man. *Science,* **157,** 501 (1967).

61. Wassermann, H. P., Human pigmentation and environmental adaptation. *Arch. Environ. Health,* **11,** 691 (1965).

62. Johnson, B. E., Daniels, F., Jr., and Magnus, I. A., Response of human skin to ultraviolet light. In *Photophysiology (Current Topics)*, vol. 4 (A. C. Giese, ed.), p. 139, Academic Press, New York (1968).

63. Blois, M. S., Blum, H. F., and Loomis, W. F., Vitamin D, sunlight, and natural selection. *Science,* **159,** 652 (1968).

64. Szabó, G., Gerald, A. B., Pathak, M. A., and Fitzpatrick, T. B., Racial differences in the fate of melanosomes in human epidermis. *Nature,* **222,** 1081 (1969).

65. Edwards, E. A. and Duntley, S. Q., The pigments and color of living human skin. *Amer. J. Anat.,* **65,** 1 (1939).

66. Prunieras, M., Interactions between keratinocytes and dendritic cells. *J. Invest. Derm.,* **52,** 1 (1969).

67. Mitchell, R. E., The skin of the Australian aborigine: A light and electronmicroscopical study. *Aust. J. Derm.*, **9,** 314 (1968).
68. Kligman, A. M., The biology of the stratum corneum. In *The Epidermis* (W. Montagna and W. C. Lobitz, Jr., eds.), p. 387, Academic Press, New York (1964).
69. Kligman, A. M., Comments on the stratum corneum. In *The Biologic Effects of Ultraviolet Radiation (with Emphasis on the Skin)* (F. Urbach, ed.), p. 165, Pergamon Press, Oxford (1969).
70. Keddie, F., and Sakai, D., Morphology of the horny cells of the superficial stratum corneum: Cell membranes and melanin granules. *J. Invest. Derm.*, **44,** 135 (1965).

DISCUSSION

DR. TAKEUCHI: How would you exclude the possibility that those three molecules are controlled by three different cistrons? Do you assume any functional difference among the three isozyme molecules?

DR. QUEVEDO: Since the c allele when homozygous results in a complete loss in tyrosinase activity, I think that it is unlikely that multiple cistrons coding for tyrosinase would explain the observed results. As yet we have no information as to possible functional differences that may exist between the multiple forms of tyrosinase.

DR. BAGNARA: It is difficult for me to see, on the basis of my own work on transplantation of the neural crest cells in amphibians, how a neural crest cell is going to migrate down precisely in the dorsal ventral line. Usually neural crest cells migrate at random. An explanation on the basis of cells giving rise to epidermal cells which slide down these in turn being melanogenic and non-melanogenic seems much more amenable.

DR. QUEVEDO: Melanoblasts do not have to migrate great distances from the neural crest, owing to the small size of the embryo. Restricted lateral migrations of stem melanoblasts coupled with their division to give clones could account for the banding patterns observed by Mintz.

Regulating Function of Agouti Gene in the Mouse*

Takuji Takeuchi**

Miyagi College of Education, Sendai, Japan

I. INTRODUCTION

In the wild-type mouse, hair bulb melanocytes of the dorsal skin produce black pigment called eumelanin at the beginning of hair growth. The melanocytes subsequently produce yellow pigment called phaeomelanin; finally they return to the eumelanin synthesis. The hair grown in the wild-type coat is, therefore, characterized by a subterminal band of yellow, due to the phaeomelanin granules, with the rest of the hair showing black or brown of eumelanin granules. This characteristic is called agouti and is controlled by A gene, wild-type allele at a (non-agouti) locus.

An animal homozygous for mutant allele, a, exclusively produces eumelanin, resulting in non-agouti black or brown hair. Another mutant allele, A^y, on the other hand, exhibits a uniform yellow coat: the melanocytes of the A^y animal produce phaeomelanin only. Therefore, it seems probable that the A gene controls the shift between the two alternative systems of melanogenesis. The shift is observed in each melanocyte but the function of the A gene is not expressed within the melanocyte. Silvers and Russell[1] have shown that it is the A gene of hair follicle cells that controls the kind of pigment produced in the melanocyte. When melano-

* This work was supported by grants 4051 and 4067 from the Ministry of Education.
** Present Address: Biological Institute, Tohoku University, Sendai, Japan.

cytes were cultured apart from the original environment, they became free of the A gene control.[2]

This system, together with various mutant alleles at a locus, provides unique material for the study of genetic regulation of cell differentiation in multicellular organisms. In this paper, an attempt was made to analyze the role of the metabolites of the tyrosinase system in the agouti pattern formation, and a possible involvement of dopa in the initiation of the gene action is reported.

II.   MATERIALS AND METHODS

Materials used in this study were newborn infants of the house mouse, *Mus musculus*, of strains C3H/HeNSa (with genotype $AABBCC$) and C57BL/6 (with genotype $aaBBCC$). Pieces of skin were excised from the dorsal side of the infants whose skin had been sterilized with 75% alcohol, and were then cut into fragments of about 1 mm $\times$ 2 mm in Hanks

Table 1.   Results of culture of the explants from 2-day-old skin with genotype $AABBCC$ in culture media with and without dopa.

Exp. no.		Concentration of dopa			
		$0\,M$	$10^{-5}M$	$10^{-4}M$	$10^{-3}M$
1	No. of explants cultured	2	2	2	2
	No. of explants with yellow pigments	0	1	0	0
2	No. of explants cultured	3	4	4	4
	No. of explants with yellow pigments	0	2	1	0
3	No. of explants cultured	4	4	4	4
	No. of explants with yellow pigments	0	1	0	0
4	No. of explants cultured	4	4	4	4
	No. of explants with yellow pigments	0	2	3	0
5	No. of explants cultured	4	4	4	4
	No. of explants with yellow pigments	1	4	4	0
Total	No. of explants cultured	17	18	18	18
	No. of explants with yellow pigments	1	9	8	0

solution. Rinsed 4 times in Hanks solution, the fragments were placed on a tripod mesh standing in petri dishes. About 5 ml of culture medium, Eagle MEM supplemented with 10% bovine serum, was introduced into each dish in such a way that the surface of the medium would just cover the fragments. The explants were cultured in a CO_2 incubator for 4 days and then fixed with buffered formalin. They were then dehydrated with alcohol and cleared with xylene for a microscopic observation.

III. RESULTS

In the wild-type mouse ($AABBCC$), yellow pigments were first observed *in situ* in the hair bulbs of the dorsal skin of 4-day-old infants. No yellow pigment was detected in 2-day-old and 3-day-old mice. However, in the hair bulbs of the skin explants from 3-day-old mice cultured for 4 days, yellow pigments were observed. The explants from 2-day-old mice did not show yellow pigment following the cultivation in Eagle's medium.

In the succeeding experiments, skin explants were cultured in the medium into which dopa, 3,4-dihydroxyphenylalanine, was added to the final concentrations of 10^{-3} M, 10^{-4} M and 10^{-5} M. Yellow pigments were observed in the hair bulbs of some explants from 2-day-old skin, cultured

Table 2. Results of culture of the explants from 2-day-old skin with genotype *aaBBCC* in culture media with and without dopa.

Exp. no.		Concentration of dopa			
		0 M	$10^{-5}M$	$10^{-4}M$	$10^{-3}M$
1	No. of explants cultured	4	4	4	4
	No. of explants with yellow pigments	0	0	0	0
2	No. of explants cultured	4	4	3	4
	No. of explants with yellow pigments	0	0	0	0
3	No. of explants cultured	4	4	4	4
	No. of explants with yellow pigments	0	0	0	0
4	No. of explants cultured	4	4	4	4
	No. of explants with yellow pigments	0	0	0	0
Total	No. of explants cultured	16	16	15	16
	No. of explants with yellow pigments	0	0	0	0

in the medium containing 10^{-4} M to 10^{-5} M dopa (Table 1). The yellow pigments were seen within the melanocytes of the hair bulbs in some cases, and also in the hair shafts in other cases. In the majority of cases, only the lower part of the hair bulb melanocytes contained the yellow pigments. On the other hand, heavy darkening of the hair bulbs as well as the epidermis was seen in the explants cultured in the medium containing 10^{-3} M dopa. All control cultures in which 2-day-old skin explants were cultivated in the medium without dopa showed negative results except one case.

Skin explants from 2-day-old infants with genotype *aaBBCC* were also cultured under the same condition. In the explants from *aaBBCC* mice cultured in the dopa-containing medium, no yellow pigment was detected (Table 2): only eumelanin was observed among the explants.

IV. DISCUSSION

Until recently, little was known about the structure and process of the formation of phaeomelanin, and many of the experimental findings differ from each other. Foster[3] first reported on the basis of a manometric study that a tryptophan-oxidizing enzyme was present in the mouse skin. This finding led to the assumption that phaeomelanin might be derived from tryptophan.[4] It might be formed as the result of the oxidation of an *o*-aminophenol by dopa quinone produced by the oxidation action of tyrosinase on dopa. If so, this explains the dual function of tyrosinase in both eumelanin and phaeomelanin formation. Genetic evidence that the formation of both eumelanin and phaeomelanin is similarly affected by mutations at albino locus indicates that tyrosinase activity is a prerequisite for the phaeomelanin formation. As a matter of fact, Fitzpatrick et al.[5] have demonstrated that phaeomelanotic hair bulbs oxidize both tyrosine and dopa: pigment produced in this case was not a yellow pigment but an unnatural black pigment.

However, attempts to demonstrate directly the formation of phaeo-melanin from tryptophan[6] have been unsuccessful, while evidence indicating that tyrosine is a possible chromogen of phaeomelanin has been accumulated. Thus, incorporation of ^{14}C-labeled tyrosine into phaeo-melanotic hair bulbs has been demonstrated.[5,7] Prota and Nicolaus,[8] on the other hand, have postulated on the basis of their experimental results *in vitro* that phaeomelanin is formed by a deviation of the eumelanin pathway, involving an interaction of cysteine with one or more quinones produced from dopa oxidation.

Other observations have indicated that the kind of pigment, i.e., eumelanin or phaeomelanin, produced by hair bulb melanocytes is determined by the follicular environment, which in turn is conditioned by genes at the *a* locus. Silvers and Russell[1] reported that when genotypically black (*aa*) melanocytes migrated into hair bulbs of a skin graft which was genotypically yellow ($A^y ac^e c^e$, phenotypically nonpigmented), yellow pigment was produced. The typical agouti hair pattern resulted when genotypically yellow (A^y-) or black (*aa*) melanocytes migrated into dorsal agouti hair bulbs (Silvers, 1958). Silvers[2] also transplanted neural crest cells from genotypically yellow mouse embryos ($A^y aBB$ and $A^y abb$) to chick extraembryonic coelom, and found that the grafts differentiated into black and brown melanocytes whenever they occurred outside of hair bulbs. The typical yellow pigment was observed only within the hair bulbs.

The follicular environment which influences the hair bulb melanocyte might involve a sulfhydryl compound. Cleffmann[9] explanted self-colored black mouse skin in a glutathion-containing medium. The skin cultures showed that a portion of the hair bulbs produced yellow pigment while, in control culture, only black hair bulbs resulted. On the other hand, in agouti skin, when a sulfhydryl-antagonist iodoacetamide was employed, yellow band formation was inhibited. Cleffmann[10] also found that the yellow melanocytes changed their color immediately to black in tissue culture, but yellow pigment formation was induced by addition of gluta-thion. The concentration needed was very high with the non-agouti *a* allele but very low with the yellow A^y allele. In agouti hair, there was a change in the minimal concentration for production of phaeomelanin with the age, the threshold being lowest at the time when yellow band would normally have been formed. This might indicate that the level of the sulfhydryl compound present in hair follicle cells is the factor responsible for the shift between the two alternative melanogeneses.

In this study, skin explants from 3-day-old infants with the genotype *AA* were shown to produce yellow pigments in the hair bulbs during the cultivation, whereas no yellow pigment was observed in the skin explants from 2-day-old mice following the culture under the same condition. This result seems to indicate that a substance that would not been synthesized under the culture condition appears *in vivo* between the second and third days after birth. Therefore, assuming that the shift from eumelanin to phaeomelanin formation in the melanocyte is controlled by the *A* gene which functions in the follicular cells, the action of the *A* gene is initiated *in vivo* on the third day of infancy.

Questions arise as to how the gene action is initiated in the follicle cells. The result of this study seems to suggest that dopa or its derivative is involved in the initiation of *A* gene. In explants from 2-day-old skin with the genotype *AA*, yellow pigments were observed following the culture in the dopa-containing medium. Although there is no direct evidence to demonstrate that the yellow pigment obtained in this study is identical with

121

natural phaeomelanin, no difference was seen in their color. It seems probable that phaeomelanin formation was induced in the hair bulbs by culturing the skin in the dopa-containing medium. On the other hand, no yellow pigment was found in the explants from non-agouti (*aa*) skin after the cultivation in the same culture medium i.e., dopa-containing medium. It seems apparent that the *A* gene is involved in the dopa-induced phaeomelanin formation *in vitro*.

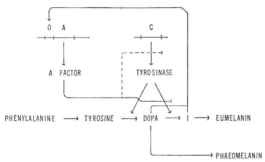

Fig. 1.   A model of the initiation and function of the *A* gene.

These results together with the experimental results accumulated lead to a tentative hypothesis on the mechanism of the formation of agouti pattern (Fig. 1). When the level of dopa (or its derivative) become high in the hair bulb melanocyte, it might diffuse into follicle cells and interact with the *A* gene to initiate its action. Although evidence supporting the presence of a site that interacts with dopa has not been obtained in this study, it is conceivable that the agouti-supressor sites, A^s, is the site in question.[11] The product of the *A* gene, which can be called *A* factor, might inhibit in some way the process from dopa to eumelanin in the melanocyte leading to the shift to phaeomelanin formation. Concerning the mode of inhibition in this case, it seems reasonable to assume an allosteric inhibition of the tyrosinase rather than repression. A question, however, remains as to the other aspect of the dual function of tyrosinase enzyme, i.e., the oxidation of both tyrosine and dopa. Further study focused upon the nature of the allosterically modified enzyme and the isozymes of tyrosinase would be required. The dopa accumulated in the melanocyte by the inhibition of the process of eumelanin formation might be utilized in the formation of phaeomelanin by combining with cysteine as assumed by Prota and Nicolaus.[8] On the other hand, a decrease in the concentration of dopa in the melanocyte might result in the termination of the phaeomelanin formation. The possibility that *A* factor does correspond to the sulfhydryl compound[9,10] and cysteine[8] cannot be rejected at present. The investigation of this possibility must be the subject of future research.

V. SUMMARY

1. In the wild-type mouse, the hair bulb melanocytes produce both eumelanin and phaeomelanin, alternatively, resulting in the so-called agouti pattern, i.e., a yellow banding of the otherwise black or brown hair. This characteristic is controlled by the A allele at a (non-agouti) locus. It seems probable that the A gene controls the shift between the two alternative pathways of melanin formation.

2. Yellow pigments were observed in the hair bulbs of the skin explants from 3-day-old mice cultured for 4 days. On the other hand, the explants from 2-day-old mice did not contain yellow pigment following the cultivation. This result seems to indicate that the gene action of A is initiated *in vivo* on the third day of infancy.

3. In the hair bulbs of the explants from 2-day-old mice with the genotype AA, yellow pigments were found following the culture in the dopa-containing medium. No yellow pigment was detected, however, in the explants from aa mice, cultured in the dopa-containing medium.

4. These observations lead to the assumption that either dopa or its derivative is involved in the initiation of the action of the A gene. When the level of dopa or its derivative becomes high in the hair bulb melanocyte, it might diffuse into follicle cells and interact with the A gene to initiate its action. The product of the A gene which is tentatively called A factor might inhibit, interacting in some way with tyrosinase, the process from dopa to eumelanin leading to phaeomelanin formation.

REFERENCES

1. Silvers, W. K. and Russell, E. S., An experimental approach to action of genes at the agouti locus in the mouse. *J. Exp. Zool.*, **130**, 199 (1955).
2. Silvers, W. K., Melanoblast differentiation secured from different mouse genotypes after transplantation to adult mouse spleen or to chick embryo coelom. *J. Exp. Zool.*, **135**, 221 (1957).
3. Foster, M., Enzymatic studies of pigment-forming ability in mouse skin. *J. Exp. Zool.*, **117**, 211 (1951).
4. Fitzpatrick, T. B. and Kukita, A., Tyrosinase activity in vertebrate melanocytes. In *Pigment Cell Biology* (M. Gordon, ed.), p. 499, Academic Press, New York (1959).
5. Fitzpatrick, T. B., Brunet, P., and Kukita, A., The nature of hair pigment. In *The Biology of Hair Growth* (W. Montagna and R. A. Ellis, eds.), p. 255, Academic Press, New York (1958).
6. Foster, M., Mammalian pigment genetics. *Advances Genet.*, **13**, 311 (1965).
7. Coleman, D. L., Effect of genic substitution on the incorporation of tyrosine into the melanin of mouse skin. *Arch. Biochem.*, **69**, 562 (1962).
8. Prota, G. and Nicolaus, A., On the biogenesis of phaeomelanins.In *Advances in Biology of Skin*, vol. 8, *The Pigmentary System* (W. Montagna and F. Hu, eds.), p. 323, Pergamon Press, Oxford (1966).

9. Cleffmann, G., Über die Beeinflussung der Wildfärbung *in vitro*. *Z. Naturforsch.*, **9b**, 701 (1954).

10. Cleffmann, G., Agouti pigment cells *in situ* and *in vitro*. *Ann. N. Y. Acad. Sci.*, **100**, 749 (1963).

11. Phillips, R. J. S., A cis-trans position effect at the *A*-locus of the house mouse. *Genetics*, **54**, 485 (1966).

DISCUSSION

DR. QUEVEDO: In preliminary results, Dr. Moyer some time ago reported that the effect of getting phaeomelanin synthesis *in vitro* by glutathion challenge actually occurred in the presence of actinomycin D, so that his conclusion was that the phaeomelanin effect was actually independent of DNA-dependent RNA synthesis, and I was wondering if this has any influence on the mechanism that you are considering here.

DR. TAKEUCHI: I do not think an alteration in the action of the *C* gene is involved in the phaeomelanin formation. But the *A* gene must acting according to this model. The effect of glutathion might be independent of the transcription of the *A* gene.

Physical Studies of the Melanins*

Marsden S. Blois, Jr.

Department of Dermatology, University of California Medical Center, San Francisco, California, and
Department of Dermatology, Stanford University School of Medicine, Stanford, California, U.S.A.

I. INTRODUCTION

The primary biological role of the melanins is usually assumed to be their pigmentary function. To the present time, no other purposes have been convincingly demonstrated for this widely occurring, unique class of bio-polymers. Because of their chemical unreactivity, they are detected, almost exclusively, through the appearance they confer upon the organisms or tissues that contain them. The color of the melanins thus relates to their only known function, affords the principal means of detecting them both in health and in disease, and has given them their name ($\mu\varepsilon\lambda\alpha\sigma$=black).

Organic substances that are black in color are relatively uncommon in nature, including the nonliving world, and, as we shall later see, attempts to account for this color (or better, lack of it) have been helpful in understanding the structure of the melanins.

The history of melanin research is unusual, as is melanin itself; in contrast with most biological compounds its starting material and principal enzyme were discovered and identified first; then the associated pathways and some of the intermediates were determined. Lastly, but incompletely, some understanding of the chemical structure of the material

* This work was supported in part by U.S. Public Health Service Grant No. CA 08064 and by the Advanced Research Project Agency through the Stanford University Center for Materials Research.

125

has been reached. This sequence is mainly accounted for by the chemical inertness, insolubility, and structural variations of the pigment which have made its direct chemical analysis so difficult. Because of this difficulty, a variety of physical techniques have been brought to bear upon the melanin problem.

For a general review of the chemistry and biochemistry of the melanins the reader is referred to Nicolaus's recent monograph.[1] In this paper I will review the results of physical techniques that have been applied to the study of the eumelanins and attempt the correlation of these results with what is known about melanin structure from chemical and biochemical studies.

II. OPTICAL SPECTROSCOPY

It is probable that melanins have been studied spectroscopically since these techniques were first applied to biological materials. In any event, absorption spectra reveal little more than the eye alone. Over visible wavelengths, melanin has a high absorbance with no absorption bands or "windows," thus accounting for the color we perceive (Fig. 1). The absorbance increases toward the shorter wavelengths in the ultraviolet as do all aromatic organic compounds, although the spectrum in the figure exaggerates this, since it is uncorrected for the effect of the Rayleigh scattering by the melanin particles. The infrared (IR) absorption (Fig. 2) spectrum shows the characteristic absorption bands expected in an aromatic organic substance, but because the material is in the solid phase, line broadening and overlapping preclude a detailed interpretation. Al-

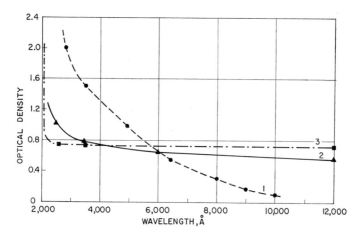

Fig. 1. Optical absorption of (1) 0.1 mg squid melanin dispersed in a 300-mg KBr pellet compared with (2) 0.3 mg charcoal, and (3) 0.1 mg graphite prepared similarly.

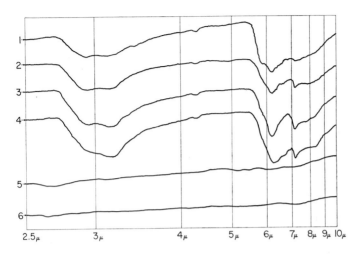

Fig. 2. Infrared absorption spectra of (1) squid melanin, (2) catechol melanin, (3) L-dopa-melanin, auto-oxidized, (4) hydroquinone melanin, auto-oxidized, (5) graphite, and (6) charcoal.

though it has been proposed that natural melanins from different species may be distinguished by means of infrared spectroscopy[2] contrary evidence also exists in that the i.r. spectra of the chemically different synthetic dopa catechol melanins are quite similar (Fig. 2).

Although detailed quantum mechanical explanations are available to account for the color of some organic compounds, the nearly uniform absorption of light over broad wavelength regions which results in blackness is poorly understood. A substance like graphite appears black because it is an electrical conductor, and the energy of the absorbed photons is used for the promotion of electrons into low-lying conduction levels and finally dissipated by the conduction process. It is presumably unnecessary that such materials be bulk conductors (in the case of graphite) so long as microscopic conductivity can occur. The example of quinhydrone may be both instructive and relevant to the melanin problem. If an equimolar solution of quinone and hydroquinone is crystallized, one obtains neither the colorless crystals of hydroquinone nor the yellow crystals of quinone, but instead the dark green, almost black, metallically lusterous crystals of quinhydrone. This has been shown to consist of two semiquinone molecules complexed together, the material being diamagnetic since the odd electron of each occupies an orbital extending over both molecules. As we shall see later, there is strong evidence that melanin also contains semiquinone radicals, and that a substantial number of these are in contact with each other in the form of π-complexes.

127

III. ELECTRON SPIN RESONANCE SPECTROSCOPY

In the first paper describing the application of electron spin resonance (ESR) to biological materials, Commoner et al.[3] reported the paramagnetism of melanin. They proposed that this may have resulted from free radicals trapped in the polymer during its synthesis and thus stabilized. The free radicals of melanin have been subsequently studied by several investigators.[4-6]

The ESR spectra of melanins are quite similar and for the most part independent of whether the melanin is of natural origin, or synthesized enzymatically by auto-oxidation. There is a dominant single, broad slightly assymetric absorption line, with a g-value near that of the free electron (g=2) which is ordinarily interpreted as arising from a free radical. As with the optical spectra, the line broadening—which in the case of ESR arises from the restriction upon molecular tumbling—limits the amount of information that may be obtained. Nevertheless, certain conclusions have emerged.

In studying a variety of aromatic free radical ions, including semi-quinones, it was found that the g-values correlated with the size of the aromatic system, and in effect with the volume of space over which the odd electron was delocalized.[7] Because of the asymmetry of the melanin absorption, its g-value could not be determined with the precision of the free radical studied in solution. However, from our measurements and the findings of other workers, the g-value of the melanin signals implied an electron delocalization over one or, at most, two aromatic rings.[5] This result argued against the hypothesis that in melanin there was a high degree of conjugation and extensive odd-electron delocalization. (Much of the speculation prior to 1960 regarding the chemical structure of melanin was centered around highly conjugated aromatic structures resulting from a regular condensation of 5,6-dihydrodioxyindole or indolequinone.)

Another conclusion that resulted from the ESR studies related to the observed concentration of unpaired electrons. It should be pointed out that quantitative data regarding this concentration are notoriously uncertain. Different workers using different spectrometers, sample tubes, etc., and different melanin preparations have reported spin density varying over a range of about 10:1. However, quantitative data may be obtained on a *relative* basis with much more confidence, and a number of such observations were carried out on the same melanin sample using the same sample tube and sample tube holder, spectrometer, etc.

It was found that as the temperature was lowered the paramagnetism increased and that this was inversely proportional to the temperature, ranging from slightly over 300°K down to liquid helium temperatures.[5]

That is to say, the Curie Law was obeyed $(\chi = \frac{K}{T})$. This finding was interesting in that it had earlier been proposed that melanin might be a semiconductor. Now, in the case of an intrinsic semiconductor, the number of unpaired (conducting) electrons is proportional to the temperature —just the opposite of the experimental result.

Another finding was the extreme stability of the free radicals in melanin —as well as of melanin itself. Melanin was heated to progressively higher temperatures, both in air and in vacuum, and after cooling, the ESR absorption was found to be unchanged. Only at temperatures of the order of 300°C, was it found that the signal began to diminish slightly, and this turned out to be due to loss of sample by sublimation. We were somewhat surprised that atmospheric oxygen did not oxidize the trapped radicals, especially at elevated temperatures, and the clear-cut failure to do so must be ascribed to the inability of molecular oxygen to diffuse into the radical-containing regions of the polymer. It was found that copper ions could apparently interact with the melanin unpaired electrons and quench its paramagnetism.[5] It was also found that the free radical signal of melanin was unaffected by treatment with oxidizing-reducing compounds such as ascorbic acid and with alkali. This stability suggested again that melanin did not contain large, highly conjugated structures, since alteration by these reagents would tend to pair off the odd electrons.

In the earlier work melanins from all sources, natural as well as synthetic, appeared to give the same broad asymmetrical ESR absorption line. Using a sample of *Ustilago* melanin which had been purified by Nicolaus and characterized by him as of the catechol type, we found the ESR spectrum to contain a weak additional line.[8] Attempts to find this additional structure in synthetic catechol melanin prepared by the alkaline autooxidation of catechol did not succeed. This problem has been solved by the use of Q-band ESR however, and Grady and Borg[6] have reported on the spectra of several natural and synthetic melanins and the effect of pH upon these spectra, particularly at alkaline pH.

It has already been mentioned that the ESR absorptions of animal melanins and the melanin produced by the auto-oxidation of dopa are indistinguishable. The synthesis of melanin in a system containing intact melanocytes has not been observed by ESR, for reasons of instrumental sensitivity, but the alkaline oxidation of dopa has been studied by this means by Wertz.[7] When the reaction is initiated, the dominant paramagnetic species present is found to be a semiquinone of dopa. As the reaction continues, other semiquinone radicals appear, and the superposition of the hyperfine structure of these species produces a complex spectrum. When the reaction has gone to completion and the insoluble, particulate melanin has been formed, all hyperfine structure is seen to have vanished and the single, broad melanin absorption remains. During

129

this reaction it seems obvious enough that there are several different semi-quinone radicals present at the same time. These are energetically capable of reacting with other diamagnetic molecules such as dopa, the starting material, or any of the intermediates of the Raper-Mason pathway (Figs. 3, 4) to form additional products. They are also capable of combining

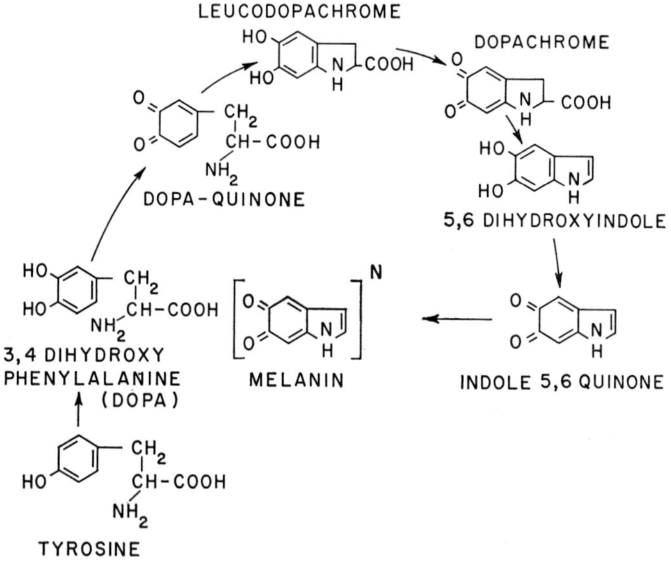

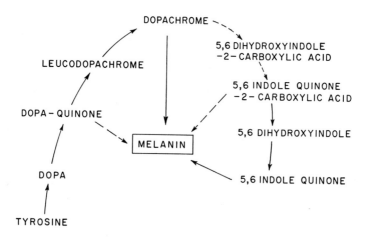

Fig. 3.   The Raper-Mason pathway of melanin synthesis.

Fig. 4.   The Raper-Mason pathway modified after Nicolaus, which suggests the direct incorporation of dopa-quinone, dopachrome, and 5,6-indole quinone-2-carboxylic acid into melanin.

130

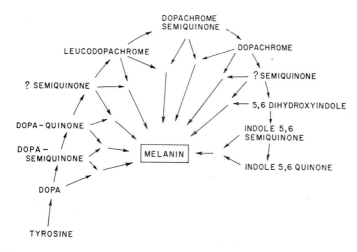

Fig. 5. The Raper-Mason pathway modified to show the probable semiquinone free radical intermediates and the multiplicity of reaction pathways.

with each other to produce new diamagnetic species (Fig. 5), which may subsequently react with yet another semiquinone free radical. It is entirely reasonable to suppose that, as the polymer grows, some of these free radicals are trapped in the growing polymer, in a random manner. For these radicals to be stably trapped, the unpaired electrons must be limited to localized regions of the polymer; otherwise they would leak out to the edges of the melanin particles and become paired off by interacting with the environment. This, as we already know, they do not do. In this *in vitro* system we have assumed that the polymerization is a random one in which several highly reactive semiquinone free radicals are involved. One may argue that well-ordered copolymerizations or stereospecific polymers may arise in the absence of enzymes, and the best evidence against such order comes to us from the X-ray diffraction evidence which we may now consider.

IV. X-RAY DIFFRACTION STUDIES

The melanins have been traditionally described as amorphous. They are not truly soluble (a number of organic solvents appear to leach out oligo-polymeric fragments which were probably not covalently bonded, and yield a colored supernatant), and crystalline forms have been neither described nor prepared. It is assumed that the melanin is highly cross-linked, but this is a supposition based upon the polyfunctional nature of the known intermediates in melanin synthesis. In any event, there has not been evidence for a crystalline or fibrous character of melanin.

131

Some years ago, after a brief attempt had been made to obtain X-ray diffraction data on melanins, we concluded that this was not a useful approach because of their amorphous nature.[5] As X-ray diffraction techniques subsequently improved — in particular automatic recording diffractometers, and the availability of powerful computing techniques for data reduction — it was decided to reinstitute the investigation of melanins with this technique. The results that will be described have been obtained by my associate, Y. T. Thathachari, over the past three years and his studies are continuing.

If one obtains an X-ray diffraction photograph from a powdered melanin sample, the principal finding is a single diffuse halo, which corresponds to a Bragg spacing of approximately 3.4Å. The exact dimensions of the spacing obtained depends, interestingly, upon the biological source of the melanin. This spacing is interpreted as arising from indole or aromatic units which are stacked upon one another to form π-complexes. It is assumed that these units are members of different polymer chains, or distant members of the same chain (Fig. 5). The existence of these stacked planer units occurs in all the melanin samples that have been studied.

A number of plant melanins were examined and the principal spacing (that between stacked, planar rings) was found to be 4.2Å while the melanins obtained from animal sources (squid and mouse melanoma) showed the principal spacing at 3.4Å. This is of interest because Piattelli et al. reported that the plant melanins are based upon catechol as the primary monomeric unit while the animal melanins are indole in type.[8] Indole is a much wider molecule than catechol, and though both are of comparable thickness it was believed there could be significant differences in the random stacking. It thus appears the concept that the π-stacking is less compact in the plant melanins than in the case of the indole melanins.[10]

With the use of automatic recording diffractometers, it has been found that these primary spacings can be determined with much more preciseness than had been expected and that not only do the classes of animal and plant melanins have different spacings, but so also do the fungal melanins (*Ustilago maydis* and *Aspergillis niger*). Furthermore, additional but very weak Bragg spacings at longer distances have been found and these are currently under investigation.

V. MECHANICAL PROPERTIES

Many biopolymers (nucleic acids and proteins) are linked by monomeric units (nucleotides or amino acids) so that each forms two covalent bonds and the resulting polymer is linear, although it may be further configured and folded by means of hydrogen bonds and van der Waals forces. It is well known that some polysaccharides are essentially linear, but since the monomers afforded the possibility of forming a third bond, there are occasional branches. Melanins consist of units that are highly polyfunctional, so that in the absence of specific enzymatic constraints (for which there is presently no evidence) one might expect to find the resulting melanin polymer to be substantially cross-linked. This might be so to a high degree if the polymer were formed by a condensation of semiquinone-type free radicals.

If melanin polymers were highly cross-linked in this manner, two properties should result. The material might be expected to be relatively insoluble—particularly if a number of different types of chemical bonds were involved—and the material should be rigid. The insolubility of melanin has already been commented upon and is very impressive.[1] The rigidity has not been adequately studied, and to this end both the compressability and the density would have to be carefully measured. We have conducted some density measurements, however, that are of interest. For example, L-dopa melanin samples prepared by alkaline oxidation have been found to have densities as high as 1.35, which is unusually high for organic polymers consisting only of C, H, O, and N. In the course of the same studies, samples of *Ustilago* melanin were found to have a density of 0.8, after compression under a pressure of 6×10^4 lb/in^2 in an attempt to collapse the presumed voids in the material. These measurements are only suggestive, but they are consistent with the concept that the melanin polymer is mechanically rigid and that it may have a high density. The latter feature is also supported by the belief that melanin may be abnormally electron dense based on the appearance of fully mature melanosomes in electron micrographs of unfixed tissue samples.

VI. ELECTRICAL PROPERTIES

Since melanins are insoluble they must be studied in the form of natural melanosomes after various purification procedures, or, in the case of synthetic melanins, as the variously sized microscopic particles which are produced. It has therefore not been possible to obtain either homogeneous continuous films or coherent solids, and electrical studies have been limited to those conducted on the dry powdered material. Despite this, attempts have been made to obtain conductivity data on the pressed powders. Potts and Au[11] report DC resistivities in the range of 5×10^{10} ohm-cm for ciliary choroidal and synthetic (enzymatically oxidized dopa) melanins. Using HCL-purified squid melanin, Thathachari found DC resistivities of approximately 4.4×10^6 ohm-cm at 100 volts.[10] He also noted that the resistivity decreased as a function of voltage, over the range of 10–100 volts, by one-third.

Appropriate AC studies on dry melanin powders have apparently not been conducted, and measurements of the dielectric constant and loss tangent as a function of frequency could possibly be of value. Parenthetically, it may be added that when the ordinary dry samples of melanin powder are introduced into the cavity of an ESR X-band spectrometer there is very little detuning of the cavity. This means that at a frequency of 9.5 GHz, melanins are very good insulators.

There is an apparent increase in the conductivity of melanins when they are illuminated,[11] as well as an apparent increase in the unpaired electron density[5] by ESR measurements, but such studies of conductivity and of photoconductivity have shed little light upon the structure of melanin and, as we shall infer below, they perhaps should not be expected to do so.

VII. DISCUSSION

Since no single physical method has led to anything approaching a complete understanding of the detailed structure of the melanins, the problem is to see how the physical data taken together relate to the chemical and biochemical evidence. Current concepts of melanin structure are heavily influenced by the analytical studies of Nicolaus[6] and the isotopic tracer studies of Hempel.[12]

Nicolaus,[9] using degradative methods and chromatographic analysis, has concluded that sepia melanin (which appears to be representative of animal melanins generally) is not a homopolymer of indole quinone, but that it consists of several different types of monomer including uncyclized aromatic compounds and pyrrolic acids in addition to indoles. Furthermore, he proposed that several types of bonds are involved including ether, peroxide, carbon-carbon, etc.

The detailed study of melanogenesis using specifically labeled precursors which Swan[13] had conducted *in vitro* was subsequently extended to the intact melanocyte by Hempel.[12] The latter has concluded that no potential bond-forming atom in dopa is consistently used or consistently not used. In other words, a given position will be a bond-forming position in some units but not in others.

In combination, the results of Nicolaus and Hempel suggest that melanin is a copolymer of considerable complexity with several types of bonds being involved. This already places melanin in a class quite different from the nucleic acids, proteins, and polysaccharides. Their evidence by itself does not answer the question whether or not melanin is a regular polymer. When we suggested[5] on the basis of the ESR studies that melanin may be a "random" polymer, it was with the idea that, if natural melanin indeed contains trapped free radicals, those free radicals were probably involved in its biosynthesis. Free radical polymerization can of course lead to the formation of quite regular polymers as in the case of several plastics of

134

technical importance, but whether or not they in fact will depends upon their chemical reactivity and the potential bond-forming sites on the molecules in the reaction mixture. The ortho-semiquinones which appear to be involved in melanogenesis are extremely reactive, and the *in vitro* study of such reaction mixtures has shown that such expected addition reactions as semiquinone pairing-off with peroxide formation, or ring attack with ether formation, did occur. Nicolaus's conclusion that such bonds appear to occur in natural melanin thus supported this hypothesis, and Hempel's finding that all potential bonding sites of dopa are involved to some extent underlined the weakness of the chemical constraints imposed when a reactive free radical intermediate collided with another molecule. The notion that melanin may be a "random" polymer should be interpreted as meaning, not that there are no chemical constraints whatsoever in its synthesis, but rather that the polyfunctional reacting species have roughly comparable probabilities of bond formation at several sites, so that a high fraction of collisions will result in bonding. Such a state of affairs would be expected to result in a polymer with a very high degree of disorder—if not random in a mathematical sense. The principal contribution of ESR to the problem of melanin structure has been to show that melanogenesis may involve free radical intermediates, from which it follows that a number of different bond types would be expected. The demonstration by Nicolaus and Hempel that several bond types are in fact involved, and the absence of any enzyme system to account for such a complex condensation mechanism, supports the hypothesis that the polymerization involves free radicals.

It is unfortunate that the one direct method of establishing molecular order—X-ray diffraction—cannot be used with the same confidence for demonstrating disorder. In the case of proteins, for example, a thoroughly denatured sample will yield an amorphous X-ray scattering diagram while a single crystal will produce a typical single-crystal pattern. The failure to obtain a crystal pattern with a protein may simply mean that the sample was inadequately prepared, not that the protein is essentially aperiodic. The amorphous X-ray diffraction data from melanin cannot be interpreted to prove that it is inherently disordered but only that means for inducing such order have not been forthcoming. Perhaps more important in the positive finding of the π-complexing in the melanins, and its occurrence in both synthetic melanin and natural melanin after various purification treatments. The data indicate that some, but not all, of the aromatic units are so associated, and a tentative interpretation has already been given.

If on the other hand one starts with the hypothesis that melanin is a highly disordered or irregular polymer, for reasons already given, then the X-ray data are fully in accord with this view. Perhaps additional data at high resolution will give additional information on the identity of nearest neighbors and shed further light on the hypothesis.

The implications of the other physical studies seem in accord with the model we have already described. Still further data are needed, in particular better data on solubility and hydrolyzability. However, if melanin is as irregular as we propose, then in principle we may expect that studies in the future—as in the past—will yield results neither as clear-cut as one might wish nor capable of unambiguous interpretation. In a sense, the difficulty lies in the fact that the obtaining of structural information implicitly requires that order be present.

VIII. ACKNOWLEDGMENTS

Thanks are due to Professor R. A. Nicolaus for his gifts of melanin preparations, to Y. T. Thathachari for many discussions on melanin structure, and to Mrs. Lina Taskovich for the painstaking preparation of samples.

REFERENCES

1. Nicolaus, R. A., *Melanins,* Herman, Paris (1968).
2. Bonner, T. G. and Duncan, A., Infra-red spectra of some melanins. *Nature,* **194,** 1078 (1962).
3. Commoner, B., Townsend, J., and Pake, G., Free radicals in biological materials. *Nature,* **174,** 689 (1954).
4. Mason, H. S., Ingram, D. J.E., and Allen, B., The free radical property of melanins. *Arch. Biochem.,* **86,** 225 (1960).
5. Blois, M. S., Zahlan, A. B., and Maling, J. E., Electron spin resonance studies on melanin. *Biophys. J.,* **4,** 471 (1964).
6. Grady, F. J. and Borg, D. C., Electron paramagnetic resonance studies on melanins. I. The effect of pH on spectra at Q band. *J. Amer. Chem. Soc.,* **90,** 2949 (1968).
7. Blois, M. S., Brown, H. W., and Maling, J. E., In *Free Radicals in Biological Systems* (M. S. Blois et al., eds.), Academic Press, New York (1961).
8. Piattelli, M., Fattorusso, E., Nicolaus, R. A., and Magno, S., The structure of melanins and melanogenesis, V. ustilago melanin. *Tetrahedron,* **21,** 3229 (1965).
9. Nicolaus, R. A., Biogenesis of melanins. *Rass. Med. Sper.,* **9,** Suppl. 1, No. 1 (1962).
10. Thathachari, Y. T. and Blois, M. S., Physical studies on melanins, II. X-ray diffraction. *Biophys. J.,* **9,** 77 (1969).
11. Potts, A. M. and Au, P. C., The photoconductivity of melanin. *Agressologie,* **9,** no. 2, 1 (1968).
12. Hempel, K., In *Structure and Control of the Melanocyte* (G. Della Porta and O. Mühlbock, eds.), Springer-Verlag, Berlin (1966).
13. Swan, G. A., Chemical studies of melanins. *Ann. N. Y. Acad. Sci.,* **100,** 1005 (1963).

DISCUSSION

DR. QUEVEDO: Would you please comment on the restrictions that might be imposed on polymerization events associated with melanin formation within a melanosome as contrasted with "free" melanin synthesis within the cells of certain organisms.

DR. BLOIS: Now I think that is a very good point in two respects. First, I suppose it can be argued that even in the absence of specific enzymes, if we have structural protein that performs its role of constraining the reacting mixture as polymerization proceeds, it is conceivable that this could introduce some kind of order. Second, biologists do not like to think of free radical reactions going on in living systems. However, we must accept the fact that we find this material [melanin] that contains trapped free radicals. One may postulate that this highly constrained system is perhaps an evolutionary response providing a means of handling a free radical reaction by walling it off inside this little reactor called a melanosome.

DR. LERNER: Can synthetic or natural melanins be arranged in monolayers with a single orientation? If so, does the orientation modify the optical properties? Do layers of synthetic or natural melanins have photoelectric properties?

DR. BLOIS: So far as the orientation of melanin is concerned, we have treated all the melanin samples by a standard purification procedure, so that one is left with these elipsoidal melanosomes, in some cases of which the protein has probably been extracted. We have never succeeded in introducing any kind of order, despite attempts to form films, and this is a criticism of the X-ray diffraction work.

DR. LERNER: How about dopa melanin?

DR. BLOIS: No, we have never succeeded in orienting dopa melanin. In the case of synthetic melanins, the particle size distribution is of course very broad, unlike the natural melanins. One obtains very tiny particles and great, big particles, which are demonstrable with the optical microscope.

DR. LERNER: And it is not possible to control that so that one gets a given size a unit and then to line that up?

DR. BLOIS: It is conceivable that one could use epitaxy or some phenomena like this, perhaps to induce orientation. I do not know, and we have certainly not accomplished it. As for the second question, what happens when these materials are illuminated, Potts and his associates at Chicago have reported on a measurement of the photo-conductivity of melanin, in which they employed natural melanin from the ciliary body and a choroid. They showed that there was an increase in the electrical conductivity of the material when it was illuminated. I think many amorphous organic materials probably would show a similar effect.

DR. FITZPATRICK: Have you any suggestions for the function of melanin other than as a light-absorbing polymer? Melanin is contained in such large quantities in parts of the body that are not exposed to light (in the lung of the frog, in the leptomeninges and the inner ear of man). Also, bats live completely in the dark, so what function might melanin have in this animal? One of my colleagues, L. B. Fred, has raised the question of whether melanin has some function other than as a light-absorber and wonders whether melanin could function in absorbing radiation other than light radiation, such as sound waves?

DR. BLOIS: I am certainly open-minded as to whether melanin may function in any way other than as a light-protecting material. Some people have proposed that it may play a physiological role in vision, but I really have no comment to make on this. I think it is yet to be proven. Others have suggested that, in addition to its pigmentary function, melanin may serve as a free radical trap. This is an attractive idea, but I think the studies that we did using ESR showed that the radicals in melanin are buried so deeply that only very tiny radicals could diffuse into the matrix and become paired off. Now it is hard to visualize UV- or radiation-produced radicals except for very small ones being able to get in. So far there is little experimental evidence to support these new proposed roles.

DR. CHAVIN: From your hypothesis, as illustrated, I noticed that tyrosine has been excluded from incorporation into melanin. Do you really mean that tyrosine cannot also be incorporated?

DR. BLOIS: I think I can only answer that in a conceptual way. If, in this reacting milieu of semiquinone free radicals, any tyrosine were present, it would be nearly as likely a target for radical attack as any other species. I happen to believe personally that tyrosine incorporation is very likely, but this remains a hypothesis.

DR. CHAVIN: We have found that (goldfish) melanin extracted from HCl skin hydrolysates when incubated with ^{14}C-labeled L-tyrosine showed a significant uptake of labeled tyrosine. Also, cold L-tyrosine plus tracer L-tyrosine-^{14}C, when subjected to 10 N HCl at 127°C for 24 hours, produced a radioactive black precipitate. It appears, therefore, that melanin may bind tyrosine in an *in vitro* situation, at least. Then, if we extrapolate to the *in vivo* situation, the melanin of each melanosome would be unique as the availability of tyrosine molecules may vary in the intracellular environment immediately surrounding the developing melanosome. Under such circumstances it may be possible that no two melanosomes would contain identical melanin polymers. Each melanin molecule, therefore, would be unique, and as a consequence the analytic problems dealing with the structure of melanin would be predictably difficult to resolve.

DR. BLOIS: Using the mouse melanoma, we have employed other radioactive quinones, including hydroquinone, somewhat along the lines of your paper yesterday, and have found that a certain amount of this, even though it is completely nonphysiologic, was taken up into the melanin polymer and apparently incorporated through co-valent bonds. Since the radioactivity could not be washed out by any technique it appears that, if an aromatic molecule could get into the melanosome and participate in this reaction mixture, it may well be indiscriminately incorporated into the pigment.

DR. CHAVIN: May I finish this up, then? Our notion is that any two molecules or any melanosomes, i.e., the melanin contained in any melanosomes, which are synthesized at different times will be different. In other words, there are probably no two melanosomes that contain exactly the same melanin polymer. This is part of our major analytic problem.

DR. BLOIS: This is what I mean by a random polymer: one assumes that in a given melanosome perhaps part or all of the melanin consists of one molecule in the sense that you can go from one atom to all other atoms through co-valent bonds. There are therefore probably no two melanosomes precisely alike in the universe. We have been intrigued for a number of years by the consequences of this assumption. For example, if there are no two identical molecules of melanin, what would be its antigenic properties? If you sensitize an animal to melanin so that he may be expected to make antibodies against it, what if you can then never find another similar molecule to challenge him with? We actually have tried over the years to find out whether melanin is antigenic or not, and we have never found any evidence that melanin is antigenic. Another interesting consequence of this idea of randomness relates to the question of how can be hydrolyzed. Can there be hydrolytic enzymes that can take it apart? Because, if there are no two melanin molecules alike, what kind of specificity properties would the enzyme have?

The Radio-Response of Malignant Melanomas Pretreated with Chlorpromazine*

Michael Cooper and Yutaka Mishima

Wayne State University School of Medicine, Detroit, and Veterans Administration Hospital, Allen Park, Michigan, U.S.A., and Department of Dermatology, Wakayama Medical University, Wakayama, Japan

I. INTRODUCTION

The radio-resistance of melanotic melanoma[1] is believed to be due to the radioprotective effect of melanin.[2,3] This protective effect is thought[3-5] to be related to the stable free radical nature of melanin,[5] which enables it to scavenge radiation-induced free radicals, thus sparing critical biological molecules.

It has recently been shown that chlorpromazine is able to bind with a high degree of selectivity to melanin,[6] and available evidence indicates that this binding is due to the formation of a charge transfer complex involving free radicals formed by chlorpromazine, and melanin.[7] In addition, the ESR signal of melanoprotein is reduced upon addition of chlorpromazine.[8]

We report here the increased radiosensitivity of malignant melanoma induced by chlorpromazine both *in vitro* and *in vivo*.[9]

II. MATERIALS AND METHODS

Fortner's melanotic melanoma no. 1,[10] obtained from J. G. Fortner in

* This research was supported by U.S. Public Health Service Research Grant No. CA-08891-02 from the National Cancer Institute and by Research Grants from Japanese Ministry of Education.

1964 and carried in our laboratory by subcutaneous implantation in Syrian (golden) hamsters, was used in this study.

A. *In vitro irradiation*

Chlorpromazine HCL (CPZ), generously supplied by Shionogi Pharmaceutical Company, was daily injected subcutaneously into 2 hamsters on the side opposite the implantation, beginning the day after tumor transplantation. The original dose was 0.25 mg per hamster (60–80 gm) and this was gradually increased to 1.0 mg over a 7-day period. At the time of tumor excision 8 days after implantation (1 day after the last CPZ injection), a total of 3.5 mg CPZ had been administered.

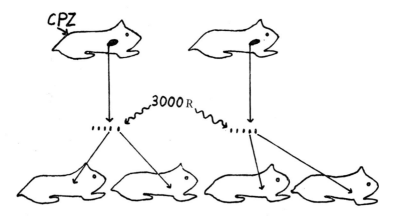

Fig. 1. *In vitro* irradiation with and without chlorpromazine, followed by melanoma implantation to determine transplantability.

Irradiation was carried out according to a modification of the method of Cobb.[2] A portion of the actively growing periphery of the tumor was removed (Fig. 1), cut into small strips approximately $1 \times 1 \times 2$ mm, and suspended in 1.0 ml saline containing 1 drop procaine penicillin-G (200,000 units/ml) and streptomycin (0.25 gm/ml) solution. The tissue strips were irradiated in a standard glass liquid scintillation vial without top, in ice, by a Dermopan type X-ray unit at 20 cm distance, step III (half-value tissue distance, 7.5 mm), with 3,000 R. One strip was then implanted subcutaneously by trocar in each of 15 hamsters. As controls, 40 hamsters received similar unirradiated implants from 6 non-CPZ-treated donors; 10 hamsters received similarly irradiated implants except that no CPZ was given to the donor animals; and 5 hamsters received unirradiated implants derived from CPZ-treated hamsters. The interval between removal of the tumor and the last implantation was less than 1 hour.

B. *In vivo irradiation*

Hamsters were implanted by trocar with a $1 \times 1 \times 2$ mm piece of melanoma tissue derived from an untreated melanoma. CPZ was injected beginning the day after transplantation. Eight controls and 9 CPZ-injected hamstars were used. The original dose was 0.25 mg and was increased by 0.25 mg every other day until a daily dose of 1.25 mg was reached, which was then continued throughout the experiment. Fourteen days after implantation the tumors were irradiated by a Dermopan type X-ray unit at 10 cm distance, step IV (half value distance, 12.5 mm), with 3,000 R. Additional single doses of 1,000 R were given until the total dose was 5,000 or 6,000 R given within 15 days. The melanomas were measured with calipers at various intervals, and at the conclusion of the experiment the irradiated melanomas were removed and their volume measured by water displacement.

III. RESULTS AND DISCUSSION

A. *In vitro irradiation*

Of 40 control hamsters receiving similar unirradiated implants from 6 donor animals, and 5 control hamsters receiving unirradiated implants derived from CPZ-treated donors, all developed tumors (Table 1). Of 15 hamsters receiving irradiated melanoma implants which had been donated by hamsters receiving CPZ, only 5 developed melanomas. Of 10 control hamsters receiving irradiated melanoma implants from donors that did not receive CPZ, all 10 developed melanomas (statistical significance: $P=0.06$). Those 5 CPZ implants that did grow, grew at a slower rate, averaging 3.3 cm in maximum diameter after 34 days as compared to 5.0 cm in the controls ($P<0.01$).

Table 1. Transplantability of *in vitro* irradiated and unirradiated tissue strips of Fortner's melanotic melanoma with and without chlorpromazine pretreatment.

Number of hamsters	Chlorpromazine pretreatment	Developing melanoma (%)
A. Unirradiated		
40	−	100
5	+	100
B. 3,000 R prior to implantation		
10	−	100
15	+	33
		($P=0.06$)

B. *In vivo irradiation*

Table 2 shows the decrease in melanoma cross-sectional area 32 days after the beginning of irradiation. Control hamsters that did not receive

Table 2. Melanoma cross-section regression 32 days after the beginning of *in vivo* irradiation with and without chlorpromazine treatment.

Chlorpromazine treatment	Decrease in size of melanoma cross section
—	1%
+	61%

chlorpromazine showed an average decrease of less than 1%, while the hamsters receiving chlorpromazine showed an average decrease of 61%. Figure 2 shows representative tumors which were treated with 6,000 R.

CPZ (−)   CPZ (+)

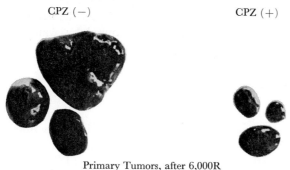

Primary Tumors, after 6,000R

Fig. 2.   Melanomas after *in vivo* irradiation with and without chlorpromazine treatment.

In one experiment using a total radiation dose of 5,000 R, the non-CPZ group averaged 1.6 cm³, as measured by water displacement, compared to 1.0 cm³ for the CPZ-treated group (Table 3). In another experiment, after 6,000 R, the control group averaged 5.0 cm³ compared to 0.13 cm³ for the CPZ-treated group (Table 3).

Table 3.   Average melanoma volume after *in vivo* irradiation with and without chlorpromazine treatment.

Exp. no.	Irradiation dose	Days since implantation	Chlorpromazine treatment	Average melanoma volume (cm³)
I	5,000 R	34	—	1.60
I	5,000 R	34	+	1.00
II	6,000 R	46	—	5.00
II	6,000 R	46	+	0.13

The incidence of metastasis in human melanomas has not been found to be increased by radiation doses of 500 R.[11] Preliminary experiments revealed that this dose does increase metasatasis in Fortner's hamster melanoma. Futhermore, no difference in tumor size was seen between

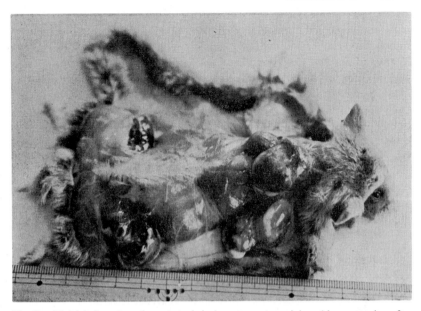

Fig. 3. Multiple lymph node metastasis in hamster not receiving chlorpromazine after 6,000 R *in vivo* total irradiation.

Table 4. Incidence of individual organ metastasis after *in vivo* irradiation with and with out chlorpromazine treatment.

Organ	Metastasis		
	−	+	++
A. Untreated			
Lymph nodes	50%	25%	25%
Lungs	25	63	12
Kidneys	12	38	50
Liver	88	12	0
Others	50	50	0
B. Chlorpromazine treated			
Lymph nodes	67%	33%	0%
Lungs	56	44	0
Kidneys	89	11	0
Liver	89	11	0
Others	78	22	0

chlorpromazine-treated animals and controls. Because of the rapid growth rate of Fortner's melanoma, the higher 3,000 R initial dose was employed. As expected, after this irradiation we observed widespread metastasis to the lymph nodes in non-chlorpromazine-treated animals (Fig. 3). However, it was found that the incidence of metastasis to the lymph nodes and viscera can be suppressed by the pre-administration of chlorpromazine.

Table 4 shows the incidence and degree of metastasis after irradiation with and without chlorpromazine. The outlined areas, which indicate individual organ metastasis of 50% or greater, are mostly seen in the metastasis-positive area for the non-chlorpromazine-treated group. In contrast, the chlorpromazine-treated group shows the outlined majority in the metastasis-negative area.

In conclusion, chlorpromazine distinctly increases the radiosensitivity of melanotic malignant melanoma, as observed by marked tumor regression, decreased transplantability, and decreased metastasis after irradiation.

REFERENCES

1. Ackerman, L. V. and Regato, O. A., *Cancer: Diagnosis, Treatment, and Prognosis* (3d ed.) St. Louis, C. V. Mosby (1962).
2. Cobb, J. P., Effect of *in vitro* X-irradiation on pigmented and pale slices of cloudman S-91 mouse melanoma as measured by subsequent proliferation *in vivo*. *J. Nat. Cancer Inst.*, **17**, 657 (1956).
3. Seiji, M. and Itakura, H., Enzyme inactivation by ultraviolet irradiation and protective effect of melanin *in vitro*. *J. Invest. Derm.*, **47**, 507 (1966).
4. Greenberg, S. S. and Kopac, M. J., Studies of gene action and melanogenic enzyme activity in melanomatous fishes. In *The Pigment Cell: Molecular, Biological, and Clinical Aspects* (V. Riley et al., eds.), p. 887, N. Y. Academy of Sciences, New York (1963).
5. Mason, H. S., Ingram, D. J. E., and Allen B., The free radical property of melanins. *Arch. Biochem.*, **86**, 225 (1960).
6. Blois, M. S., On chlorpromazine binding *in vivo*. *J. Invest. Derm.*, **45**, 475 (1965).
7. Potts, A. M., The reaction of uveal pigment *in vitro* with polycyclic compounds. *Invest. Ophthal.*, **3**, 405 (1964).
8. Bolt, A. G. and Forrest, I. S., 152nd Amer. Chem. Soc. Meeting (1966).
9. Cooper, M., and Mishima, Y., Effect of chlorpromazine on the radio-sensitivity of malignant melanoma. 2nd meeting, Jap. Soc. Invest. Dermatol., Osaka, May 24, 1969.
10. Fortner, J. G., Mohy, A. G., and Schrodt, G. R., Transplantable tumors of the Syrian (golden) hamster. I. Tumors of the alimentary tract, endocrine glands and melanomas. *Cancer Res.*, **21**, 161 (1961).
11. Sahatchiev, A. A., Raichev, R. D., and Ivonova, M., Clinico-experimental studies on radiosensitivity of malignant melanoma. In *Structure and Control of the Melanocyte* (G. Della Porta and O. Mühlbock, eds.), p. 338, Springer-Verlag, Berlin (1966).

DISCUSSION

DR. BLOIS: I would like to congratulate Dr. Cooper and Prof. Mishima on their beautiful experiment. It may be of some interest to add that we have been led to conduct a similar experiment using chloroquine, but with a completely different rationale in mind. As I understand the present work, it was based upon the concept that the chlorpromazine worked to pair off some of the melanin free radicals and thus to reduce its supposed radiation-protective effect. Our idea was that, because of its melanin-binding effect, chloroquine (or, for that matter, chlorpromazine) must be present in higher concentration in the melanoma cells than in the other body cells. Then, when the ionizing radiation is directed against the tumor, these chlorinated molecules used undergo radiolysis and produce chlorine free radiclas which are toxic and would enhance the radiation-damage effect.

Our own experiments are incomplete at this writing, but there is already a suggestion that the chloroquine may be radiosensitizing against pigmented melanoma.

DR. QUEVEDO: Thank you for your excellent paper. Do you plan to make paired comparisons of the responses of amelanotic and melanotic varieties of malignant melanoma to X-irradiation so as to determine the precise contribution of chlorpromazine-melanin interactions in the observed increased tumor radiosensitivity?

DR. COOPER: Yes, such studies are in progress.

DR. FITZPATRICK: These very interesting studies do not explain the radio-resistance of nonpigmented malignant melanoma in which there is little or no melanin present. Therefore it may be worthwhile to look for another explanation for your results, rather than the reduction of the free radical absorbing properties of melanin by chlorpromazine.

DR. COOPER: I do not think your explanation would apply directly to our experiment, but it may have some relationship.

The Photobiology of Melanin Pigmentation in Human Skin*

Madhukar A. Pathak, Yoshiaki Hori, George Szabó, and Thomas B. Fitzpatrick

Department of Dermatology, Harvard Medical School, and the Massachusetts General Hospital, Boston, Massachusetts, U.S.A.

I. INTRODUCTION

Melanin pigmentation, which follows exposure of skin to solar radiation or to ultraviolet (UV) light from artificial sources, is known to involve two distinct photobiological processes. The first is described as immediate pigment-darkening (IPD), or tanning, or direct pigmentation. The second, described as new pigment formation, melanogenesis, primary melanization, or indirect pigmentation, is a complex process that involves: (1) an erythemal response (sunburn, or vasodilation); (2) increase in the number of functioning melanocytes; (3) an increase in tyrosinase activity; (4) an increase in the number of melanized melanosomes; and (5) an increase in the transfer and redistribution of melanized melanosomes in the keratinocytes.

In this paper we shall cover only the newer concepts concerning the photobiology of melanogenesis. For additional information about the processes of IPD and melanogenesis, the reader is referred to some of our other publications.[1-5]

* This work was supported by Public Health Service Research Grants Nos. RO1 CA-05003, CA-05010, and CA-05401 from the National Cancer Institute.

149

II. THE IMMEDIATE PIGMENT-DARKENING REACTION (IPD)

The IPD process begins without any latent period and is best observed in a pigmented individual undergoing sun exposure. The reaction can also be elicited in the previously exposed (tanned) areas of fair-skinned individuals. The skin begins to darken within 5 to 10 minutes of midday summer sun exposure. The darkening of skin becomes maximal after 1 hour of irradiation. The darkened area, if left covered, fades rapidly for the first 10 to 15 minutes, and thereafter the color usually gets lighter gradually, so that after 3 to 4 hours the irradiated areas are barely detectable except when observed under Wood's ultraviolet light. Sometimes, however, after prolonged exposure to solar radiation (over 2 hours), residual pigmentation may be visible for as long as 24 to 36 hours.

The IPD reaction is related to two important events: (1) the migration and redistribution of already existing melanosomes from basal cells to the superficial keratinocytes; and (2) an oxidation reaction which involves the generation of unstable, semiquinone-like free radicals in the melanin polymer present in melanosomes. There is no change in the population of melanocytes immediately after irradiation.[1,2] Observations concerning these two events are briefly described below.

A. *Electron microscopic observations of IPD reaction*

Two pigmented subjects (one medium-colored Negro and one lightly pigmented Mexican) were exposed to midday solar radiation for 15 minutes in the month of April in Florence, Arizona. In addition, one Negro subject was exposed to solar radiation for 60 minutes in the month of June in Boston. Thiersch grafts were obtained from unexposed control sites and from exposed test sites after 15 and 60 minutes of sun exposure. Immediately after excision, the tissues were fixed in a 2.5% glutaraldehyde, 4% paraformaldehyde mixture in 0.1 M sodium cacodylate buffer, pH 7.4, at room temperature for 1.5 hours and then washed in 0.1 M sodium cacodylate buffer, pH 7.4. The tissues were postfixed in 1.33% osmium tetroxide in collidine buffer, pH 7.4, at 4°C for 2 hours, then dehydrated in graded ethanol solutions and propylene oxide, and embedded in Epon 812. The sections were cut on a Porter-Blum Ultramicrotome MT-2 stained with uranyl acetate and lead citrate, and examined on a Siemens Elmiskop I.

The pattern of melanosome distribution in the basal cell region of human skin in which IPD was induced by solar irradiation was observed by electron microscopy to differ markedly from the pattern found in unirradiated control sites. In the unirradiated site, melanosomes were present in the basal cells. They were scattered at random throughout the cytoplasm of the cells, with some tendency to cluster in the perinuclear area (Fig. 1). In the biopsies from the irradiated area obtained within 15 minutes after inducing an IPD reaction, there was an apparent increase in the number of melanosomes, not only in basal cells but also in suprabasal cells (Fig. 2). The majority of melanosomes in the exposed skin were in the perinuclear halo, with the result that the nucleus was virtually surrounded by a dense ring of melanosomes. These findings suggest that either there was a rapid transfer and redistribution of pigmented melanosomes through the dendrites to the malpighian cells or there occurred a change in the electron density of the melanosomes, i.e., the relatively less electron-dense melanosomes had become more dense as the result of an oxidation reaction. If a migration or transfer of melanosomes had indeed taken place, then there should have been a decrease in the melanosome

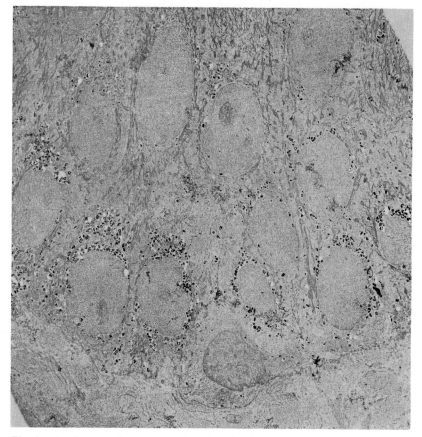

Fig. 1.   An electron micrograph of the unirradiated skin. The melanosomes were distributed at random throughout the cytoplasm of the cells, with some tendency to cluster in the perinuclear area.

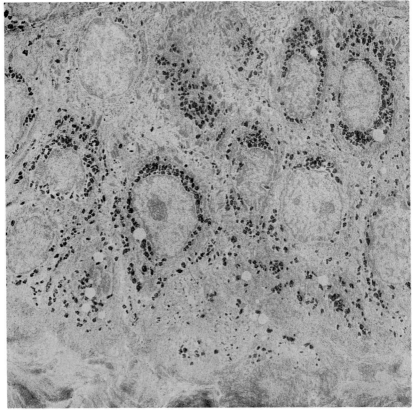

Fig. 2.   An electron micrograph of the irradiated skin showing the melanosomal changes. Biopsy (Thiersch graft from the back region) from the irradiated area was obtained immediately after exposure to solar radiation. The basal and suprabasal cells appeared to have a corona of melanosomes around nuclei of the keratinocytes.

content of the melanocytes after induction of the IPD reaction. No such change in the melanocytes could be found in the electron micrographs examined. It appears that, under the influence of photic energy, the granules not only are redistributed in the suprabasal malpighian cells but also undergo changes in the optical density. In the unstained sections obtained from the light-exposed areas, when compared to sections from the unexposed sites, we could detect these granules very readily. In presenting these observations, it must be stated that to give quantitative data on the number of melanosomes in the melanocytes and malpighian cells is at present impractical; the sampling problems, particularly when very small-sized pieces of skin are studied by electron microscopy, are evident. All we can stress in this presentation is that, when compared to the adjacent control biopsies in the subjects studied, the IPD reaction appears to cause definite changes in the distribution and number of melanosomes.

B. *Biophysical nature of IPD reaction*

It has been speculated that IPD represents an oxidation reaction of melanin pigment already present in the skin in a slightly less colored, reduced state, and that the darkening of the pigment is a reversible process in which the brown color represents a comparatively reduced state of the melanin polymer and the dark, black color represents an oxidized state of the polymer.

In an effort to elucidate the biophysical nature of this immediate pigment-darkening reaction, we have examined several specimens of fair-skinned Caucasian and pigmented Negro skin. Electron spin resonance (ESR) spectra were recorded in each specimen before and after irradiation by means of a Varian X-band spectrometer and 100 kc modulation. While at 77°K, each specimen was irradiated with ultraviolet or visible light (320 to 700 mμ) emitted by a 1,600-watt xenon lamp with appropriate filters.[6]

Human skin, when examined with an ESR spectrometer, was found to exhibit a stable ESR signal before irradiation only if it was pigmented (Fig. 3). A single, almost symmetrical absorption peak of g-value=2.003 and line width 5 to 6 gauss at 77°K is associated with unpaired electrons (free radicals) present in the melanin. The nonpigmented, so-called "white" human skin did not exhibit a stable free-radical signal of this type. Irradiation of pigmented skin (wavelengths 320 to 400 mμ and 400 mμ to 4.0 μ)

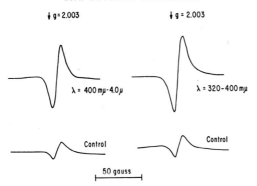

E.S.R. SPECTRA AFTER IRRADIATION OF PIGMENTED HUMAN SKIN
WITH DIFFERENT WAVELENGTHS

Fig. 3. Electron spin resonance (ESR) spectra obtained from pigmented skin before (control) and after exposure to long-wave ultraviolet and visible radiation. Irradiation of pigmented skin was found to enhance the intrinsic melanin, free-radical signal.

was found to enhance the intrinsic melanin signal (Fig. 3). The increase in unpaired spins in the irradiated skin melanin is due to generation of semiquinone-like free radicals in the melanin polymer. No other types of radicals were produced in either the nonpigmented or the pigmented skin when the skin was exposed to long-wavelength ultraviolet and visible radiation (320 to 700 mμ). Note that the action spectrum for IPD reaction is also in this range.[1,2] The observation that long-wave ultraviolet and visible radiation enhances only the free-radical signal of intrinsic melanin suggests that IPD is an oxidation reaction that involves the generation of unstable semiquinone-like free radicals in melanin. One must bear in mind, however, that the free radicals generated in this oxidation reaction are very unstable and are associated with the component of the IPD reaction that decays rapidly after cessation of irradiation. Exposed areas of skin that continue to remain hyperpigmented for 3 to 4 hours after the cessation of radiation are the result of the redistribution of melanosomes in the basal, malpighian, and granular cells of the epidermis, as discussed earlier. These melanosomes may contain oxidized melanin that is generally dark.

III. MELANOGENESIS BY UV IRRADIATION

The process of melanin pigmentation involves the production, transfer, and distribution of melanosomes. The number of melanosomes produced and the rate of their transfer to malpighian cells are regulated primarily by genetic factors. Genes control the morphology of melanocytes and melanosomes, and also the color (black, brown, yellow-red, etc.) of the melanin polymer. Genetic factors also determine the fine structure of melanosomes, the level of tyrosinase activity, the polymerization process of indole-quinone monomers, and the length and size of the melanocytic dendrites that deliver the melanosomes to malpighian cells. Solar radiation profoundly influences this process of melanogenesis. The degree of melanogenesis that follows exposure of human skin to solar radiation varies with the total dose of radiation received and the severity of epidermal-cell damage. The dose of radiation that causes only mild damage promotes pigmentation more effectively than the dose that causes severe damage. When the previously unexposed regions of the body have been exposed to a single but severely damaging dose of radiation (e.g., a fair-skinned individual receiving 45 and 90 minutes of midsummer, midday exposure that is approximately equal to 3 and 6 times, respectively, the minimal erythema dose), the skin in the first 48 hours shows damage in all epidermal layers. When there is moderate to severe damage, the epidermis goes through the following stages: (1) individual cell dyskeratosis (24 hours); (2) epidermal-cell disorganization (48 hours); (3) arborization of melanocytic dendrites (48 to 72 hours); (4) epidermal-cell regeneration and desquamation of damaged melanocytes (72 hours); (5) late epidermal regeneration and melanocytic hypertrophy (96 hours), during which the perikaryon increases in size, and the dendrites become much fully outlined by the presence of melanosomes and appear longer and more anastomotic; (6) cornification and continued melanocytic hypertrophy, manifested by increased numbers of melanocytes and increased tyrosinase activity (6 to 15 days); (7) epidermal thickening and diminished melanocytic activity (10 to 31 days).

When the epidermal damage does not involve all epidermal layers (e.g., in deeply pigmented Negroes and Caucasians), the changes are similar to those listed above, but differ in the following respects: the stage of epidermal disorganization does not occur, the epidermis showing only multiple foci of dyskeratosis without general disorganization of the cell

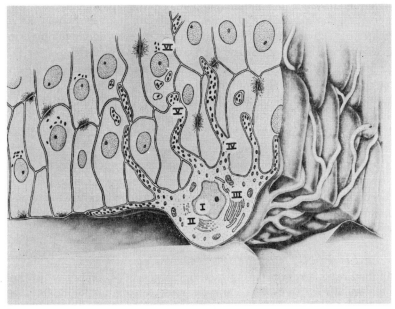

Fig. 4. Biological control of melanin pigmentation in human skin stimulated by solar radiation at various levels of organization of the melanocyte. The augmentation of melanin pigmentation by solar radiation involves a close interaction of melanocytes and keratinocytes and appears to occur at the levels shown in this figure.

Level I: Melanocyte distribution, Level II: Melanosome production, Level III: Tyrosinase biosynthesis, Level IV: Tyrosine–melanin biosynthesis, Level V: Transfer of melanosomes to keratinocytes, Level VI: Oxidative darkening of early, intermediate and late melanosomes.

(Figure adapted from T. B. Fitzpatrick and A. S. Breathnack, 1963, Das epidermale Melanin-Einheit-System, *Derm. Wschr.*, **147**, 481–89.)

layers. The numerical increase of melanocytes is evident within 72 hours after irradiation and is accompanied by dendritic arborization and hypertrophy of melanocytes. Enhanced tyrosinase activity in melanocytes is evident within 48 to 72 hours. The skin becomes visibly hyperpigmented within this same length of time.

Thus, it is obvious that the augmentation of melanin pigmentation in human skin stimulated by solar radiation involves a close interaction of melanocytes and keratinocytes and is characterized by the formation of new melanosomes within the melanocytes and their transfer and distribution within an increased population of keratinocytes. This augmentation of melanin pigmentation, illustrated in Fig. 4, appears to occur at the following levels:

Level I: An increase in the number of functional melanocytes as the result of proliferation or activation of melanocytes, or both. Hypertrophy of the melanocytes and increased arborization of melanocytic dendrites may accompany this increase.

Level II: An increase in the number of melanosomes (early, intermediate, and fully melanized), both in the melanocytes and in malpighian cells, due to increased synthesis of new melanosomes in melanocytes.

Level III: An increase in tyrosinase activity due to the synthesis of new tyrosinase in proliferating melanocytes.

Level IV: Activation of tyrosinase as the result of the direct effect of radiation on tyrosinase-inhibiting sulfhydryl compounds.

Level V: An increase in the transfer of melanosomes as the result of increased turnover of keratinocytes. This process may exhibit an increase in the number and size of melanosome complexes in certain races (e.g., Caucasians) and also increased arborization of melanocytic dendrites.

These concepts concerning the biological control of melanin pigmentation at various levels of organization of the melanocyte are briefly discussed below.

A. *Level I: Effect of ultraviolet light on melanocyte population*

Published reports regarding the increased melanin pigmentation following UV radiation suggest that a numerical increase of functioning melanocytes is responsible for enhanced pigment production.[3,4,7-18] The melanocytes reveal an increase in size of nucleus and perikaryon; the dendrites become more fully outlined by the presence of melanosomes and appear longer and much more branched (Fig. 5).

This increase in the number of functioning melanocytes, as determined by histochemical techniques involving incubation in dopa solution, could be the result of either: (1) proliferation of melanocytes, or (2) activation of "dormant" dopa-negative melanocytes as a result of enzymic (tyrosinase) activation in preexisting but inactive melanocytes.[18]

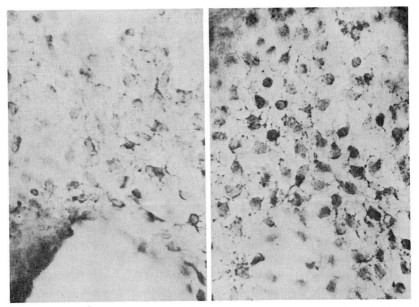

Fig. 5. Human skin epidermal melanocytes before (left) and after exposure (right) to solar radiation. The average number of melanocytes per mm² area of the unexposed abdominal skin was 830±40, whereas the melanocyte count from the biopsy obtained from the adjacent site at 72 hours following an exposure to solar radiation was 1,440±65.

The existence of enzymically inactive melanocytes has been repeatedly mentioned by several investigators. It is believed that there are melanocytes which are both "enzymically inactive" (i.e., they exhibit minimal or no tyrosinase activity) and structurally unrecognizable (i.e., they do not synthesize melanosomes). These so-called dormant melanocytes are presumed to be activated as a result of exposure of skin to UV radiation.[17]

There is no doubt that a numerical increase of functional melanocytes occurs following irradiation, but what is not known is whether the increase in numbers of melanocytes after UV stimulation derives from increased melanogenic activity within existing dormant melanocytes, as well as from proliferation of melanocytes. Melanocytes have been observed to undergo mitosis,[11,19-21] but the regularity with which this process occurs has not been established. This event is seen only rarely because of the difficulty of detecting melanocytes in the process of division and because of the loss of their cytological characteristics during mitosis. At the height of mitotic activity of epidermal cells, which usually occurs between 48 and 72 hours after UV irradiation, melanocyte mitosis is rarely seen. In addition, the search for this mitotic activity in melanocytes is further complicated by

157

the uncertainty of identifying these cells; during division they do not seem to have characteristic dendritic processes.

B. Level II: Effect of ultraviolet light on melanosome formation

In all races, the number of fully melanized melanosomes increases in the melanocytes as well as in the malpighian cells as a result of UV irradiation (Figs. 6, 7). In studying the cytology of racial differences in human skin color, Szabó et al.[3, 23] observed that UV radiation has a stimulating effect on the rate of melanosome formation in melanocytes and on the rate of melanization of melanosomes. When compared to unirradiated skin, the irradiated skin of Caucasians showed more melanized melanosomes in the melanocytes, whereas the Negroid skin showed a greater number of melanosomes that were in earlier stages of melanization. In all races (Caucasians, Negroes, American Indians, Mongoloids), the number of melanosomes inside the malpighian cells was found to be much higher after irradiation than before.

It is now well established that, in human epidermal keratinocytes, mature melanosomes are dispersed singly (e.g., in Negroes and Australian aborigines) or as a complex of two or more melanosomes (e.g., in Cauca-

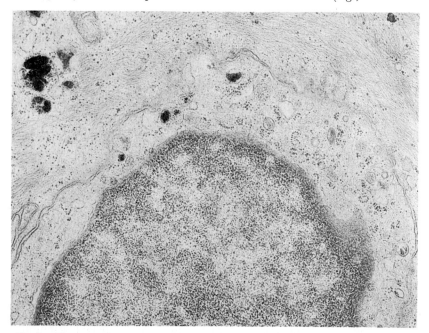

Fig. 6. Electron micrograph of a melanocyte of a Caucasian before exposure to UV radiation. Only a part of a melanocyte surrounded by a portion of a keratinocyte is shown. There are very few fully melanized melanosomes in the melanocyte. Melanosome complexes in the keratinocyte can also be seen in the left corner.

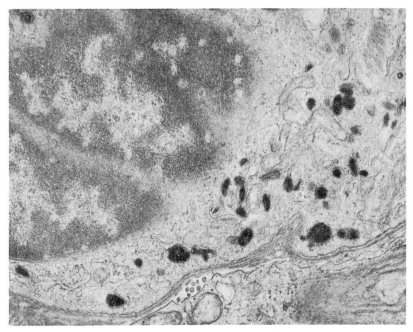

Fig. 7.   Part of a melanocyte of the same Caucasian after ultraviolet exposure. It is full of melanized melanosomes. Melanosomes in early stages of formation are also present.

sians and Mongoloids). The melanosomes, either singly or in a complex of two or more, are always surrounded by a limiting membrane.[23, 24] After UV irradiation, this characteristic grouping of melanosomes persists in Caucasians and Mongoloids. In addition, UV radiation induces larger complexes with more melanosomes. Instead of a melanosome complex of 2 or 3 melanosomes surrounded by a membrane, 5, 10, or more melanosomes may be clustered within a single membrane. These changes in melanosome complex are more evident in the basal cell regions. In the more superficial layers of the epidermis, the melanosome complexes are not so numerous, and in keratinizing cells just below the horny layer, single melanosomes may be observed in Caucasians, owing to the disintegration of cellular content.

Thus, the color of the human skin stimulated by UV radiation is influenced by: (1) the increased production of melanosomes by the melanocytes, and (2) the increased transfer and distribution of melanosomes in the keratinocytes. In dark (e.g., Negro) skin these melanosomes are individually dispersed inside the malpighian cells, and they are present in the entire epidermis, including the stratum corneum, whereas in lighter skin (e.g., that of fair-skinned Caucasians) there is an aggregation of melanosomes—2, 3, or more being inside a single encompassing mem-

159

brane—and they are evident in the basal cell regions only. In the more superficial layers of the epidermis, these melanosome complexes are not numerous and may even be absent.

Thus, the hyperpigmentation stimulated by UV radiation in all races is due to the increased production of melanosomes by the melanocytes, and their increased transfer and distribution in the keratinocytes. The intensity of visible coloration of the skin is related to both the number and the location of melanosomes in the epidermis.

C. Level III: Effect of UV irradiation on tyrosinase activity

It is often said that UV irradiation of skin causes an increase in tyrosinase activity. In 1949 and 1950, Fitzpatrick et al.[25, 26] postulated that tyrosinase in human skin exists in a partially inhibited form. They observed that tyrosinase in normal unirradiated melanocytes was unable to oxidize tyrosine to form melanin, whereas, when the irradiated human skin biopsies were incubated with tyrosine, the melanocytes were able to form new melanin. The mechanism by which ultraviolet light activates the tyrosinase reaction and evokes hyperpigmentation of skin was postulated by these investigators to include: (1) oxidation of tyrosine to dopa (3,4-dihydroxyphenylalanine) by direct photochemical action, and (2) the acceleration of tyrosine→dopa→melanin reaction by the presence of trace amounts of dopa thus formed.

Although this hypothesis is interesting, and one must consider the possibility that ultraviolet light acts in vivo to bring about the oxidation of tyrosine to dopa in a manner similar to the in vitro photochemical reactions,[25, 26] one must realize that the photochemical formation of dopa requires unusually large doses of UV irradiation. Under physiological conditions, this photochemical reaction seems unlikely.

There is a great need to differentiate between the primary and secondary changes that can take place in melanocytes after irradiation. The primary changes are those that occur during the process of absorption of radiant energy by molecules; the secondary changes are those that result from cell damage and are interrelated with the repair process which sets in after the cells are damaged. It is highly unlikely that UV irradiation will directly activate tyrosinase, at least in the primary process of light absorption. Irradiation of several enzymes has revealed that ultraviolet light causes their inactivation. Absorbed light energy migrates through a protein (i.e., enzyme) molecule until it reaches a "weak link" and causes damage at that point. Such a weak link has been supposed to be a group of disulfide and hydrogen bonds necessary for enzyme activity.[27, 28] It is therefore essential to bear in mind that specific damage, such as disruption of a specific group of easily broken bonds that are essential for maintenance of biological activity (either hydrogen or disulfide bonds, or both, e.g., cysteine, glutathione), can take place in the tyrosinase present in the melanocytes and can result in the inactivation of the enzyme. Such an inactivation of mammalian tyrosinase does occur in the *in vitro* system when the purified enzyme solution is irradiated with ultraviolet light of wavelengths from 260 to 320 mμ (Burnett and Pathak, unpublished observations, 1968). In the *in vivo* system, however, this type of inactivation is less likely to be observed because the melanocyte, by its very nature, has a chromophore, melanin, that is capable of protecting the cell against damaging effects of radiation. The presence of melanin in the form of a melanoprotein can effectively absorb, scatter, and attenuate the UV radiation[2] and protect the melanocyte and its content against inactivation or death.[29] It is apparent, therefore, that secondary changes are responsible for the differences in tyrosinase activity in the melanocytic system after UV irradiation. It would seem more likely that the increased melanogenic activity is derived mainly from an increase in tyrosinase activity due to the synthesis of new tyrosinase in the proliferating melanocytes and an increase in the rate of melanogenesis in melanocytes that are already present and engaged in the synthesis of melanin.

D. *Level IV: Effect of UV irradiation on tyrosinase activation*

The speculation that the inhibition of melanin formation is a function of sulfhydryl compounds in the epidermis is based on the hypothesis that both the substrate (tyrosine) and the enzyme (tyrosinase) in melanocytes are unable to react fully because of the inhibitory action of certain sulfhydryl-containing compounds present in the epidermal cells. The oxidation of the sulfhydryl-containing compounds by such pigmentogenic stimuli as ultraviolet light, X-ray, heat, and inflammatory skin diseases releases the tyrosinase, thereby facilitating the enzyme reaction with the substrate and allowing pigmentation to take place.

The concept that the regulation of melanogenesis may depend upon the release of inhibition of tyrosinase by sulfhydryl compounds has evolved gradually. First, it was observed that crude aqueous extracts of rabbit or guinea pig skin could inhibit the oxidation of tyrosine by tyrosinase.[30-32] Then Rothman et al.[33] extended these observations and were able to demonstrate this inhibitory effect on the oxidation of tyrosine by tyrosinase in the aqueous extracts of human epidermis. Rothman found for human skin, as Ginsberg had found for skin of guinea pigs, that a factor is present which inhibits melanin formation *in vitro*. He identified this as a water-extractable, dialyzable, sulfhydryl-containing compound. In subsequent publications, Rothman and Flesch[34, 35] have presented evidence that the inhibitory substance present in water extracts of human epidermis indeed contains sulfhydryl groups because (1) this inhibition was abolished by compounds that combine with sulfhydryl groups, and (2) there was an immediate decrease in sulfhydryl concentration in rabbit skin upon exposure to UV radiation. Flesch[35] also observed that the reduced glutathione is thirteen times more effective in inhibiting dopa oxidation than the most effective inhibitor, cysteine.

Several other investigators have postulated the existence of tyrosinase inhibitor as an important factor in regulating melanogenesis. Riley et al.[36] and Hirsch[37] reported an inhibitor of dopa-oxidase activity in mouse melanoma. Satoh and Mishima[38] also observed the presence of a tyrosinase inhibitor in amelanotic and melanotic malignant melanoma of the Syrian golden hamster. They concluded, however, that the substance, although dialyzable, was not a sulfhydryl compound. Recently, Chian and Wilgram[39] also demonstrated the existence of a tyrosinase inhibitor (molecular weight less than 5,000; extracted from the albino strain of the S91 mouse melanoma, and the S91 pigmented strain, the Harding–Passey, and B16 mouse melanomas) that could inhibit the soluble tyrosinase. This inhibitor was found to be inactivated by ultraviolet light; therefore, these investigators speculated that the induction of tanning after exposure to ultraviolet light could be the result of the inactivation of the inhibitor. They believed that this tyrosinase inhibitor not only inhibits the fully functioning enzyme but also acts as a repressor of the gene that regulates the synthesis of tyrosinase. Inactivation of the repressor by ultraviolet light could lead to an increased synthesis of tyrosinase followed by an increased number of melanosomes.

Recently, Halprin and Ohkawara[40] added additional support to this intriguing concept by presenting experimental evidence that the tripeptide reduced glutathione (γ-glutamylcysteinylglycine) appears to be the inhibitory substance that controls the activity of the enzyme tyrosinase. These workers measured the concentration of reduced glutathione and the activity of glutathione reductase in Negro and fair-skinned Caucasian subjects before and after exposure to ultraviolet light. Glutathione-reductase activity and the content of reduced glutathione were found to be lower in Negro skin than in Caucasian. The concentration of reduced glutathione in vivo was approximately 100 times the minimum concentration necessary to cause inhibition of melanin formation. Furthermore, hyperpigmentation following exposure to ultraviolet light was preceded by a significant drop in glutathione-reductase activity, combined with a fall in the content of reduced glutathione and a rise in oxidized glutathione. Halprin and Ohkawara therefore believed that one of the mechanisms responsible for post-ultraviolet hyperpigmentation is manifested by a fall in glutathione-reductase activity and a decrease in reduced-glutathione content that consequently lead to favorable conditions for melanogenesis. It is interesting to note that other investigators, as far back as 1953, also observed that hyperpigmented skin in postinflammatory conditions has about half the sulfhydryl content of the surrounding normal skin, and the vitiliginous skin has a sulfhydryl content about twice as high.[41]

Although tanning of human skin requires direct exposure to solar radiation, the exact sequence of the regulatory factors that program the performance of the epidermal melanocytes in UV induced melanogenesis remains speculative. It is conceivable that the levels of the sulfhydryl compounds, such as glutathione and cysteine, may not only regulate the functional state of tyrosinase but also act as repressors of the genes that regulate the synthesis of tyrosinase. When the repressor is destroyed or modified through a process such as oxidation, it may lead to an increased synthesis of new enzyme in melanocytes engaged in synthesis of tyrosinase.

There are two mechanisms by which the amount of melanin formed can be regulated by the enzyme tyrosinase. One involves changes in activity of the already synthesized enzyme, and the other involves changes in the quantity of the enzyme present through variation of its rate of synthesis. These regulatory mechanisms, one involving the control of tyrosinase action, and, the other, control of tyrosinase formation by the removal of the repressor, must be kept in mind in speculating on the role of sulfhydryl compounds in melanogenesis stimulated by solar radiation. Inhibition of tyrosinase by such agents as reduced glutathione and cysteine may likely be brought about from direct interaction of the sulfhydryl group at a site on the enzyme molecule that results in a reduced affinity for the substrate (tyrosine) at the active center of the enzyme. The increased availability of oxygen as a result of vasodilation after sunburn may also exercise a direct effect on the levels of sulfhydryl groups; it may oxidize reduced glutathione and favor release of enzymes to full activity. Repression is, on the contrary, indirect in its mechanism of action and involves the biosynthesis of protein (enzyme). This difference between the inhibition and repression of tyrosinase by these inhibitors can be understood only when mechanisms of regulation of tyrosinase activity are studied.

E. *Level V: Transfer of melanosomes to keratinocytes*

In visualizing the process of melanin pigmentation stimulated by solar radiation, it would be misleading to stress only the alteration in the melanocytic system. While it is true that the melanocyte is the sole producer of melanosomes in the epidermis, hyperpigmentation of the skin is due to the distribution of the melanosomes in the malpighian cells and is brought about by the transfer of melanosomes by melanocytic dendrites.

Ultraviolet irradiation can either stimulate the transfer of melanosomes to the proliferating keratinocytes by the process of arborization of melanocytic dendrites, or it can inhibit the transfer of melanosomes by causing intercellular edema. No investigation of the relation of variations in the size and length of dendrites in human melanocytes of different races of man has been made; it seems, however, that some variations in color reflect the inability of melanocytes to transfer melanosomes to epidermal keratinocytes because the dendrites of the melanocytes are short and stubby. Quevedo and Smith,[12] in a study of the effect of UV radiation on the plantar skin of D and Ln mouse genotypes, noted that tanning is due to an increase in melanin-granule (melanosome) synthesis in epidermal melanocytes and the transfer of these granules to the epidermal cells. These investigators found that the melanocytes of the plantar skin of "dd" genotypes of mice possess a few fine dendritic processes. These areas did not become hyperpigmented after exposure to ultraviolet light, presumably because of the poor development of their dendrites and failure to transfer melanosomes to the epidermal cells. Although melanogenesis was stimulated, melanin pigmentation of epidermal cells did not ensue. Melanocytes with short, stubby dendrites similar to those observed in the dilute strain of mouse mentioned above have been observed by Mitchell[42] in human skin that had been exposed to prolonged solar radiation. She found that the melanocytes appear round because their dendrites were "only short stumps." The adjacent malpighian cells associated with these deformed melanocytes have very few melanosomes, suggesting that transfer of melanin granules has been blocked.

It is a common experience to observe profound arborization of melanocytic dendrites in biopsies obtained after UV irradiation or in conditions associated with postinflammatory hyperpigmentation.[43] It seems that ultraviolet light activates an integrated mechanism by which melanosomes that are synthesized at an accelerated rate in melanocytes are delivered to malpighian cells by dendritic processes. It is interesting to note that the population of epidermal melanocytes and the melanocyte:keratinocyte ratio are constant and characteristic for each body region.[44,45] For example, the melanocyte distribution in normal human skin (unexposed) is considered to be approximately 1 melanocyte per 10 keratinocytes in the basal layer of the epidermis of arm, ear, and leg. It is also known that UV radiation induces hyperplasia. Melanocytes can maintain contact and deliver melanosomes to this increased population of keratinocytes only through their dendritic processes.

IV. SUMMARY

The augmentation of melanin pigmentation in human skin by solar radiation involves a close interaction of melanocytes and keratinocytes and is characterized by the formation of new melanosomes within the melanocytes and their transfer and distribution within an increased population of keratinocytes. This augmentation of melanin pigmentation reveals: (1)

an immediate pigment-darkening oxidation reaction in the melanin present in already existing melanosomes and involving the formation of semiquinone-like free radicals; perinuclear redistribution of melanosomes in the basal and suprabasal cells; (2) an increase in the number of functional melanocytes as the result of proliferation and/or activation of melanocytes; (3) hypertrophy of melanocytes and increased arborization of their dendrites; (4) an increase in the number of melanosomes in all stages (early, intermediate, and fully melanized), both in the melanocytes and malpighian cells, due to increased synthesis of new melanosomes in melanocytes; (5) an increase in the number and size of melanosome complexes in certain races; (6) an increase in the transfer of melanosomes as a result of increased turnover of keratinocytes; (7) an increase in tyrosinase activity due to the synthesis of new tyrosinase in proliferating melanocytes; and (8) activation of tyrosinase as the result of the direct effect of radiation on tyrosinase-inhibiting sulfhydryl compounds.

REFERENCES

1. Pathak, M. A., Photobiology of melanogenesis: Biophysical aspects. In *Advances in Biology of Skin*, vol. 8, *The Pigmentary System* (W. Montagna and F. Hu, eds.), p. 397, Pergamon Press, Oxford (1967).

2. Pathak, M. A. and Stratton, K., Effects of ultraviolet and visible radiation and the production of free radicals in skin. In *The Biologic Effects of Ultraviolet Radiation. Proceedings of the First International Conference* (F. Urbach, ed.), p. 207, Pergamon Press, Oxford (1969).

3. Szabó, G. S., Photobiology of melanogenesis: Cytological aspects with special reference to differences in racial coloration. In *Advances in Biology of Skin*, vol. 8, *The Pigmentary System* (W. Montagna and F. Hu, eds.), p. 379, Pergamon Press, Oxford (1967).

4. Szabó, G. S., Current state of pigment research with special reference to macromolecular aspects. In *Biology of the Skin and Hair Growth* (A. G. Lyne and B. F. Short, eds.), p. 705, Angus and Robertson, Sydney (1965).

5. Fitzpatrick, T. B., Mammalian melanin biosynthesis. *Trans. St. John Hosp. Derm., Soc.*, **51**, 1 (1965).

6. Pathak, M. A. and Stratton, K., A study of the free radicals in human skin before and after exposure to light. *Arch. Biochem.*, **123**, 468 (1968).

7. Bloch, B., The problem of pigment formation. *Amer. J. Med. Sci.*, **177**, 608 (1929).

8. Peck, S. M., Pigment (melanin) studies of the human skin after application of thorium X. With special reference to the origin and function of dendritic cells. *Arch. Derm.*, **21**, 916 (1930).

9. Becker, S. W., Dermatological investigations of melanin pigmentation. In *Biology of Melanomas* (R. W. Miner, ed.), **4**, 82, New York Academy of Science Special Publication, New York (1948).

10. Fitzpatrick, T. B., Becker, S. W. Jr., Lerner, A. B., and Montgomery, H., Tyrosinase in human skin: Demonstration of its presence and of its role in human melanin formation. *Science*, **112**, 223 (1950).

11. Becker, S. W., Jr., Fitzpatrick, T. B., and Montgomery, H., Human melanogenesis: Cytology and histology of pigment cells (melanodendrocytes). *Arch. Derm.*, **65**, 511 (1952).

12. Quevedo, W. C. Jr. and Smith, J. A., Studies on radiation-induced tanning of skin. *Ann. N. Y. Acad. Sci.*, **100**, 364 (1963).

13. Bischitz, P. G. and Snell, R. S., A study of the melanocytes and melanin in the skin of the male guinea-pig. *J. Anat.*, **93**, 233 (1959).

14. Snell, R. S., The effect of ultraviolet irradiation on melanogenesis. *J. Invest. Derm.*, **40**, 127 (1963).

15. Quevedo, W. C. Jr., Szabó, G., Virks, J., and Sinesi, S. J., Melanocyte populations in UV-irradiated human skin. *J. Invest. Derm.*, **45**, 295 (1965).

16. Papa, C. M. and Kligman, A. M., The behavior of the melanocytes in inflammation. *J. Invest. Derm.*, **45**, 465 (1965).

17. Mishima, Y. and Widlan, S., Enzymically active and inactive melanocyte populations and ultraviolet irradiation: Combined dopa-premelanin reaction and electron microscopy. *J. Invest. Derm.*, **49**, 273 (1967).

18. Pathak, M. A., Sinesi, S. J., and Szabó, G., The effect of a single dose of ultraviolet radiation on epidermal melanocytes. *J. Invest. Derm.*, **45**, 520 (1965).

19. Billingham, R. E., Dendritic cells. *J. Anat.*, **8**, 93 (1948).

20. Billingham, R. E., and Silvers, W. K., The melanocytes of mammals. *Quart. Rev. Biol.*, **35**, 1 (1960).

21. Giacometti, L. and Allegra, F., The effect of wounding upon the uptake of ^{3}H-thymidine by guinea-pig epidermal melanocytes. In *Advances in Biology of Skin*, vol. 8, *The Pigmentary System* (W. Montagna and F. Hu, eds.), p. 89, Pergamon Press, Oxford (1967).

22. Clark, W. H., Jr., Pathak, M. A., Szabó, G., Bretton, R., Fitzpatrick, T. B., and El-Mofty, A. M., The nature of melanin pigmentation induced by furocoumarins (psoralens). In *Book of Abstracts*, Fifth International Congress on Photobiology, Hanover, N. H., p. 57 (1968).

23. Szabó, G. S., Gerald, A. B., Pathak, M. A., and Fitzpatrick, T. B., Racial differences in the fate of melanosomes in human epidermis. *Nature*, **222**, 1081 (1969).

24. Hori, Y., Toda, K., Pathak, M. A., Clark, W. H., Jr., and Fitzpatrick, T. B., A fine structure study of the human epidermal melanosome complex and its acid phosphatase activity. *J. Ultrastruct. Res.*, **25**, 109 (1968).

25. Fitzpatrick, T. B., Becker, W., Jr., Lerner, A. B., and Montgomery, H., Tyrosinase in human skin: Demonstration of its presence and of its role in human melanin formation. *Science*, **112**, 1 (1950).

26. Fitzpatrick, T. B., Lerner, A. B., Calkins, E., and Summerson, W. H., On the mechanism of melanin formation by the action of ultraviolet radiation. *J. Invest. Derm.*, **12**, 7 (1949).

27. McLaren, A. D. and Shugar, D., *Photochemistry of Proteins and Nucleic Acides*. Macmillan, New York (1964).

28. Augenstein, L. G. and Riley, P., The inactivation of enzymes by ultraviolet light. IV. The nature and involvement of cystine disruption. *Photochem. Photobiol.*, **3**, 353 (1964).

29. Kitano, Y. and Hu, F., The effects of ultraviolet light on mammalian pigment cells *in vitro*. *J. Invest. Derm.*, **52**, 25 (1969).

30. Schaff, F., Manometrische Vergleichsuntersuchungen mit Pressäften aus Weisser und pigmentierter Meerschweinhaut. *Arch. Derm. Syph.*, **176**, 646 (1938).

31. Schaumann, K. and Danneel, R., Physiologie der Kälteschwärzung beim Russenkanincher. *Biol. Zentralbl.*, **58**, 242 (1938).

32. Ginsberg, B., The effects of the major genes controlling coat color in the guinea-pig on the dopa oxidase activity of skin extracts. *Genetics*, **29**, 176 (1944).

33. Rothman, S., Krysa, H. F., and Smiljanic, A. M., Inhibitory action of human epidermis on melanin formation. *Proc. Soc. Exp. Biol. Med.*, **62**, 208 (1946).

34. Flesch, P. and Rothman, S., Role of sulfhydryl compounds in pigmentation. *Science*, **180**, 505 (1948).

35. Flesch, P., Inhibotory action of extracts of mammalian skin on pigment formation. *Proc. Soc. Exp. Biol. Med.*, **70**, 136 (1949).

36. Riley, V., Hobby, G., and Burk, D., In *Pigment Cell Growth* (M. Gordon, ed.), p. 231, Academic Press, New York (1953).

37. Hirsch, H. M., Inhibition of melanogenesis by tissues and the control of intracellular autooxidants. In *Pigment Cell Biology* (M. Gordon, ed.), p. 327, Academic Press, New York (1959).

38. Satoh, G. J. Z. and Mishima, Y., Tyrosinase inhibitor in Fortner's amelanotic and melanotic malignant melanoma. *J. Invest. Derm.*, **48**, 301 (1967).

39. Chian, L. T. Y. and Wilgram, G. F., Tyrosinase inhibition: Its role in suntanning and in albinism. *Science*, **155**, 198 (1967).

40. Halprin, K. M. and Ohkawara, A., Human pigmentation: The role of glutathione. In *Advances in Biology of Skin*, vol. 8, *The Pigmentary System* (W. Montagna and F. Hu, eds.), p. 241, Pergamon Press, Oxford (1967).

41. Van Scott, E. J., Rothman, S., and Greene, C. R., Studies on the sulfhydryl content of the skin. *J. Invest. Derm.*, **20**, 111 (1953).

42. Mitchell, R. E., The effect of prolonged solar radiation on melanocytes of the human epidermis. *J. Invest. Derm.*, **41**, 199 (1963).

43. Pinkus, H., Staricco, R. J., Kropp, P. J., and Fan, J., The symbiosis of melanocytes and human epidermis under normal and abnormal condition. In *Pigment Cell Biology* (M. Gordon, ed.), p. 127, Academic Press, New York (1959).

44. Szabó, G., Quantitative histological investigations on the melanocyte system of the human epidermis. In *Pigment Cell Biology* (M. Gordon, ed.), p. 99, Academic Press, New York (1959).

45. Szabó, G., The regional anatomy of the human integument with special reference to the distribution of hair follicles, sweat glands and melanocytes. *Phil. Trans. Roy. Soc.*, *Ser. B*, **252**, 447 (1967).

DISCUSSION

DR. BAGNARA: I know very little about this subject, but I would like to ask how the binding of psoralens to DNA and RNA is related to the photosensitization reaction in the skin.

DR. PATHAK: Photosensitization of skin by an exogenous agent, either applied topically or given orally, can be defined as an event occurring as a result of absorption of light that is either lethal to the cell or highly damaging to the functional state of the cell. Cutaneous cells when exposed only to long-wave ultraviolet radiation (320–400 mμ) are not damaged; likewise, psoralen molecules in the absence of UV radiation do not produce any cellular damage. Only when the cutaneous cells are exposed to long-wave ultraviolet radiation in the presence of psoralen can one observe the event of photosensitization. To understand this photosensitization at a molecular level, you have to understand first what ultraviolet light (290–320 mμ), which causes sunburn reaction in the skin, does to the cell. We have to know where the absorption of light occurs and what is the target of the absorbed energy. The bacterial chemists and physiologists have provided us with very good evidence that the target of the absorbed ultraviolet radiation is usually the DNA of the cell. When bacteria are exposed to UV radiation of wavelengths shorter than 320 mμ, one finds that the major change occurs in the DNA of the cell. There is formation of thymine dimers. The formation of this 3,4-cyclobutane-type pyrimidine dimer appears to be lethal to bacteria. When you study the molecular changes in epidermal cells exposed to various wavelengths, it turns out that the 290–320 mμ wavelengths that cause cutaneous sunburn reaction also produce a similar type of thymine dimers. One also observes the formation of thymine dimers after irradiation at 250 mμ wavelength, but we shall not worry about the effect of these wavelengths because, for all practical purposes, the 250 mμ wavelength is not present in our solar spectrum at the surface of the earth. What is interesting is the fact that wavelengths of 320, 340, or 360 mμ do not evoke any dimerization of thymine and do not yield 3,4-cyclobutane-type pyrimidine dimers in cutaneous cells. However, when psoralen molecules are simultaneously present, the same wavelengths will give the 3,4-cyclobutane-type of a photoadduct of psoralen and thymine. When this occurs, we believe that the cutaneous cells are photosensitized and damaged in the same manner as the bacterial cells are damaged.

Dr. Lerner: Because pigment darkening occurs rapidly and fades fast, can one view the skin as having photoreceptors? Perhaps melanin undergoes a reversible photoelectric reaction.

Dr. Pathak: I think that melanin undergoes a reversible photoelectric reaction. I do feel that neurophysiologists and people expert in electrical conductance must undertake experiments to determine changes in the charge-transfer properties of melanosomes and melanocytes, not only before irradiation, but also during and after irradiation. Studies of variations in cyclic AMP levels during exposure to light should also be undertaken. I think these studies are badly needed to explain some of the changes in the immediate pigment-darkening reactions that we see under *in vivo* conditions when people are exposed to solar radiation.

Dr. Bagnara: In reference to Dr. Lerner's point, I would like to comment further, especially in the light of Dr. Blois's presentation. We know of three clear cases wherein light has a direct effect on chromatophore response. The first relates to amphibian melanophores in tissue culture, and the other two concern tail darkening. In *Xenopus* tadpoles, individual tail melanophores are directly sensitive to light. One can focus a fine beam of light on specific melanophore processes and the reaction is propagated from that point to the whole melanophore. The third example of this is the demonstration of the tail-darkening reaction in the larva of *Agalychnis dachnicolor,* the Mexican frog I spoke of yesterday. At the ultrastructural level the tail melanophores possess no unusual structure that might be a photoreceptor. I can only conclude, therefore, that the melanosomes themselves must be the light receptor.

DR. PATHAK: With respect to Dr. Bagnara's comments I should like to add that long-wave ultraviolet radiation penetrates deep into the dermo-epidermal junction. This kind of radiation promotes pigment-darkening reaction much better than the short ultraviolet radiation of wavelengths 290–320 mμ. It is possible that there is a kind of a receptor at the dermo-epidermal junction that is preferentially absorbing radiation of 320–600 mμ and evoking photocatalytic changes in the transfer and redistribution of melanosomes.

DR. BLOIS: With respect to Dr. Lerner's comment on light sensing by the skin there seem to be at least two possibilities. One is the photoconductivity of melanin as reported by Potts and Au. The other is a thermal effect arising from the presence of intense and nonspecific light absorption by melanosomes, embedded in a largely translucent and turbid medium. One might expect the melanosomes to undergo a modest temperature rise which would produce thermal gradients within the melanocyte, which in turn might alter chemical reaction rates.

DR. PATHAK: Dr. Blois, I should however like to add to your comments that the infrared radiation does not evoke this kind of a color change which is manifested by an immediate pigment-darkening reaction. I do believe that the temperature gradient will alter the rate of the pigment-darkening reaction.

DR. FITZPATRICK: I would like to go back to Dr. Bagnara's question concerning the relation of these findings to increased photosensitivity with psoralens; Dr. Pathak did not mention that, at least *in vitro,* it has been shown that the nonphotosensitizing psoralens do not bind to DNA.

Fine Structure of Pigment Granules in the Human Hair Bulb: Ultrastructure of Pigment Granules*

Kowichi Jimbow and Atsushi Kukita

Department of Dermatology, Sapporo Medical College, Sapporo, Japan

I. INTRODUCTION

Pigment granules in the hair follicles are elaborated within a specialized cell, the melanocyte, located in the hair bulb and actively engaged in melanogenesis. Since the pioneer investigation of Birbeck et al.,[1] a number of studies have been reported on the ultrastructural characteristics of pigment granules in the hair melanocyte and their possible developmental changes in the elaboration of mature pigment granules.[2,3] These studies described pigment granules in the early stage (premelanosomes) as hollow vacuoles which are formed from the enlargement or fusion of the Golgi vesicles and which contain tenuous materials in the form of folded fibrils or incomplete lamellae. These inner structures in the vacuoles are rapidly thickened and defined by the progressive deposition of melanin on their surfaces. The mature pigment granules (melanosomes) thus formed are dense bodies without any internal structures. They are then transferred into surrounding keratinocytes.

In spite of the above proposed growth processes from immature to mature pigment granules, there are still unsolved problems and divergent views on their fine structures. The attention of the present study is

---

* This study was supported by grants from the Ministry of Education, Japan, the Far East Basic Research Fund of Sears, Roebuck and Co., Inc., Chicago, Illinois, and Japan O'Leary Inc. Fund for Pigment Research.

171

directed toward confirming the inner structures of premelanosomes and demonstrating the internal structures of mature melanosomes which have been considered to be structureless.

The terms "premelanosome" and "melanosome" are used as suggested by Fitzpatrick et al., who reported the results of the International Nomenclature Committee.[4]

II. MATERIALS AND METHODS

Observations were carried out on the melanocytes in the hair bulb of normal human black hair follicles of the scalp. Active hair follicles (anagen phase) were removed under local or general anesthesia and dissected under a stereoscopic microscope. These specimens were immediately fixed in 2% osmium tetroxide buffered with phosphate for 2 hours, or in 4% glutaraldehyde buffered with cacodylate and then 1% osmium tetroxide buffered with phosphate for 2 hours. They were dehydrated in ethanol solutions and embedded in epoxy resins. Longitudinal sections were made by means of a Porter-Blum ultramicrotome MT-2 equipped with a Du-Pont diamond knife. These sections were stained with uranyl acetate and lead citrate, and examined in a Hitachi HS-8 electron microscope under an accelerating voltage of 50 kV.

III. OBSERVATIONS AND DISCUSSION

A. *Melanocytes*

Melanocytes in the hair bulb can be readily distinguished from surrounding keratinocytes by virtue of the absence of desmosomes and tonofilaments, and the presence of pigment granules in various developmental stages. Most melanocytes distend their cytoplasm longitudinally in the hair follicles. In their cytoplasm, the well-developed rough-surfaced endoplasmic reticulum and Golgi apparatus are prominent. Centrioles are rarely observable. The filamentous structures found occasionally in the epidermal melanocytes can hardly be seen. Premelanosomes are mostly distributed proximal to the nucleus, particularly near the Golgi apparatus. Melanosomes are scattered in the periphery away from the nucleus or in the cytoplasmic processes (Fig. 1).

B. *Premelanosomes*

1. Earliest form and distribution

Near the Golgi apparatus, large spherical vacuoles about $1.0\,\mu$ in diameter are present. The outer membranes defining the vacuoles have a unit membraneous structure.[2] They contain amorphous structures or structures

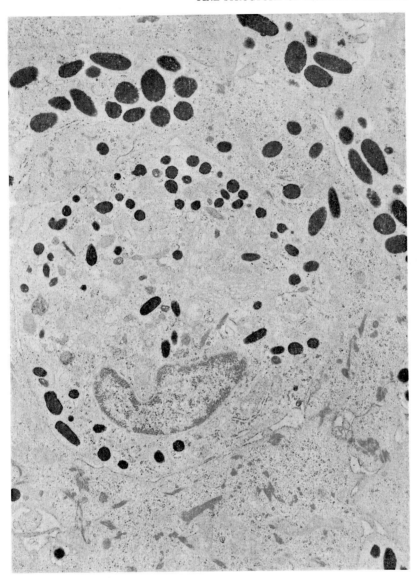

Fig. 1.　Melanocyte among the keratinocytes. In the cytoplasm, well-developed rough-surfaced endoplasmic reticulum and Golgi apparatus are observable. While premelanosomes are distributed in the center of the cytoplasm, mature melanosomes are scattered in the periphery away from the nucleus or in the cytoplasmic processes. ×6,700.

resembling a string of beads folded in a complicated fashion (Figs. 2, 3). A small number of them possess, in addition to the amorphous structures, lamellar or filamentous structures with a regular striation pattern

173

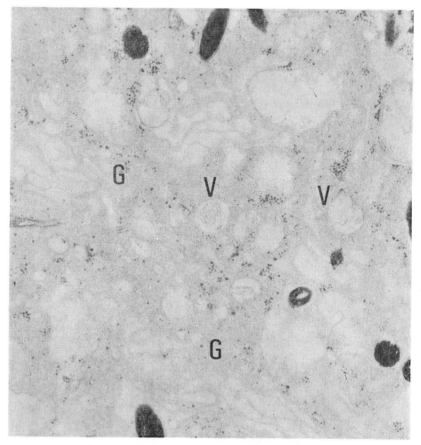

Fig. 2.   Near the Golgi apparatus, spherical vacuoles containing amorphous structures or structures resembling a string of beads folded in a complicated fashion are present. G: Golgi apparatus; V: premelanosome. ×23,000.

of fine densities or cross striations along their length (Fig. 4). Their regular striations show the same periodicities as those in premelanosomes in the more advanced stage of maturation. Therefore the earliest form of premelanosomes in human black hairs may be speculated to be the large spherical vacuoles. Their inner structures consist of amorphous structures of protein matrices which later change into filamentous or lamellar structures with a regular striation pattern. The specific topographic relation between the large vacuoles and Golgi apparatus suggests that premelanosomes may be evolved from the Golgi apparatus. However it is not certain whether they arise from the enlargement of small vesicles that pinch off from the cisternae of Golgi system or instead are contained in

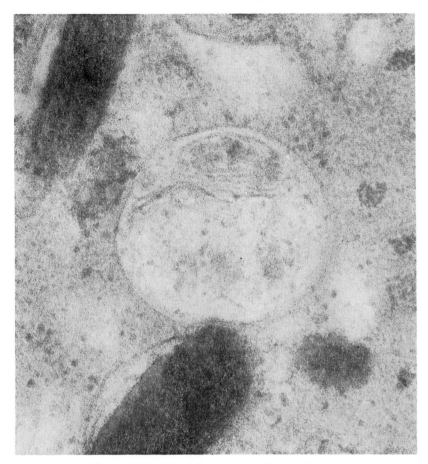

Fig. 3. The earliest premelanosome. The outer membrane defining the vacuole has a unit membranous structure. Note that the amorphous structure in the vacuole is changing into filamentous or lamellar structures, but in this stage they do not show the regular striation pattern seen in the more developed premelanosome. ×98,500.

and develop within a tubular smooth endoplasmic reticulum which is connected with Golgi apparatus during melanogenesis.[3,5-8]

The above findings suggest that the form and distribution of the earliest premelanosomes are essentially similar to those in the melanocytes of guinea pig black hairs, human skin and mucous membrane, and the retinal pigment epithelium of human fetus and chick embryo.[10-13] But their earliest form is different from those in human melanoma cells, in which the earliest premelanosomes take the form of long and slender vacuoles.[9] Their distribution differs significantly from the retinal pigment epithelium of fetal mice and S-91 mouse melanoma cells, where the

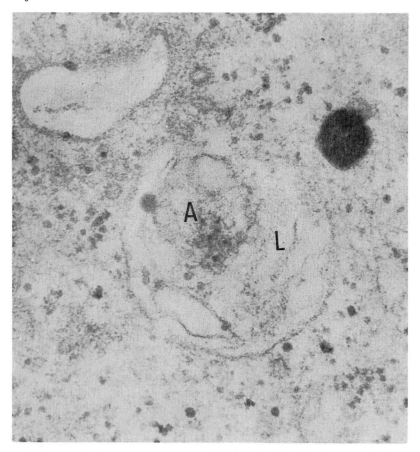

Fig. 4. Premelanosome in a more advanced stage of maturation than that in Fig. 3. Note that the lamellar or filamentous structures reveal regular striation patterns composed of dark lines. In this stage the amorphous structures still coexist. Some of the filamentous structures have changed into lamellar structures. A: amorphous structure; L: lamellar structure. ×98,000.

earliest premelanosomes are reported to be distributed throughout the cytoplasm, even in the dendritic processes, and to be synthesized even in the total absence of any structures resembling Golgi apparatus.[14, 15]

2. Structures in developmental stages

As maturation progresses, the spherical vacuoles of the earliest premelanosomes become elongated, and their inner structures arrange themselves with regularity (Figs. 5, 8). While morphological sequences from early to late premelanosomes are essentially similar in the various melanocytes of different origin, there is still disagreement on the ultrastructure of inner structures in the earlier stage of melanin synthesis.

176

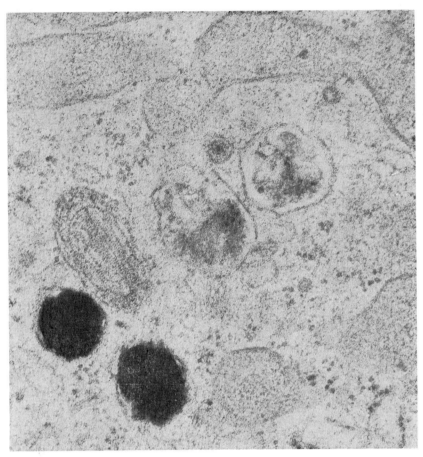

Fig. 5. Three premelanosomes in the early stage of maturation are visible. Note that spherical vacuoles change into elongated ones, and their inner structures arrange themselves with regularity. The inner structures of the premelanosome in the most advanced stage of maturation among these three premelanosomes show a regular striation pattern. ×7,800.

Moyer, Drochmans, and Schroeder reported in their studies of pigment epithelium of fetal mice and human skin and mucous membrane that the inner structure of premelanosomes consists of fibrils which lates aggregate in parallel lines to form a latticelike sheet.[15-17] On the other hand, Birbeck and Breathnach et al. proposed that premelanosomes in human black hairs and retinal pigment epithelium of human fetus are composed of either the folds of several concentric cylindrical membranes or a single membrane wrapped in a spiral.[1,12]

In the present study of human black hairs, the longitudinal sectioning of premelanosomes shows that their inner structures can be classified into

177

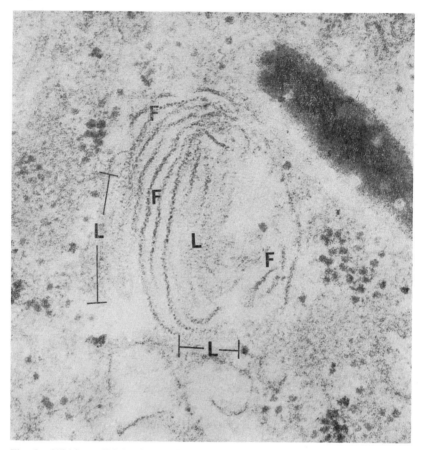

Fig. 6. Within a slightly elongated vacuole, filamentous and lamellar structures are observable. Note that each filament becomes wider and is altered into a lamella as it changes its direction and twists. L: lamellar structure; F: filamentous structure. ×108,000.

two types: filamentous and lamellar structures. Both structures exhibit the same regular striation pattern with dark lines that measure approximately 35 Å in diameter and are 90 Å apart.[3] The filamentous structures are frequently observable along the corner of the vacuoles, running paralle, to their long axis, or slightly curved. The lamellar structures are generally present in their centers. These structures appear to join at the poles forming an oval-shaped structure (Figs. 5, 8). On occasion some of the filamentous structures are found to become wider and to be altered into the lamellar structures as they change their directions and twist (Fig. 6). Cross sections of these premelanosomes provide further information on the structural relationship between the filamentous and lamellar struc-

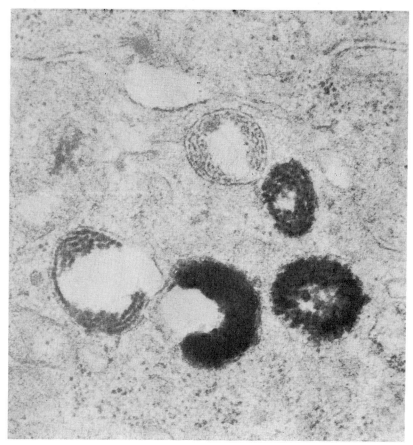

Fig. 7. Cross sections of premelanosomes in the various stages of maturation. The cross-sectioned inner structures are revealed as dark lines of variable length. Some of these dark lines appear to fuse laterally to form a wider one. ×55,200.

tures. In cross section, the inner structures are observed as concentrically arranged dark lines of various lengths (Fig. 7). On the basis of these findings it may be possible to suspect that the inner structures of early premelanosomes in human black hairs consist of lamellar structures. In oblique or perpendicular section, they appear as filaments; their parallel sections appear as lamellae.

The nature of striation patterns along the lamellae is interpreted as representing the alignment of macromolecules of proteins. Drochmans stated that they are helically arranged filaments along the long axis of inner structures.[18] Their photographic examination also confirmed that their configurations appear as a regular, tightly coiled spiral.[17] Birbeck suggested that these helical thread appearances are derived from the array

179

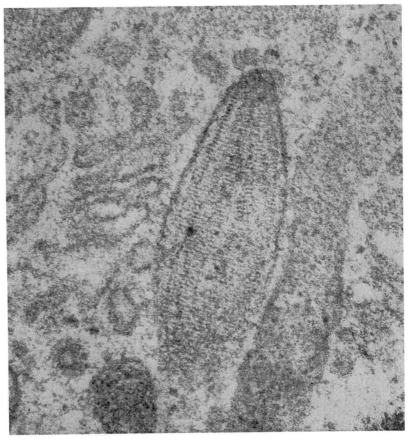

Fig. 8. An elongated premelanosome showing a regular striation pattern composed of dark lines. While lamellar structures are located in the center of the premelanosome, filamentous structures are observed in its peripheral corner. They join at the poles. × 104,000.

of tyrosinase molecules.[2] Seiji assumed that such protein molecules consist of not only tyrosinase but also various proteins including other enzymes and constitutional protein moieties.[9]

Melanin synthesis begins with the deposition of fine granular, osmiophilic substances on the protein matrix of lamellar structures. As this continues, the lamellar structures become electron dense and thickened (Figs. 9, 10). Some of them are found to fuse laterally, forming a wider one. The condensation or fusion of these inner structures and the activation of tyrosinase are considered to be processes which have some factors in common and which start at the same time.[18] Melanin deposition

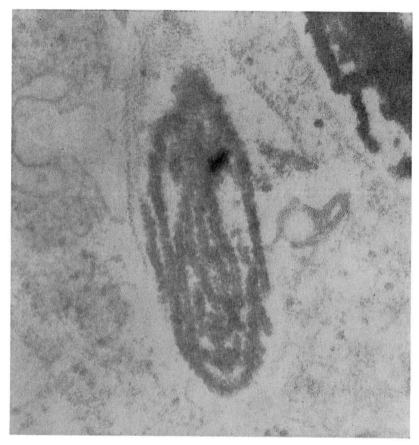

Fig. 9. As melanin synthesis begins, the inner structures of the premelanosome become electron dense and thickened, and the characteristic striation pattern seen in the more immature premelanosomes cannot be observed. ×105,800.

appears to last until the spaces between the inner structures become filled (Figs. 11, 12).

C. *Melanosomes*

Fully developed melanosomes have been described as electron-dense, amorphous bodies. However, these amorphous bodies are reported to reveal faint internal structure in the sections stained with phosphotungstic acid before embedding the fixed blocks.[16] Recently it was also suggested that premelanosomal structures are observable if care is taken in printing the electron micrographs.[3] In the present study, it has been clearly demonstrated that ultrasectioning (300–400 Å in thickness) enables the

181

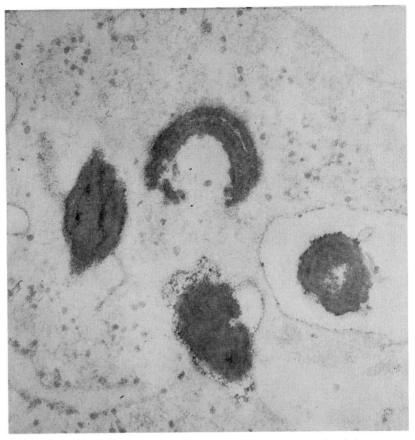

Fig. 10. Late premelanosomes. Melanin is deposited on both sides of their inner structures. ×79,200.

discernment of previously unknown inner structures of mature melanosomes (Figs. 13, 14).

Two components can be distinguished: the dense cortical shell or region, and the less dense central core. The central core consists of regularly folded, lamellar structures disposed in the amorphous matrix. The appearance of these lamellar structures varies according to the direction in which granules are cut. In longitudinal sections they appear as light parallel lines. In cross sections they appear as light concentric lines. Their shape and disposition may correspond to those of lamellar structures in the latest premelanosomes. The cortical shell is present below the outer membrane of melanosomes and encloses the central core. It is composed of fine granular, osmio-philic substances.

In addition to the above two substructures, round electron-lucent

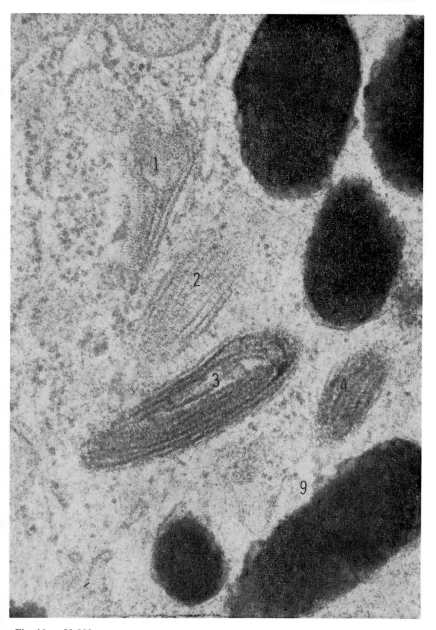

Fig. 11. ×52,800.

Figs. 11 and 12. Morphological sequences from premelanosomes to mature melanosomes. The numbers printed on these figures represent the possible sequence of developmental changes in the pigment granules. In the early stage of maturation, the inner structures of premelanosomes reveal a regular striation pattern along their length,

183

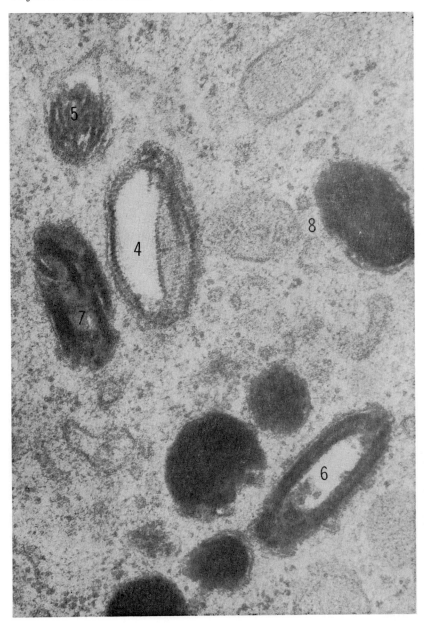

Fig. 12. ×27,000.

but in the later stage they become electron dense and thickened, and such a regular striation pattern cannot be seen. Mature melanosomes appear as electron-dense amorphous bodies.

184

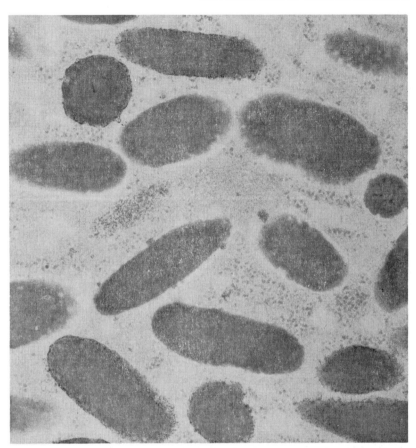

Fig. 13.   Ultrasections of fully developed melanosomes. The presence of internal structures is clearly observable. ×46,000.

structures observed previously in the melanosomes of dermal melanocytosis are resolved in the cortical shell and central core.[20] Unstained
sections fixed with osmium tetroxide alone show that they are surrounded
by fine granular substances of cortical shell (Fig. 15). Serial sectioning
suggests that they are globular shaped and measure up to about 400 Å in
diameter. Most of them appear to be attached to the outer layer of the
central core and are protruding into the cortical shell. These globular
bodies are also found in the melanosomes transferred into the surrounding keratinocytes, and in rare instances in the premelanosomes in which
melanin is almost completely deposited (Fig. 17). The ultrastructural
characteristics of fully developed melanosomes described above are
represented schematically in Fig. 18.

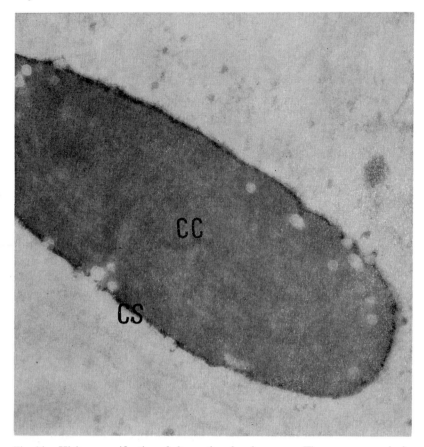

Fig. 14. Higher magnification of ultrasectioned melanosomes. They are composed of a dense cortical shell with fine granular, osmiophilic substances and a less dense central core with premelanosomal structures appearing as light lines. Note the presence of electron-lucent, round structures in the cortical shell and central core. Around the cortical shell the outer membrane of the melanosome can be seen. CS: cortical shell; CC: central core. ×109,500.

Mottaz and Zelickson reported that some melanosomes of very dark brown hairs display a small amount of the "moth-eaten" effect, which is not noted in any of the lighter brown hairs.[3] The morphological characteristics of the effect termed "moth-eaten" and the electron-lucent, globular bodies of the present study seem to be identical. The study of melanosomes of human skin of the vulvar regions showed that granular, dense structures which are arranged in a regular pattern and more or less aggregated in units 30–50 mμ in diameter may be distinguished within the cortical shell.[16] Drochmans explained these structures as melanin particles. The

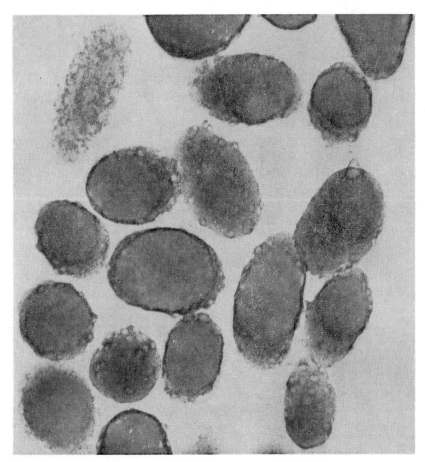

Fig. 15. Unstained sections of mature melanosomes. The presence of round, electron-lucent structures is clearly demonstrated. Most of them appear to be attached to the outer layer of central core and to be protruding into the cortical shell. They are surrounded by fine granular, osmiophilic substances of cortical shell. ×46,000.

globular bodies observed in the present study are, however, electron-lucent and are surrounded by fine granular, osmiophilic substances. Therefore, the globular bodies or the effects termed "moth-eaten" may be different from the melanin particles of Drochmans.

In the present study it is still not clear whether these globular bodies are present only in fully developed melanosomes or were already present in premelanosomes in the masked state and become apparent as melanization is completed. Much further study is needed to elucidate the nature and significance of these structures.

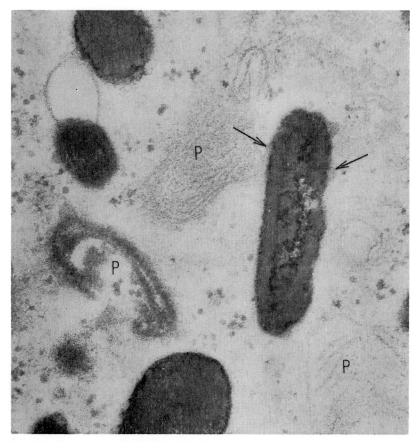

Fig. 16. Electron-lucent globular bodies seen in the mature melanosomes are noted in a premelanosome in which melanin is almost completely deposited (arrows). However, they are not noted in the earlier premelanosomes (P) surrounding this premelanosome. ×79,200.

IV. SUMMARY

Developmental changes of pigment granules in human black hairs were studied by means of an electron microscope. The earliest premelanosomes were assumed to be the large spherical vacuoles located near the Golgi apparatus and containing amorphous protein matrices. As maturation progressed, these spherical vacuoles became elongated, and their inner structures arranged themselves with regularity, showing a striation pattern along their length. The inner structures in the early stage of maturation were observed to be composed of several lamellar structures which appeared as filaments or lamellae depending on the plane of sectioning. Their cross sections appeared as dark lines of various lengths.

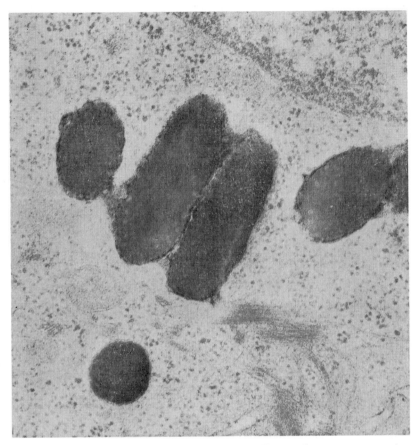

Fig. 17. Melanosomes transferred into keratinocytes. These melanosomes possess electron-lucent, round structures. ×900,000.

Melanization processes in premelanosomes seem to be essentially
similar in the various tissues. Melanin synthesis began with the deposition
of fine granular substances of melanin on both sides of lamellar structures
and continued to the extent that they obscured these inner structures. In
these processes, the lamellar structures were occasionally found to fuse
laterally, forming a wider one disposed concentrically in the vacuoles.

It has been clearly demonstrated that ultrasectioning of fully developed
melanosomes enables the discernment of previously obscured inner
structures, which included the dense cortical shell composed of fine granular, osmiophilic substances, and the less dense central core with the
printing of premelanosomal structures. The present study further confirmed that the "moth-eaten" effect described by Mottaz and Zelickson

189

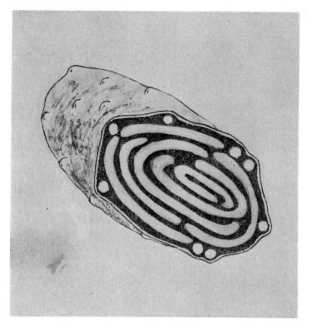

Fig. 18.   Schematic representation of a fully developed melanosome. Note the disposition of electron-lucent, globular bodies.

consists of globular-shaped bodies; these bodies appear to be attached to the outer layer of central core and to be protruding into cortical shell. However, their nature and significance are still uncertain.

REFERENCES

1. Birbeck, M. S. C., Mercer, E. H., and Barnicot, N. A., The structure and formation of pigment granules in human hair. *Exp. Cell Res.*, **10,** 505 (1956).
2. Birbeck, M. S. C., Electron microscopy of melanocytes: The fine structure of hair-bulb premelanosomes. *Ann. N. Y. Acad. Sci.*, **100,** 540 (1963).
3. Mottaz, J. H. and Zelickson, A. S., Ultrastructure of hair pigment. In *Advances in Biology of Skin*, vol. 9, *Hair Growth* (W. Montagna and R. L. Dobson, eds.), p. 471, Pergamon Press, Oxford (1969).
4. Fitzpatrick, T. B., Quevedo, W. C., Levene, A. L., McGovern, V. J., Mishima, Y. and Oettle, A. G., Terminology of vertebrate melanin-containing cells., *Science,* **152,** 88 (1965).
5. Birbeck, M. S. C., Electron microscopy of melanocytes. *Brit. Med. Bull.*, **10,** 220, (1962).
6. Hirone, T., Electron microscopic studies on the subcellular origin and ultrastructure of melanin granules in the retinal pigment epithelium of the chick embryo. *Jap. J. Derm. Ser. B,* **74,** 164 (1964).

7. Horiki, N., Electron microscopic study of dendritic cells of human epidermis. *Hifu* [*Skin*], **8,** 233 (1966).

8. Novikoff, A. B., Albala, A., and Biempica, L., Ultrastructural and cytochemical observations on B-16 and Harding-Passey mouse melanomas; The origin of pre-melanosomes and compound melanosomes. *J. Histochem. Cytochem.,* **16,** 299 (1968).

9. Maul, G. G., Golgi-melanosome relationship in human melanoma *in vitro. J. Ultrastruct. Res.,* **26,** 163 (1969).

10. Parakkal, P. F., The fine structure of melanocytes in the hair follicles of the guinea-pig. In *Advances in Biology of Skin,* vol. 8, *The Pigmentary System* (W. Montagna and F. Hu, eds.), p. 179, Pergamon Press, Oxford (1967).

11. Squir, C. A. and Waterhouse, J. P., The ultrastructure of the melanocyte in human gingival epithelium. *Arch. Oral Biol.,* **12,** 119 (1967).

12. Breathnach, A. S. and Wyllie, L. M.-A., Ultrastructure of retinal pigment epithelium of the human fetus. *J. Ultrastruct. Res.,* **16,** 584 (1966).

13. Toda, K. and Fitzpatrick, T. B., Ultrastructural and biochemical studies of the formation of melanosomes in the embryonic chick retinal pigment epithelium. Presented at the Seventh International Pigment Cell Conference, Sept. 2–6, 1969, Seattle, Washington (in press).

14. Demopoulos, H. B. and Yuen, T. G. H., Fine structure of melanosomes in S-91 mouse melanomas. **50,** 559 (1967).

15. Moyer, F. H., Genetic variations in the fine structure and ontogeny of mouse mela-nin granules. *Amer. Zool.,* **6,** 43 (1966).

16. Drochmans, P., The fine structure of melanin granules (the early, mature, and com-pound forms). In *Structure and Control of the Melanocyte* (G. Della Porta and O. Mühl-bock, eds.), p. 90, Spring-Verlag. Berlin (1966).

17. Schroeder, H. E., Melanin-containing organelles in cells of the human gingiva. *J. Periodont. Res.,* **4,** 1 (1969).

18. Drochmans, P., Ultrastructure of melanin granules. In *Advance in Biology of Skin,* vol. 8, *The Pigmentary System* (W. Montagna and F. Hu, eds.), P. 169, Pergamon Press, Oxford (1967).

19. Seiji, M., Melanogenesis. In *Ultrastructure of Normal and Abnormal Skin* (A. S. Zelick-son, ed.), p. 183, Lea and Febiger, Philadelphia (1967).

20. Kukita, A., Sato, S., Jimbow, K., and Kamino, T., A study of fine structure of melanosome using a 200-kV electron microscope. *Technical Data Sheet,* No. 11 (1969).

DISCUSSION

DR. KUKITA: I would like to ask the doctors in this seminar about the nature of the electron-lucent bodies we found in the fully melanized melanosome.

DR. FUJII: Do you have some idea of the chemical nature of fine-granular osmio-philic granules?

DR. JIMBOW: I think the fine granular substances of the cortical shell may represent the melanin attached to the outer membrane of melanosomes.

DR. SEIJI: Let me say a few words in response to Dr. Kukita's question.

When I examined the retinal pigmented epithelia of chick embryos, 17 or 20 days old, which are fully melanized, I found small, electronless opaque particles, which I think are exactly the same as the ones you mentioned.

DR. PATHAK: A number of people, including Dr. Szabó in our laboratory, have observed these electron-lucent substructures in melanosomes, but none of us has commented on or stressed the existence of these electron-lucent substructures, as you have pointed out today. Their nature remains to be investigated, and it would be interesting to know whether similar electron-lucent substructures are observed in melanosomes in their native state, without prefixing with gluteraldehyde and osmium or other counter-staining techniques.

DR. BLOIS: I would like to ask Dr. Jimbow, or perhaps someone in the audience, if he has had any experience with taking electron micrographs of melanosomes that have been fixed only with something like gluderalde-hyde and never exposed to osmium. I have heard over the years that melanosomes are normally electron dense if no heavy metal comes into contact with them. Is there any comment on that?

DR. JIMBOW: We have never observed sections fixed with glutaraldehyde alone.

DR. HIRONE: I have examined melanotic melanomas *de novo* and their metastatic lesions in regional lymph nodes. The same round bodies Dr. Jimbow has described here were found in the cells of metastatic melanomas, but were not found in the cells of primary lesions. The fixatives used were glutaraldehyde and osmium tetroxide.

DR. QUEVEDO: Dr. Brumbaugh has observed globular bodies within melanosomes of chicken melanocytes, which, I believe, he concluded make a contribution to matrix structure.

DR. MISHIMA: I would like to congratulate Dr. Jimbow and Professor Kukita on their beautiful demonstration. I also saw 400 A-sized electron-lucent substructures in my materials, for example, within melanosomes of blue nevus cells. I would like to ask two questions now:
1. Is there any regularity in the distribution of the electron-lucent substructures?
2. Do you think a premelanosome is primarily composed of spiral structures containing subunit particles?

192

DR. JIMBOW: I did not find any regularity in the distribution of the electron-lucent bodies. We have now been studying the relation between these electron-lucent bodies and the inner structure of premelanosomes, but these bodies are hardly observable in premelanosomes.

DR. FUJII: Recently, we have studied with an electron microscope very thin sections, cut with a diamond knife, of the melanophores of the goby, *Chasmichthys gulosus*. Sometimes we found crystalloid patterns in the matrix of melanosomes. However, no electron-lucent globular structures like those described by Drs. Jimbow and Kukita were found. The matrix was rather homogeneous.

DR. HIRONE: In the early premelanosomes of cells of melanotic melanoma *de novo*, I found several helical fibrils with small granules, 35–45 Å in diameter, occupying the medial and lateral apogees along the helix at definitive intervals. In advanced premelanosomes, the fibrils were laterally associated and cross-linked, forming membranes with cross-striation, just as Dr. Jimbow has described. So I interpret the lamellar structures to be membranes resulting from cross-linkage of fibrils.

Behavior of Melanosomes in Melanocytes and Keratinocytes of Japanese Skin and Black Hair

Shoichi Kinebuchi, Tatsuji Kobori, and Yoshiaki Hori*

Department of Dermatology, Tokyo Teishin Hospital, and
Faculty of Medicine, University of Tokyo,* Tokyo, Japan

I. INTRODUCTION

Recent observations with electron microscopy have revealed that in Caucasian skin, two or three melanosomes are frequently aggregated within a limiting membrane in epidermal keratinocytes, whereas in Negro skin a single melanosomes is usually observed within a limiting membrane. This evidence may indicate an important cytogenic factor controlling skin color.[1]

We wish to report on the behavior of melanosomes in melanocytes and keratinocytes of Japanese skin and in hair bulbs in the same individuals.

II. MATERIALS AND METHODS

Specimens were taken from the hairy skin of 4 normal young male Japanese. The hair follicles were then separated from the surrounding dermis. For routine electron microscopy, hair bulbs and skin specimens were fixed for 3 hours in 2.5% glutaraldehyde in 0.1 M Na-cacodylate buffer, pH 7.4. After initial fixation, the specimens were washed in buffer and subsequently fixed for 3 hours at 4°C in 1% O_8O_4 solution, at pH 7.4, in 0.1 M Na-cacodylate buffer. Routine dehydration and embedding in Epon 812 were then carried out.

The direction of the ultrathin sections of hair bulb was exactly oriented.

195

Staining of ultrathin sections was carried out with uranyl acetate and lead citrate.

Specimens of red hair bulbs obtained were fixed and observed following the methods described above.

III. RESULTS

The melanocytes of hair bulbs are larger than those of epidermis, and the dendritic processes of the hair melanocyte are longer than those of the epidermal melanocyte. The epidermal melanocytes are attached to the basement membrane along a large part of the cell membrane, whereas hair melanocytes are attached to the basement membrane of the dermal papilla with a very small part of the cell membrane, and the greater part of the cell body of the hair melanocyte is located between the hair keratinocytes (Figs. 1–3). The cytoplasm of the hair melanocyte contains well-developed Golgi apparatus, many melanosomes in various stages of melanization, and other organelles (Fig. 4). The hair melanocytes contain many more melanosomes than the epidermal melanocytes, and most of the melanosomes in hair melanocytes are in stage IV of melanization (Figs. 3, 4). The dendritic processes of the hair melanocytes are full of stage IV melanosomes (Fig. 5).

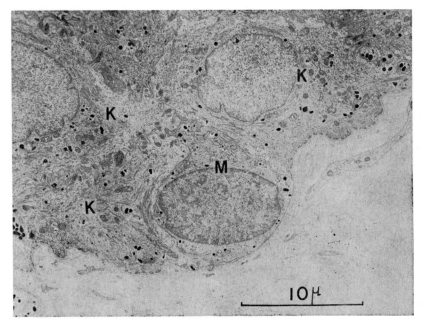

Fig. 1.   Epidermal keratinocytes (K) and a melanocyte (M).

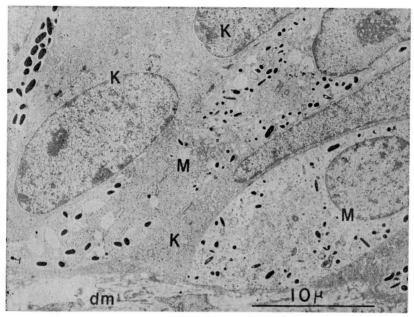

Fig. 2. Hair melanocytes (M) and keratinocytes (K); dm: dermal papilla.

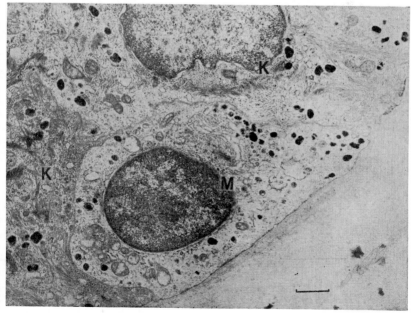

Fig. 3. Epidermal melanocyte (M) and keratinocyte (K).

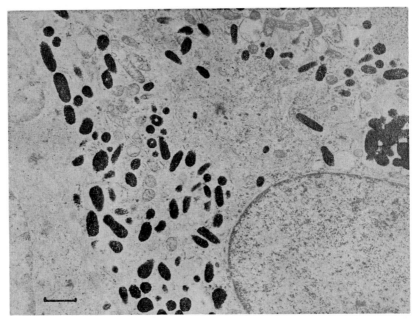

Fig. 4. Cytoplasm of hair melanocyte. It contains well-developed Golgi apparatus, many melanosomes, and other organelles. Melanosomes in the hair melanocyte are larger than those in the epidermal melanocyte.

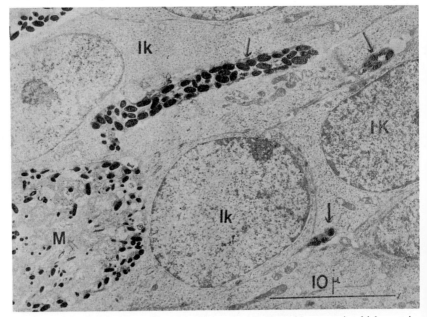

Fig. 5. Melanocyte (M) and its dendritic processes (indicated by arrows), which contain many fully melanized melanosomes, between immature keratinocytes (IK).

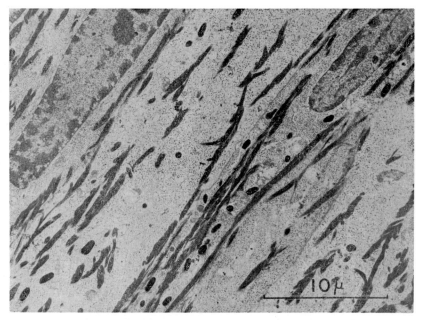

Fig. 6.   Cortex of the hair. Hair keratinocytes have many tonofilaments and fully mela-nized melanosomes. Each melanosome exists singly.

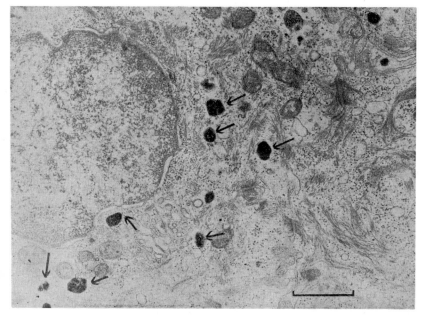

Fig. 7.   Epidermal keratinocyte. Many melanosome complexes (indicated by arrows) are seen in the cytoplasm of the keratinocyte.

199

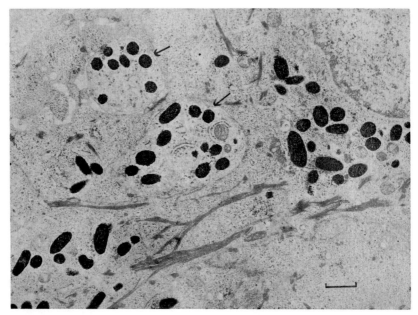

Fig. 8. Cortex of the hair. Dendritic processes (indicated by arrows) of the melanocyte are seen in cytoplasm of the hair keratinocyte. They contain many fully melanized melanosomes.

In the cytoplasm of hair keratinocytes, most of the melanosomes are singly dispersed (Fig. 6). The fully melanized melanosomes are usually singly dispersed or in the form of two or more melanosomes (melanosome complex) in epidermal keratinocytes (Fig. 7). The melanosomes, either singly or in the form of a complex, are surrounded by a limiting membrane (Fig. 7). Dendritic processes of melanocytes can often be found in the cytoplasm of hair keratinocytes; they are full of stage-IV melanosomes (Fig. 8). Almost no melanosome complexes are recognized in the hair keratinocytes, although many melanosomes aggregate in the hair keratinocytes, and each melanosome is surrounded by a single unit membrane (Fig. 8). Immature (undifferentiated) keratinocytes are observed in the lower part of the hair matrix or near the dermal papilla and have few or no tonofilaments and poorly developed desmosomes. Although some dendritic processes of melanocytes are observed between the immature keratinocytes, no melanosomes are observed in the cytoplasm of immature keratinocytes (Figs. 9, 10).

In the hair cuticle, no melanosomes can be seen (Fig. 11). In the cytoplasm of medullary cells, a small number of melanosomes and trichohyaline granules are found (Fig. 12).

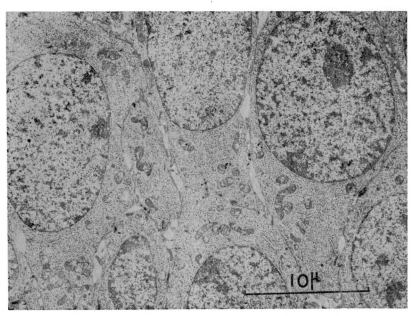

Fig. 9.   Immature (undifferentiated) keratinocyte of the hair matrix. These cells have very poor tonofilaments and undeveloped desmosomes.

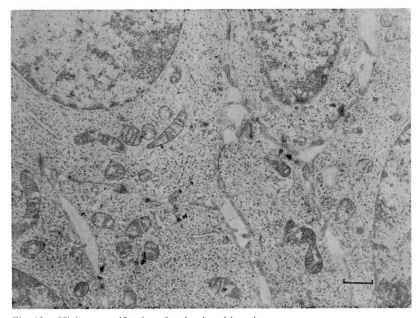

Fig. 10.   Higher magnification of undeveloped keratinocyte.

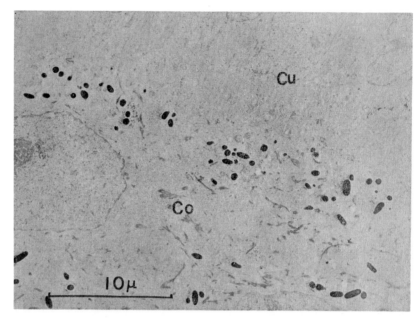

Fig. 11. Keratinocytes of the cortex (Co) have many melanosomes, while keratinocytes of hair cuticle (Cu) have none.

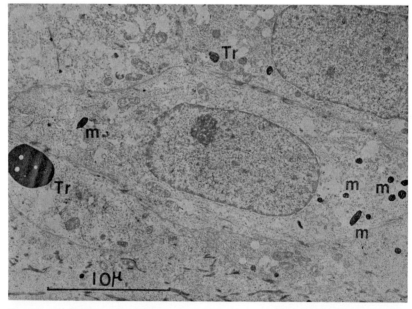

Fig. 12. Medullary cells. These cells have a few tonofilaments and trichohyaline granules (Tr) and melanosomes (m).

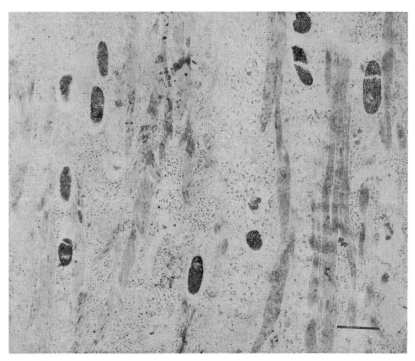

Fig. 13.   Keratinocytes of hair cortex. Melanosomes are seen singly.

The hair melanosomes are two to four times larger than epidermal mela-nosomes (Figs. 13, 14). In red hair, almost all of the melanosomes in the hair keratinocyte form melanosome complexes (Figs. 15, 16). Most of the melanosomes in the melanosome complexes consist of destructed melano-somes and amorphous substances. In normal Japanese skin, melanosomes are seldom found in the upper part of the epidermis, while, in normal Japanese hair, melanosomes still remain in the cytoplasm of well-kera-tinized keratinocytes.

IV.   DISCUSSION

According to Mishima[3] and Pinkus,[2] the hair melanocyte is considered to have migrated from the epidermis. The hair melanocyte should be gene-tically the same as the epidermal melanocyte. It must be considered, of course, that the epidermal melanocyte and hair melanocytes are geneti-cally controlled in different ways, once they locate in the epidermis or in the epidermis or in the hair. Thus, it may be assumed that environmental factors can control melanosome formation and melanin synthesis. These environmental factors may be keratinocytes or dermal elements which are close to the melanocytes.

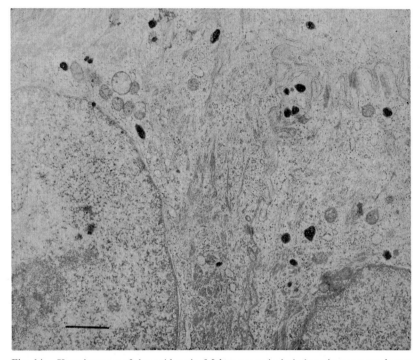

Fig. 14.   Keratinocytes of the epidermis. Melanosomes in hair keratinocytes are larger than those of epidermal keratinocytes.

The melanocytes of the hair matrix are located around the dermal papilla. The nature of dermal papilla in the hair bulb is different from that of the papillary layer in the dermis. The nature of the keratinocytes that surround the melanocytes is also different in the hair and the epidermis. Mitotic figures of hair keratinocytes are frequently seen, and mitotic activity of the hair keratinocyte is probably stronger than that of the epidermal keratinocyte.

It was reported by Hori et al. that melanosomes in Negro skin are larger than those in Caucasian skin. It was also reported by Mottaz et al.[4] that the state of melanosomes in the cytoplasm of hair melanocytes is different in black, blond, and red hair. This may be thought to depend upon genetic factors. But, when the differences in size and behavior of epidermal and hair melanosomes in the same individual are discussed, environmental factors rather than genetic factors should be emphasized. Synthesis of melanin and melanosome formation may be regulated by environmental and genetic factors. If it were true that hair melanocytes had migrated from the epidermis in the embryonic stage, hair and epidermal melanocytes should have the same origin. The difference in size between hair and epidermal melanosomes may be due to merely environmental factors.

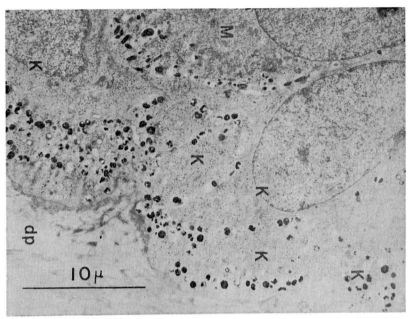

Fig. 15. Dermal papilla (dp), keratinocytes (K), and melanocyte (M) in red hair. Immature keratinocytes have many melanosome complexes.

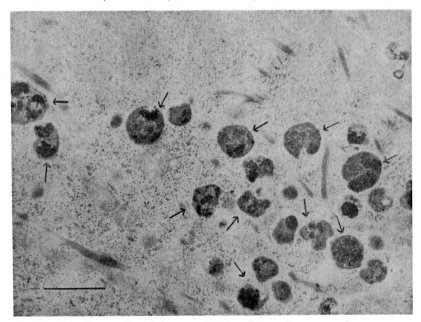

Fig. 16. Keratinocyte of red hair cortex. The keratinocyte has many melanosome complexes (indicated by arrows). Most melanosomes in the melanosome complexes are broken.

205

In the cytoplasm of the Caucasian keratinocyte, two or three melanosomes aggregate and form a melanosome complex, while in the cytoplasm of the Negro keratinocyte, the melanosomes are usually dispersed singly. The relation of morphology of melanosomes between Caucasian and Negro epidermal keratinocytes may be induced to that between Japanese epidermal and hair keratinocytes.

Immature (undifferentiated) keratinocytes do not have melanosomes in their cytoplasm in normal Japanese hair bulb, but in red hair, immature hair bulb keratinocytes have melanosome complexes in their cytoplasm. These findings suggest that there is some genetical regulation in the form of transfer of melanosomes from melanocyte to keratinocyte, and there is increased phagocytic activity in keratinocytes of red hair bulbs.

In the cytoplasm of keratinocytes of Japanese hair bulb, dendritic processes of melanocytes are often found. This fact, however, does not imply phagocytosis of the dendritic processes of melanocytes by keratinocytes.

V. SUMMARY

Normal young male Japanese skin and hair bulbs were investigated by electron microscopy. The size and melanosome formations of the epidermal melanocytes differ from those in hair melanocytes. Hair melanocytes are larger than epidermal melanocytes and produce many more melanosomes than do epidermal melanocytes. The size and behavior of melanosomes in epidermis also differ from those in hair in the same individual. Hair melanosomes are two to four times larger than epidermal melanosomes. The melanosomes in the hair keratinocytes are dispersed singly, whereas most of the melanosomes in the epidermal keratinocytes form melanosome complexes. It may be assumed that these differences depend mainly upon environmental factors. Red hair bulbs were also investigated.

VI. ACKNOWLEDGMENTS

We wish to thank Prof. T. B. Fitzpatrick, who kindly gave us the opportunity to conduct this investigation.

REFERENCES

1. Hori, Y., Toda, K., Pathak, M. A., Clark, W. H., Jr., and Fitzpatrick, T. B., A fine-structure study of the human epidermal melanosome complex and its acid phosphatase activity. *J. Ultrastruct. Res.*, **25,** 109 (1968).
2. Pinkus, H., Embryology of hair. In *The Biology of Hair Growth* (W. Montagna and R. A. Ellis, eds.), p. 1, Academic Press, New York (1958).
3. Mishima, Y. and Widlan, S., Embryonic development of melanocyte in human hair and epidermis. *J. Invest. Derm.*, **46,** 263 (1966).

4. Mottaz, J. H. and Zelickson, A. S., Melanin transfer: Possible phagocytotic process. *J. Invest. Derm.*, **49,** 605 (1967).

DISCUSSION

DR. PATHAK: I have two questions to ask:
1. I would like to know whether in your studies you observed any phosphatase reaction indicative of hydrolytic activity.
2. Did you study the hair bulbs of gray hair and, if so, did you observe any significant differences in the number and size of their melanosome complexes?

DR. KOBORI: Well, we are now conducting acid-phosphatase reactions and so far all melanosomes, both single and in complexes, show a positive reaction. But no conclusions have yet been made. Grey hair is a very difficult sample to collect, but we are now in the process of studying it.

DR. SEIJI: I think the differences among Caucasions, Negroes, and Asiatics in the distribution of the melanosomes in keratinocytes is a very interesting problem. This difference might be due to a different transfer mechanism. I wonder, Dr. Pathak, if you have ever examined under an electron microscope the hair bulb of a Caucasian—how these melanosomes are located.

DR. PATHAK: This study concerning the hair bulb melanosomes is still in progress. Melanosome complexes are indeed present in the hair bulbs of Caucasian subjects. Like the melanosome complexes present in the keratinocytes of the epidermis, the melanosome complexes of the hair bulb can be easily recognized in the dermal papillae of the hair in the anagen phase. I do believe that the presence or absence of recognizable melanosome complexes is a very important feature in controlling the degree of hair color. If you have melanosome complexes having hydrolytic activity (e. g., acid phosphatase), these complexes will render a different hue or color to the hair. The hair will be lighter in color. On the contrary, if the melanosomes are not in the form of membrane-limited complexes, but are single and dispersed randomly in the keratinocytes, they will tend to give a darker hue or color to the hair.

DR. QUEVEDO: The observation of melanosome complexes in epidermal keratinocytes and single melanosomes in hair keratinocytes may reflect a shorter period of "melanocyte-keratinocyte contact" in the hair bulb than in the epidermis.

DR. FITZPATRICK: It is possible to speculate, as Dr. Quevedo has first mentioned, that the presence of a solitary melanosome in the keratinocyte of the Japanese hair bulb may be related to the rate of proliferation of keratinocytes in the hair. It would appear that the Japanese hair bulb follows the same pattern as the Negro epidermis: large, fully melanized melanosomes in the melanocytes, and solitary melanosomes in the keratinocytes. I do not know if any studies have ever been done on the keratinocyte turnover in the Negro epidermis as compared to other races; this would be very interesting in view of the results reported today.

DR. KOBORI: I have no definite conclusions, but the state of melanosomes in the keratinocytes appears to contribute to hair color. But the observations concerning melanosomes in red hair which Dr. Kinebuchi made in Boston were very interesting ones. In red hair, you can see many melanosome complexes in the keratinocytes. The appearance of melanosomes singularly or in complexes seems to contribute to the color of the hair.

DR. SEIJI: I wonder, Dr. Pathak, if you have observed Caucasian skin after irradiation, when there is some hyperpigmentation; in this case are the melanosomes located in the keratinocyte in complexes or in an individual state?

DR. PATHAK: Such melanosomes are found in the form of larger complexes; post irradiation these complexes contain a greater number of melanosomes than prior to irradiation. I personally feel that the hydrolytic activity associated with these membrane-bound complexes plays a significant role in regulating the intensity of skin color. As you examine a cross-section of the epidermis of a Caucasian subject, you are surprized to see that the basal cells are loaded with melanosomes. As you go through the suprabasal cells toward the granular layer, you find very few complexes present, and you begin to wonder what happened to those melanosome complexes that were abundant in the basal cells. The answer is that the formation of melanosome complexes is one of nature's ways of digesting melanosomes. Up to now, we always believed that melanosomes are indestructible. This is no longer a valid assumption. Melanin present in the melanosomes may be intractable to digestion, but melanosomes can be broken down by these hydrolytic enzymes. This appears to be one of the modes of regulating color. In my skin, which is fairly well pigmented, I can see these melanosome complexes way up to the granular layer, but in a fair-skinned individual, I do not find these complexes above the suprabasal cells. So one has to assume that these complexes are being digested or that something is happening to these melanosomes.

Melanin Synthesis in Cultured Melanoma Cells[*]

Atsushi Oikawa, Masami Nohara, and Michie Nakayasu

Biochemistry Division, National Cancer Center Research Institute, Tokyo, Japan

I. INTRODUCTION

The development of melanin in melanocytes is controlled at various stages in a series of reactions leading to formation of melanosomes.[1] Among the different types of amelanotic melanoma cells, the occurrence of melanosomes lacking tyrosinase activity has been reported.[2-4] Another type of regulation, which has been suggested, is control of the level of a specific inhibitor(s) of tyrosinase.[5,6] For studies on the control of melanin synthesis, a method of culturing a stable line of cells forming melanin is required.

Two cultured cell lines of mouse melanoma B16 were established by Hu and Lesney.[7] From one of these, highly pigmented and non-pigmented cell lines were isolated by colony-cloning from agar plate cultures by Claunch et al.[8] The cell line C_3B is one of the most stable lines for pigment formation, although the degree of pigmentation varies from culture to culture, and sometimes pigment is lost completely. However, pigmented clones can be regained by selecting pigmented colonies on a soft agar plate culture.[8] During studies to establish stable pigmented cell lines, we observed an interesting effect of the cultivation temperature on pigmentation. Our results are described here.

---

* This work was supported in part by a grant from the Japanese Ministry of Education.

II. MATERIALS AND METHODS

A. *Cells*

The cell line C$_3$B isolated by Claunch et al.[8] was used. Cells were maintained as monolayer cultures in TD15 culture flasks with a rubber stopper, containing 5 ml of Eagle's minimum essential medium supplemented with 20% calf serum. A mixture of polymixin, neomycin and penicillin[7] or a solution of kanamycin alone was added to prevent bacterial contamination. Cells were subcultured once a week. The size of inoculum was 5 to 10 × 10^5 cells per flask.

B. *Cell culture*

Volumes of 1 ml culture medium containing 1 to 2 × 10^5 dispersed cells were placed in Leighton tubes with a surface area of 5 cm^2. Incubation was carried out at the temperatures indicated in the results. The medium was changed every day or every other day to assure the best proliferation of cells. The cell number was counted with a hemocytometer after trypsinization.

C. *Assay of tyrosinase activity in situ—melanin synthesis*

The methods described in this and the next section, D, were essentially those of Pomerantz[9] with some modifications for use with cultured cells.[10]

Cells were grown in 1 ml of culture medium containing 2 μc/ml of L-tyrosine-3,5-^{3}H. The labeled medium was changed every day and, after used, medium was stored in a freezer ($-15°$C) for the assays described below.

To 0.5 ml of thawed medium, 0.1 ml of 1 M TCA (trichloroacetic acid) and about 50 mg of Norit A were added to adsorb the unreacted tyrosine and its reaction products. The medium was mixed well and, after standing for 30 minutes at room temperature, the suspension was filtered through a Millipore membrane-HA and the radioactivity of 0.1 ml of the filtrate was counted in Diotol under a liquid scintillation spectrometer.

D. *Assay of tyrosinase activity in the cell-free extract*

Cells were trypsinized, washed with isotonic saline, and packed tightly by centrifugation. They were homogenized in distilled water, usually at a concentration of 3 to 6 × 10^5 cells per ml. The assay mixture contained 0.5 μmole of L-tyrosine-3,5-^{3}H (10 μc), 0.05 μmole of L-dopa (3,4-dihydroxyphenylalanine), 25 μg of chloramphenicol and cell homogenate as enzyme, in 0.5 ml of 80 mM sodium phosphate buffer (pH 6.8). After incubation for 5 to 15 hours at 38°C, the reaction mixture was treated in the identical manner as for the assay *in situ*. The assay was carried out in duplicate at two concentrations of enzyme, which assured a linear relation of the extent of reaction to the amount of enzyme.

E. *Calculations*

Tyrosinase activity (T) is expressed as the amount of tyrosine converted by the enzyme at a specified temperature in a specified time, where T_0 is the amount of tyrosine initially contained in the culture media; cpm_t, the

$$T = 2 \times \frac{cpm_e - cpm_{-e}}{cpm_t - cpm_0} \times T_0$$

total radioactivity; and cpm_0, the radioactivity without incubation after treatment with Norit A. Cpm_e and cpm_{-e} are the radio-activities after Norit A treatment of culture media incubated with and without cells, respectively, in the assay *in situ*, or those of reaction mixture incubated with and without cell homogenate, respectively, in the assay of the cell-free extract. The radioactivity found on incubation without cell homogenate, cpm_{-e}, was the same as that found on incubation with boiled cell homogenate.

According to Pomerantz,[11] only one of the two tritium atoms in the tyrosine molecule is released as water by tyrosinase. Hence, the radio-activity released by the enzyme ($cpm_e - cpm_{-e}$) represents half of that in tyrosine before hydroxylation, so that there is a factor of 2 in the equation. If all the tyrosine molecules which have been oxidized to dopa proceed through the whole series of reactions to melanin, the tritium atom at position 7 in the dihydroindol ring of a dopachrome (position 5 in tyrosine) may also be released as water on condensation forming melanin. Thus the factor 2 in the equation should be replaced by a factor close to unity. However, as the true value of the factor has not been determined, we tentatively adopted the value 2 in this work.

The presence of reactions other than that of tyrosinase resulting in release of tritium was ruled out by the fact that no radioactivity was found when a cell-free extract of an amelanotic melanoma cell line (C_2W) was assayed for tyrosinase, or when diethyldithiocarbamate, a potent tyrosinase inhibitor, was added to reaction mixture containing active enzyme.[10]

The culture of C_2W cells gave no tyrosinase activity *in situ*. The tritium released from tyrosine in the C_3B cell culture is quantitatively related to the amount of melanin synthesized.[10]

F. *Cell fractionation*

Cells were trypsinized, washed with isotonic saline, and homogenized in 0.25 M sucrose. The homogenate was fractionated by differential centrifugation. The precipitates obtained on centrifugation at $800 \times g$ for 10 minutes, $10,300 \times g$ for 20 minutes and $105,000 \times g$ for 60 minutes were designated as the "nuclear," "melanosomal-mitochondrial," and "microsomal" fractions, respectively, and the final supernatant as the "cell sap" fraction. These fractions were used for tyrosinase assay.

G. Determination of protein
Protein was determined by Lowry's method with bovine serum albumin as standard.

III. RESULTS

A. Cell proliferation
The growth curve of cultured melanoma cells, line C₃B, is shown in Fig. 1. The growth patterns of cells at 34.5°C and 38°C were practically the same, and the doubling time was approximately 24 hours at both temperatures. At below 34.5°C cells did not become firmly attached to the glass surface and grew more slowly. At 40°C many cells died at an early stage of growth.

At the plateau of the growth curve the cell number was about 1 to 3×10^6 per Leighton tube.

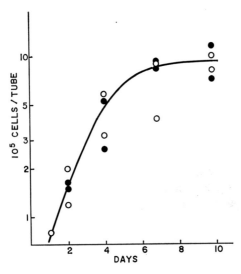

Fig. 1. Growth curve of C₃B melanoma cells cultured at 34.5°C and 38°C. 2.1×10^5 cells were inoculated per tube, and incubated at 38°C. One day after inoculation, half the cultures were transferred to 34.5°C. Two cultures at each temperature were used each time for cell counts. Open circles, 38°C; closed circles, 34.5°C.

B. Tyrosinase activity in situ
When cells were tryspinized and dispersed, many heavily pigmented cells were damaged and died by the second day in the new culture. Surviving cells were faintly pigmented. The total activity and the specific tyrosinase activity of the culture began to increase rapidly when cells regained their proliferative activity (Fig. 2).

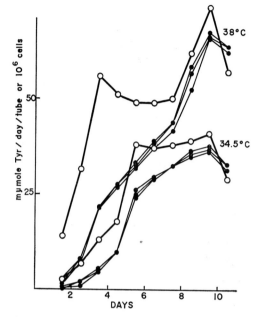

Fig. 2. Tyrosinase activity in cultures. Cultures were made as in Fig. 1. ●—●, tyrosinase activities *in situ* of 3 independent cultures at the respective temperatures; ○—○, tyrosinase activity *in situ* per 10⁶ cells, at the respective temperatures.

The specific activity of tyrosinase of cells cultured at 38°C was definitely higher than that of cells cultured at 34.5°C, especially in the early phase of growth. The tyrosinase activity of a cell-free extract assayed at 38°C was 5 to 16% higher than that assayed at 34.5°C for the same enzyme preparation, but this difference does not explain the big difference in tyrosinase activity *in situ* between cultures incubated at 34.5°C and 38°C.

Table 1. Effect of cultivation temperature on tyrosinase activity of cell-free extracts.

| Exp. no. | Age of culture | Cells cultured at | | A/B |
| | | 34.5°C (A) | 38°C (B) | |
	(days)	(μmoles/mg protein/5 hrs)		
1	5	1.39	0.35	4.0
2	5	0.91	0.29	3.1
	7	0.52	0.25	2.1
		0.54	0.24	2.3
3	5	1.0	0.5	2.0
4	5	0.51	0.12	4.3

C. *Tyrosinase activity of cell-free extracts*

The specific activity of tyrosinase in a homogenate of cells cultured for 5 days at 38°C was compared with that of cells cultured at 34.5°C. Con-

213

trary to expectation, the results in Table 1 clearly show that the cell-free extract from the culture at 34.5°C was 2 to 4 times more active than that from the culture at 38°C. Homogenates of cells cultured for 7 days at the two temperatures also showed a similar relationship.

D. Tyrosinase activity of subcellular fractions

Results (Table 2) show the distribution of tyrosinase in subcellular fractions of cells cultured at 35°C and 38°C. It was found that the increased tyrosinase activity in cells grown at 35°C was due to equal increases in the tyrosinase activities in all 3 subcellular fractions: the "melanosomal-mitochondrial," "microsomal," and "cell sap" fractions. The "nuclear" fractions were heavily contaminated with particulate cell components and cell debris, so that comparison of "nuclear" fractions was meaningless.

Table 2. Tyrosinase activity of cell fractions prepared from two cultures incubated at different temperatures.

Cell fraction	Specific activity* Cells cultured at			Total activity** Cells cultured at	
	35°C (A)	38°C (B)	A/B	35°C	38°C
Homogenate	0.31	0.22	1.4		
Nuclei	0.29	0.21	1.3	0.044 (16.5)	0.043 (21.4)
Melanosomes (mitochondria)	1.02	0.46	2.2	0.173 (65)	0.121 (61)
Microsomes	0.22	0.095	2.3	0.039 (14.5)	0.028 (14)
Cell sap	0.073	0.034	2.1	0.012 (4.5)	0.007 (3.5)

* μmole tyrosine converted/mg protein/12 hours at 38°C.
** μmole tyrosine converted/12 hours at 38°C by a cell fraction prepared from a cell homogenate containing 1 mg protein. Figures in parentheses are percentages of the activity in each fraction.

E. Change of activity on shifting the incubation temperature

Two groups of 4 cultures were incubated in Leighton tubes, at 38°C (H), and 34.5°C (L) respectively for 4 days. Then the medium in all the culture tubes was replaced with labeled medium and 2 tubes from each group were incubated at the other temperature, resulting in 4 subgroups of 2 cultures each, (H), (H–L), (L), and (L–H). Incubation was continued for 20 hours more. During this incubation, the medium in all cultures was replaced by fresh, labeled medium after 5 and 15 hours. The media were stored for assay of in situ tyrosinase activity. After incubation, the cells in the 2 tubes of each subgroup were combined and washed for use in tyrosinase assay of the cell-free extract. Results of the tyrosinase assay in situ and those of the cell-free extracts of the 4 subgroups are shown in Table 3. Tyrosinase activities in situ in the first 5 hours of incubation were very low for some unknown reason(s), but the activities in situ in the second

214

period (5–15 hours) of incubation were similar to those in the third period (15–20 hours) in all 4 subgroups, respectively.

Table 3. Tyrosinase activity in cultures after temperature shifts.

Temperature*	Activity in situ			Activity in cell-free extract**
	1st 5 hrs.	2nd and 3rd 5 hrs.	4th 5 hrs.	
	(mμ moles/culture/5 hours)			(μmole/mg protein/5 hours)
H	5.2	30.7	27.3	0.5
	9.0	27.4	27.5	
H–L	(−2.1)	7.9	8.1	0.6
	(−3.1)	9.1	8.8	
L	(−1.4)	11.0	13.2	1.0
	(−0.8)	10.5	11.2	
L–H	4.3	20.1	22.6	0.4
	2.8	19.8	22.7	

* H=38°C, L=34.5°C.
** Assayed 20 hours after temperature shift. For details of procedure, see text.

On increasing the incubation temperature, the *in situ* activity of the subgroup (L–H) increased to close to that of subgroup (H). On decreasing the temperature the *in situ* activity of subgroup (H–L) decreased to that of subgroup (L). Increase in temperature resulted in decrease in activity of the cell-free extract after 20 hours incubation to the level of that of subgroup (H). However, there was little increase in activity on decreasing the temperature.

IV. DISCUSSION

The pigmentation characteristics of cell line C_3B are of mixed type,[10] in which tyrosinase activity *in situ* increases in the exponential growth phase and then again in the late stationary phase after a plateau of activity (Fig. 2). As mentioned in section E of "Methods," this *in situ* tyrosinase activity is a measure of the melanin-synthesizing activity.[10] Actually, when cultures grown at 34.5°C and 38°C were compared by eye, the difference in color of the cell sheets was clearly visible.

The tyrosinase activity *in situ* in cultures at 38°C is higher than that in cultures at 34.5°C, and the reversed relationship of the tyrosinase activities of cell-free extracts prepared from cultures at the two temperatures is also clear. Possible explanations for this are (1) the half-life of tyrosinase may be longer at the lower temperature, resulting in a new steady state at a higher tyrosinase level; (2) tyrosinase synthesis may be enhanced at the lower temperature; and (3) deposits of melanin are smaller at the lower temperature, as observed experimentally, possibly resulting in an increased

215

population of premature melanosomes. These possibilities require testing. The third hypothesis seems the least likely, because, if it were true, the increase in melanosomal tyrosinase activity should be higher than the activities in the microsomal and cell sap fractions but this was not the case (Table 2).

As shown in Table 3, tyrosinase activity in homogenate of the subgroup of which the temperature was raised (L–H) reached the level of subgroup H and that of the subgroup of which the temperature was lowered (H–L) increased only slightly and did not reach the level of subgroup L within 20 hours. These facts favor the first rather than the second possibility.

Table 4. Tyrosinase activity of cells cultured at 34.5°C and 38°C for 5 days.

| | Cultured at | | A/B |
| | 34.5°C (A) | 38°C (B) | |
	(μmole/mg protein/5 hours)		
Cell-free extract (a)	0.95±0.36*	0.31±0.16*	3.0
In culture (b)	0.0074**	0.023**	0.32
b/a	0.0078	0.073	

* Average and standard deviation calculated from values in Table 1.
** Calculated from the experiment shown in Fig. 2. Corresponding values for cultures at 34.5°C on the 4th and 6th day were 0.0058 and 0.0156, respectively; those for the culture at 38°C were 0.0254 and 0.0223, respectively.

Table 4 summarizes results of two tyrosinase assays for cultures incubated at 34.5°C and 38°C. In the culture at 38°C the activity *in situ* was about 7% of the activity of a cell-free extract for the cells and was less than 1% in the culture at 34.5°C. This lower value for the fraction of active enzyme in the culture at 34.5°C implies the presence of a regulatory mechanism acting on the reaction at an enzymic level, which inhibits the reaction more at 34.5°C than at 38°C and is apparently independent of regulation of the amount of tyrosinase measured in the cell-free extract. This view is also supported by the fact that the response of tyrosinase activity *in situ* to the new incubation temperature is almost completed within 10 hours, while the activity of the cell-free extract of the culture in which the temperature wad decreased (H–L) does not reach the higher level (Table 3). More detailed studies on this are in progress.

V. SUMMARY

Cultured cells of mouse melanoma B16 grow at the same rate at 34.5°C and 38°C. The rate of melanin synthesis increases in the exponential phase of growth and reaches a plateau when cultures reach the stationary phase.

216

Cultures at 38°C synthesize more melanin and reach a plateau earlier than cultures at 34.5°C. On the other hand, growth at 34.5°C is more favorable for tyrosinase activity in the cell-free extract than the higher temperature (38°C). This elevated tyrosinase activity at the lower temperature represented increased activity of all cell fractions with activity, namely, melanosomal, microsomal, and soluble fractions.

When the temperature was raised from 34.5°C to 38°C during culture, the rate of melanin synthesis increased and tyrosinase activity in the cell-free extract decreased to the levels of the culture at 38°C, within one day. When the cultivation temperature was lowered from 38°C to 34.5°C, the rate of melanin synthesis decreased to the level of the culture at the lower temperature, but little change in tyrosinase activity of the cell-free extract was observed.

Some possible mechanisms for regulation of tyrosinase activity were discussed.

REFERENCES

1. Quevedo, W. G., Jr., Genetic regulation of pigmentation in mammals. *These proceedings* (1970).
2. Hu, F., Swedo, J. L., and Watson, J. H. L., Cytological variations of B-16 melanoma cells. In *Advances in Biology of Skin,* vol. 8, *The Pigmentary System* (W. Montagna and F. Hu, eds.), p. 549, Pergamon Press, Oxford (1967).
3. Seiji, M., Kikuchi, A., and Komiya, T., Molecular biology of malignant melanomas. *Saishin Igaku,* **23,** 555 (1968).
4. Mishima, Y., Subcellular activities in the ontogeny of nevocytic and melanocytic melanomas. In *Advances in Biology of Skin,* vol 8, *The Pigmentary System* (W. Montagna and F. Hu, eds.), p. 509, Pergamon Press, Oxford (1967).
5. Chian, L. T. Y. and Wilgram, G. F., Tyrosinase inhibition: Its role in suntanning and in albinism. *Science,* **155,** 198 (1967).
6. Satoh, G. J. Z. and Mishima, Y., Tyrosinase inhibitor in Fortner's amelanotic and melanotic malignant melanoma. *J. Invest. Derm.,* **48,** 301 (1967).
7. Hu, F. and Lesney, P. F., The isolation and cytology of two pigment cell strains from B-16 mouse melanomas. *Cancer Res.,* **24,** 1634 (1964).
8. Claunch, C., Oikawa, A., Tchen, T. T., and Hu, F., Biochemical studies on certain "pigmented" and "non-pigmented" strains of melanoma cells. In *Advances in Biology of Skin,* vol. 8, *The Pigmentary System* (W. Montagna and F. Hu, eds.), p. 479, Pergamon Press, Oxford (1967).
9. Pomerantz, S. H., Tyrosine hydroxylation by mammalian tyrosinase: An improved method of assay. *Biochem. Biophys. Res. Commun.,* **16,** 188 (1964).
10. Oikawa, A., Claunch, C., and Tchen, T. T., unpublished observations.
11. Pomerantz, S. H. and Warner, M. C., 3,4-Dihydroxy-L-phenylalanine as the tyrosinase cofactor: Occurrence in melanoma and binding constant. *J. Biol. Chem.,* **242,** 5308 (1967).

Discussion

DR. QUEVEDO: It is possible that the higher rate of melanin synthesis observed in melanoma cells cultured at 38°C reflects cell selection for increased melanosome assembly from structural and enzymic proteins. Your demonstration of increased tyrosinase activity in homogenated samples of cells cultured at 34°C seems to agree with this view. In addition, Whittaker has proposed that, during rapid growth of chicken retinal melanocytes *in vitro*, the loss of pigment results from the failure of "tyrosinase messenger RNA" to compete successfully with general "growth messenger RNA" for translation on cytoplasmic ribosomes.

DR. OIKAWA: As I showed in Table 3, interconversion of tyrosinase activity *in situ* between two cultivation temperatures occurs within about 10 hours. It is unlikely that the selection of a cell population favorable to increased tyrosinase activity could occur during this time. The higher tyrosinase activity in homogenates of cells cultured at 34°C may be explained in terms of the prolonged half-life of the enzyme at that temperature.

As for the mRNA theory, we have no evidence for our cells. However, in view of the fact that the decrease in tyrosinase activity *in situ* following the lowering of cultivation temperature is accompanied by a slight change of the enzyme activity of cell homogenate (Table 3), it seems likely that the regulatory mechanism for tyrosinase activity *in situ* works independently of the total amount of tyrosinase, which is more than 10 times the acting enzyme (Table 4). The mechanism may include a kind of inhibitor, which cannot act in the cell-free system.

The cell-line used in the present experiment showed a high rate of melanin synthesis at the exponential phase of growth (Fig. 2). In another cell line, not presented here, the rate of melanin synthesis was highest at the late exponential phase and then declined. These results could not be explained by Whittaker's proposal.

DR. FITZPATRICK: When you have a white tumor, is this due to presence of melanosomes and the presence of an inhibitor? I think Dr. Hori did a study recently on a human melanoma.

DR. HORI: When amelanotic melanoma cells which did not contain any melanosomes and were dopa-negative were cultured in the tissue culture medium, those cells became dopa-positive and synthesized melanosomes. Have you examined the activity of hydrolytic enzymes of cultured cells? I feel that when melanoma cells get old melanin synthesis stops and melanin degradation becomes more prominent.

DR. OIKAWA: I have not assayed hydrolytic enzymes of cultured cells. Our results seem to agree with your observation in that the rapidly growing cells synthesize more melanin than the cells at the stationary phase. The second increase in tyrosinase activity *in situ* after a plateau was observed in the late stage of culture (Fig. 2), but at this stage cells might be degenerative; in other words, in an old culture, any controlling mechanism would be inactivated and masked tyrosinase would be activated.

DR. HU: The formation of individual pigmented colonies in "roller tube" culture is perhaps due to the reduction in the size of inoculum, since only a portion of the inoculum finally gets attached to the glass surface and develops into colonies. This may be comparable to Cahn and Cahn's finding that retinal pigment cells lose their pigment in mass culture and regain the pigment in closing conditions. Do the white cell tumors produced by injection of white cell line remain white after serial transplantation? If this tumor remains white after several transplantation generations, this will distinguish a true white tumor from a temporarily white tumor. Ultrastructure of white cell line (different clonal lines, but the same original cell line, from Dr. Oikawa) shows no typical premelanosomes, but there are structures similar to premelanosomes without the typical internal structures. These I consider to be abnormal premelanosomes in both cultured cells and tumors.

DR. OIKAWA: I have not done such a serial transplantation.

DR. SEIJI: Collaborating with Dr. Oikawa, I took some electron microscopic pictures of tissue-cultured melanocytes. There were a lot of melanosomes in the melanocytes. They were very irregular in shape and size. Some of them looked like combined ones, consisting of two or three melanosomes surrounded by a unit membrane. Part of the melanosomes showed myelin-like structures.

DR. JIMBOW: I would like to ask Dr. Seiji if the amorphous bodies or structures coexisting with premelanosomes and surrounded by a unit membrane might be derived from lysosomal degradation.

DR. SEIJI: I imagine so.

DR. LERNER: Several years ago the thermostat in the animal house at Yale University broke down. The animal house overheated and most of the animals died. A few surviving hamsters with melanomas had the tumors transplanted. The new melanoma that appeared was light in color. Have you kept animals at low and high temperature to determine the effect of temperature *in vivo* on melanoma color?

DR. OIKAWA: No, I have not.

The Membrane System and Melanin Formation in Melanocytes*

Makoto Seiji,** Hideko Itakura, and Kazuhiko Miyazaki

Department of Dermatology, Tokyo Medical and Dental University School of Medicine, Tokyo, Japan

I. INTRODUCTION

The melanosome concept of melanogenesis was proposed in 1963, based on various electron microscopic and biochemical evidence accumulated.[1] Since the experimental studies that are to be reported herein concern some of the problems encountered in the various steps of melanosome formation, the melanosome concept is described briefly to provide the background for these problems.

Tyrosinase, the enzyme responsible for melanin formation in the melanocyte, is synthesized in ribosomes; enzyme molecules synthesized are transferred across the limiting membrane of the network and deposited in its interstices. In the form of granules or in solution, the enzyme protein molecules move through the channels of the endoplasmic reticulum (rough-surface membranes) to the Golgi area to be separated there into quanta, each surrounded by its own membranous envelope (smooth-surface membranes), that is, vesicles; within each envelope, the tyrosinase molecules assume an ordered pattern, after which melanin biosynthesis

* This work was supported by a research grant from the Ministry of Education, Japan, and in part by a grant from West Wood Pharmaceuticals, Buffalo, New York, and from the Japan Olili Pigment Research Fund, Osaka.
** Present Address: Department of Dermatology, Tohoku University School of Medicine, Sendai, Japan.

begins and the particle is known as a melanosome. As melanin accumulates on the internal network, the density of the enclosed particles increases until the interstices of the internal network have been filled in. Eventually, the final product, in which melanization is completed and no tyrosinase activity may be detected, is the mature melanosome.

The first problem described herein is the status of tyrosinase in the membrane system, such as the rough endoplasmic reticulum and vesicles. The transfer mechanism of tyrosinase synthesized in the ribosomes, from the rough endoplasmic reticulum through the membrane system to the melanosomes, has not yet been clarified. As a first step for studying such a dynamic mechanism, the status of tyrosinase in the rough-surface membranes and the smooth-surface membranes was examined. The second problem concerns the relationship between tyrosinase activity and melanin formation. At the final stage of melanosome formation, melanin accumulates on the melanosomes; during this period tyrosinase activity of the melanosome has been shown to decrease, and a reciprocal relationship between tyrosinase activity and melanization of melanosomes was found to be present.[2] In the present study, using soluble tyrosinase instead of melanosomes, the inhibitory mechanism of tyrosinase during melanization is studied *in vitro*.

II. MATERIALS AND METHODS

A. *Protease*
Trypsin was purchased from Sigma Chemical Company, St. Louis, Missouri, and used without further purification.

B. *Preparation of the smooth-surface membrane fraction*
Harding–Passey mouse melanomas, which had been serially transplanted in Swiss strain mice, were excised when they were about 1.0–1.5 cm in diameter and were promptly homogenized in nine volumes of 0.25 M sucrose at 0°C with a Potter glass Teflon homogenizer to give a 10% homogenate (w/v). All subsequent procedures were carried out at below 4°C. The homogenate was centrifuged at $11,000 \times g$ for 10 minutes, and the resulting supernatant was centrifuged at $105,000 \times g$ for 60 minutes. The sediment (the small granule fraction) was resuspended in 1.32 M sucrose, and the smooth-surface membrane fraction was isolated by the Rothschild method,[3] with a slight modification; the small granule suspension was placed at the bottom of the centrifuged tube instead of the "mitochondrial supernatant." The smooth-surface membrane fraction was also separated from the small granule fraction by the method of Dallner et al.[4] A 7 ml portion of the small granule suspension in 0.34 M sucrose containing 0.01 M MgCl$_2$ was layered over 4.5 ml of 1.5 M

sucrose containing 0.01 M MgCl$_2$, and centrifuged at 105,000 $\times g$ for 60 minutes. The opaque fraction layer over 1.5 M sucrose solution was carefully separated from the tightly packed pellet of rough-surface membranes, transferred to a new tube, and diluted with distilled water to reduce the sucrose concentration to 0.25 M. The diluted suspension was centrifuged again at 105,000 $\times g$ for 60 minutes to precipitate the smooth-surface membranes as a pellet. A 5 ml portion of the smooth-surface membrane suspension in 0.8 M sucrose solution was layered over the discontinuous density gradient tube which consists of 2 ml of 1.32 M and 3 ml of 1.23 M sucrose solution. The 0.8 M sucrose layer was carefully collected, diluted, and centrifuged again to precipitate the smooth-surface membrane fraction as a pellet.

C. *Preparation of the smooth-surface membrane fraction from the rough-surface membrane fraction*

Deoxycholate (DOC) in 0.15 M sucrose, 0.025 M KCl, 0.01 M MgCl$_2$, 0.035 M Tris buffer, pH 7.8, was mixed with the rough-surface membrane fraction and isolated from the small granule fraction by the methods described above, for a final concentration of 0.4%. Four ml of the mixture was layered over 5 ml of 0.3 M sucrose, 0.035 M Tris buffer, pH 7.6, containing 0.6 M KCl and 0.01 M MgCl$_2$, and centrifuged at 105,000 $\times g$ for 60 minutes. The opaque layer corresponding to 0.15 M sucrose layer was carefully collected and lyophilized, redissolved in 1.5 ml of cold water, and placed on a Sephadex G-50 column, which was then eluted with 0.05 M Tris-HCl buffer, pH 7.2. Two-milliliter fractions of the eluates were collected. Protein and tyrosinase activity in the eluates were measured by absorption at 280 mμ and by the colorimetric method,[5] respectively. The protein fractions with tyrosinase activity were combined and centrifuged at 105,000 $\times g$ for one hour; this resulted in the sedimentation of a considerable amount of membranous material, which is designated as RS in this paper.

D. *Proteolytic digestion of RS and smooth-surface membranes treated with DOC*

A mixture containing RS (about 3.5 mg protein per ml), 0.02 M Tris-HCl buffer, pH 7.4, and a suitable concentration of trypsin was incubated at 0°C. After incubation for a desired period of time, the mixture was centrifuged at 105,000 $\times g$ for 60 minutes. The amounts of various components of the membranes thus released into the supernatant were determined. The extent of solubilization of a component was expressed as the percentage of that component not precipitated under this centrifugal condition relative to that present in the original incubation mixture. In estimating the solubilization of protein, correction was made for the

223

amount of protease used, which was determined by the method described by Lowry et al.[6] In experiments in which solubilization of enzyme was studied, measurements were also made on the whole incubated mixture to evaluate the inactivation of the enzymes caused by the protease treatment.

E. *Analytical procedures*

Protein was determined according to the method of Lowry et al.,[6] with the use of bovine serum albumin as standard. For analysis of phospholipids and RNA using a modification of the Schneider method,[7] up to 5% cold perchloric acid was added to the sample, and the mixture was centrifuged. The precipitate was extracted with ethanol and then with ethanol-diethylether (3:1 v/v). The combined extracts were digested and analyzed for phosphorus by the method of Fiske and Subbarow.[8] The phospholipid content was obtained by multiplying the amount of phosphorus by 25. The residue of ethanol-ether extraction was treated with 6% perchloric acid at 90°C for 15 minutes and RNA was determined from the phosphorus content of the RNA fraction. Ribonucleic acid content was obtained by multiplying the amount of phosphorus by 11.

F. *Enzyme assay*

Tyrosinase activity (E.C. 1.10.3.1., *o*-diphenol: oxygen oxidoreductase) was estimated manometrically by measuring the oxygen consumption with L-dopa as substrate in $M/10$ phosphate buffer, pH 6.8, and was also estimated by the colorimetric method of Shimao.[5] The optical density was determined with the Klett–Summerson photoelectric colorimeter ($\Delta E =$ scale reading). Both ATPase (E.C. 3.6.1.3.) and NADH-cytochrome C reductase activities were determined by the methods described by Ernster et al.[9] Acid phosphatase (E.C. 3.1.3.2.) and esterase (E.C. 3.1.1.3.) activities were estimated by the methods described by Gianetto and De Duve,[10] and Nachlas and Seligman,[11] respectively.

G. *Preparation of soluble tyrosinase*

Soluble tyrosinase was prepared from either the melanosome fraction or the smooth-surface membrane fraction by the method described by Seiji et al.[12] using trypsin digestion followed by gel filtration and purification with DEAE.

H. *Effect of incubation in L-dopa on the activity of soluble tyrosinase*

The *in vitro* experiment designed to test the relationship between artificial melanization and changes of tyrosinase activity is essentially the same as that used by Seiji et al.[2] for the melanosomes isolated from Harding-Passey mouse melanoma. Figure 5 shows the experimental procedure.

224

Soluble tyrosinase A was incubated with a desired amount of L-dopa. When the oxygen consumption had reached its maximum after more than 3.5 hours of incubation, the enzyme activity of the soluble tyroassine A was measured, using 1 mg of L-dopa as a substrate. Simultaneously, soluble tyrosinase B was incubated under identical conditions, but without dopa, so that the tyrosinase activity of the two soluble tyrosinases A and B could be compared.

I. Morphological methods

For electron microscopy, suitable aliquots of each fraction were centrifuged at $105,000 \times g$ for one hour. The sediment thus obtained was fixed at 0°C with 2.5% glutaraldehyde in phosphate buffer at pH 7.2 for two hours and then in 1% osmium tetroxide solution for two hours. The sediment was then dedydrated in a graded series of ethyl alcohol and embeded in Epon 812. The specimens were sectioned with a Porter–Blum Ultrathin microtome and studied with the Hitachi electron microscope HU 11 B.

III. RESULTS

A. Chemical composition of microsomes, smooth-surface membranes, rough-surface membranes, and RS

Table 1 shows the chemical properties of microsomes, smooth-surface membranes, rough-surface membranes, and RS isolated from Harding–Passey mouse melanoma. The biochemical compositions of RS were compared with those of the smooth- and rough-surface membranes. Phospholipid content of RS was three times as high as that of rough-surface membranes and about half of that of smooth-surface membranes. On the other hand, RNA content was 1.7 times as high as that of smooth-surface

Table 1. Chemical composition of various cell particles isolated from Harding–Passey mouse melanoma.

	μg-PL/mg-pr.	μg-RNA/mg-pr.	Tyrosinase ΔE/10min/mg-pr.	ATPase μg-Pi/mg-pr.
Microsome	243	33.1	72	232
Smooth	556	11.1	287	1,083
Rough	175	54	22	326
RS	276	10.5	75	194

	Acid phase μg-Pi/mg-pr.	Esterase μg-β-naphthol/mg-pr.	DPNH-Cyt. C Reductase O.D./5 min/mg-pr.
Microsome	11	374	5.63
Smooth	43	59.5	3.46
Rough	8.2	1,179	8.10
RS	26	375	4.95

membranes and about half of that of rough-surface membranes. Tyrosinase activity of RS was one-fourth of that of the smooth-surface membranes and three times as high as that of rough-surface membranes. Esterase and NADH-cytochrome C reductase activities were found to be much higher in the rough-surface membranes than in the smooth-surface membranes; on the other hand, the smooth-surface membranes were rich in acid phosphatase and ATPase activities.

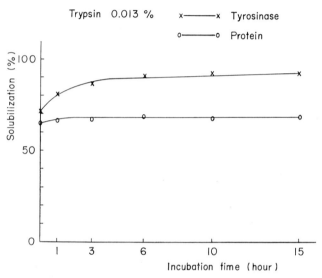

Fig. 1. Time course of solubilization of protein and tyrosinase from RS by the action of trypsin. RS, isolated from the rough-surfaced membranes with DOC (2 mg/ml), was incubated with 400 μg of trypsin at 0°C. After incubation for the indicated period of time, the mixture was centrifuged at 105,000 × g for 60 minutes, and the amount of protein and tyrosinase solubilized in the supernatant was determined. The control was incubated in the absence of added protease.

B. *Digestion of RS with trypsin*

As shown in Fig. 1, trypsin resulted in progressive solubilization of protein from RS. The proteolytic action, however, became limited when about 70% of the total protein had been solubilized, regardless of the protease concentration used. The time required to reach this limit was dependent on the protease concentration; at a concentration of about 200 μg trypsin per mg membranes protein, this limit could be reached within six hours at 0°C. The recovery of tyrosinase activity was fairly good; tyrosinase appeared to be unaffected by protease digestion. Tyrosinase activity, about 90% in all, had been solubilized. The apparent solubilization of more than 60% of the protein without incubation (time: 0) was due to the action of the added protease during centrifugal recovery of the mem-

226

branes from the suspension (0°C, 60 minutes). Figure 1 also shows that about 40% of the protein was recovered in the supernatant even when the membranes were incubated in the absence of added protease. This result appeared to represent the detachment of adsorbed cytoplasmic proteins rather than the action of contaminating proteases, because this solubilization was not significantly time dependent.

Table 2. Solubilization of phospholipids and protein from RS by trypsin.

RS trypsin digest	Distribution (%)	
	PL	Protein
Sup.	21.1	68.5
Ppt.	78.9	31.5

It is likely that the limitation is a reflection of the presence in the membranes of a certain barrier against the proteolytic activity. Table 2 shows the solubilization of protein and phospholipid of RS with trypsin. As can be seen, protein, about 70% in all, was recovered in the supernatant fraction and 21% of total phospholipid of RS was also found to be present in the supernatant. It was concluded that the proteolytic attack, limited by a structural barrier, was accompanied by selective release of tyrosinase and also some release of phospholipid, from the RS.

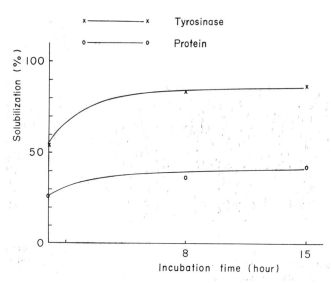

Fig. 2.   Time course of solubilization of protein and tyrosinase from the smooth-surfaced membranes treated with DOC. The control was incubated in the absence of added protease.

C. *Proteolytic digestion of smooth-surface membranes treated with DOC*

RS has been prepared with DOC by detaching ribosomes from rough-surface membranes. Since DOC treatment might alter the nature of the membranes of the rough-surface membrane, the effect of DOC on the membranes was subjected to study using smooth-surface membranes.

The smooth-surface membranes were treated with DOC in exactly the same way as RS was prepared from the rough-surface membranes. The smooth-surface membranes thus treated were digested with trypsin according to the same procedures by which RS was digested. Figure 2 shows

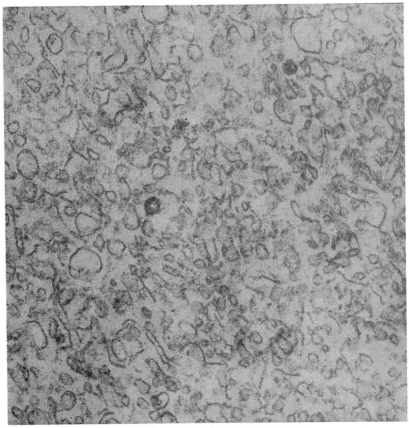

Fig. 3.   Section of the residue of the RS in pellet form, isolated with DOC treatment from the rough-surfaced membranes. Pellet fixed with 2.5% glutaraldehyde in 0.1 M phosphate buffer, pH 7.2, postfixed in 1% OsO_4 in the same buffer. Epon embedding sections doubly stained with uranyl acetate and lead citrate. The pellet consists mainly of RS which has no ribosome. The vesicles retain their unit membranes appearance and are symmetrical. Note that a majority of RS was organized in closed vesicles but some membrane fragments with free margins also occurred.

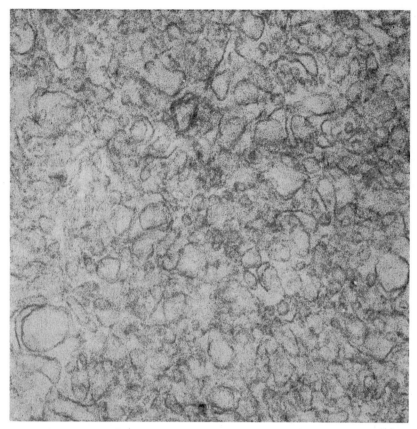

Fig. 4. Electron micrograph of trypsin-digested RS, comprised of closed membranes limited by unit membrane as they originally are.

the experimental results thus obtained. The solubilization of protein from the smooth-surface membranes treated with DOC progressed slowly and the proteolytic action became limited when about 40% of the total protein had been solubilized. The apparent solubilization of more than 25% of the protein without incubation (time: 0) was due to the action of the added protease during centrifugal recovery of the membranes from the suspension (0°C, 60 minutes). Figure 2 also shows that about 15% of the protein was recovered in the supernatant even when the membranes were incubated in the absence of added protease. With respect to the solubilization of the tyrosinase activity, about 87% of the total tyrosinase activity had been solubilized and about 54% without incubation. About 6% of the total phospholipid of the smooth-surface membranes was found to be present in the supernatant. When the experimental results obtained

229

with the smooth-surface membranes were compared with those of the smooth-surface membranes treated with DOC, the extent of the solubili-

Table 3. The effect of incubation in dopa on the activity of soluble tyrosinase.

Exp. no.	Dopa (mg)	Original tyrosinase (μl/min.)	Incubated without dopa (μl/min.)	Incubated with dopa (μl/min.)
1	0.3	1.85	1.56	0.82
	0.6	1.85	1.56	0.70
	1.2	1.85	1.56	0.52
2	0.3	5.02	4.03	3.70
	0.6	5.02	4.03	3.16
	0.9	5.02	4.03	2.97
	1.2	5.02	4.03	2.62
	1.5	5.02	4.03	2.50
3	0.3	7.48	6.17	5.91
	0.6	7.48	6.17	4.96
	0.9	7.48	6.17	3.92
	1.2	7.48	6.17	2.97
	1.5	7.48	6.17	2.28

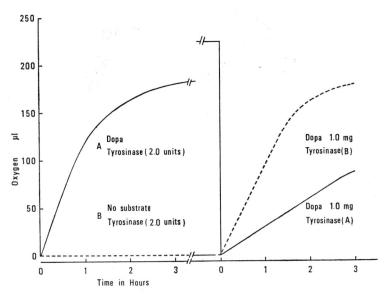

Fig. 5. Effect of dopa incubation on soluble tyrosinase isolated from Harding–Passey mouse melanoma. Tyrosinase of experimental series A was incubated with L-dopa in Warburg apparatus at 38°C with $0.1M$ phosphate buffer at pH 6.8. The control group B was incubated without dopa under the same conditions. After the oxygen consumption of the experimental group reached its maximum, 1 mg of L-dopa was added to the reaction vessels of both groups for measurement of tyrosinase activity.

zation of protein and tyrosinase from the membranes was higher in the latter than in the former; in other words, when treated with DOC, smooth-surface membranes become more susceptible to proteolytic digestion.

D. *Electron microscopic observations*

Under the electron microscope, RS appeared as membrane-bound, closed vesicles with no recognizably attached ribosomes, as shown in Fig. 3. Note that RS is organized in closed vesicles; some membrane fragments with free margins also occur. Figure 4 shows the electron micrograph of the residue of trypsin digestion, which was comprised of closed membranes limited by unit membrane as they originally were, but their width was 73 Å, which is slightly narrower than that of RS, 81 Å.

E. *Effect of incubation in L-dopa on the activity of soluble tyrosinase*

At the end of incubation the color of the reaction mixture of A was black and that of B was entirely clear and colorless. Table 3 shows a significant decrease in the reaction velocity of tyrosinase in the reaction mixture A after incubation. A linear relationship between the concentration of dopa in the incubation mixture and the reaction velocity of tyrosinase is shown in Fig. 3 (see also experiment no. 3 in Table 3). It appears to show clearly that the relationship is inverse.

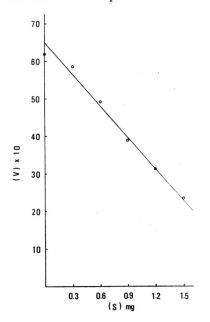

Fig. 6. Relationship between dopa concentration (S) and reaction velocity (V: rate of oxygen consumption, in microliters per minute) of tyrosinase incubated in dopa.

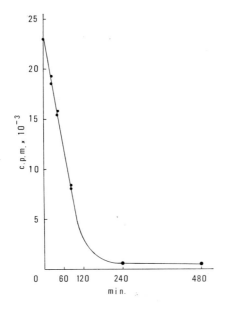

Fig. 7.   Time course of the formation of ^{14}C-melanin in the reaction mixture consisting of ^{14}C-dopa and soluble tyrosinase.

F.   In vitro melanin formation with ^{14}C-L-dopa and soluble tyrosinase

The experimental procedures were the same as those used in the previous experiment except that a small amount of ^{14}C-dopa was mixed with L-dopa. The total radioactivity of the supernatant, which was obtained from the reaction mixture by centrifugation at the end of various periods of incubation, were recorded. Figure 7 shows clearly that ^{14}C-L-dopa was oxidized and disappeared rapidly from the supernatant. In other words it appears that ^{14}C-L-dopa was oxidized and polymerized as a melanin quite rapidly and precipitated. The relationship between melanin formation and enzyme as a protein present in the reaction mixture *in vitro* was investigated in the following model experiment.

G.   Behavior of albumin in the melanin formation system in vitro

The soluble tyrosinase was incubated with L-dopa and ^{131}I-albumin *in vitro*. The radioactivity of an aliquot of the supernatant obtained from the reaction mixture by centrifugation at the end of various incubation periods was determined with a liquid scintillation counter. Figure 8 shows that ^{131}I-albumin in the supernatant disappeared rapidly after some in-

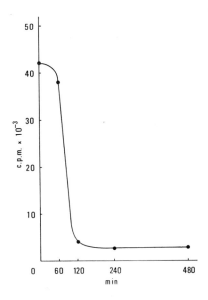

Fig. 8.   Time course of the formation of ^{131}I-albumin-melanin complex in the reaction mixture consisting of ^{131}I-albumin, dopa, and soluble tyrosinase.

duction period for about one hour as melanin formation proceeded. It can be interpreted that, as melanin formation takes place, albumin which is not involved in the enzyme reaction but is present in the reaction medium must combine with melanin and be precipitated with them.

IV.   DISCUSSION

The membrane component of the rough-surface membrane was isolated using deoxycholate treatment and density gradient centrifugation. The membranes thus isolated are called RS in the present paper. This RS can be characterized biochemically by comparing its chemical composition to those of the smooth-surface membranes and the rough-surface membranes. As shown in Table 1, the RNA content of RS was almost the same as that of the smooth-surface membranes; therefore, RS obtained in the present study was not appreciably contaminated by rough-surface membranes and free ribosomes, and was essentially homogeneous, as shown in the results of the electron microscope observations. On the other hand, the phospholipid content per mg of membrane protein of RS was much lower than that of the smooth-surface membranes; therefore protein content in the RS might be higher than in the smooth-surface membranes. Specific tyrosinase activity of RS was one-fourth of that of the smooth-surface membranes. It is indicated that the smooth-surface membranes, which probably contain most of the Golgi vesicles, are much higher in tyrosinase content than RS, which may be a channel of tyrosinase transfer. The low content of acid phosphatase, one of the lysosomal enzymes, in RS also may indicate that the smooth-surface membranes would be contaminated with Golgi vesicles. In the previous report,[12] the smooth-surface membranes were digested with proteolytic enzymes. The experimental results obtained showed that 70–80% of tyrosinase in the membranes was solubilized by proteolytic enzymes. The finding that proteolytic solubilization of protein from the smooth-surface membranes never exceeded a limit of about 30%, regardless of the specificity and concentration of the protease used, seemed to suggest the presence, in the structure of each membrane, of a barrier against protease action. And it was also speculated that a protease could attack the membranes only at the outside surface of the membranes. The attack, which liberated and decomposed the protein components located on the surface area of the membranes, eventually came to be limited by the barrier. RS was digested with trypsin to study the digestive effect of RS. The proteolytic action became limited when about 70% of the total protein had been solubilized. Tyrosinase activity, about 90% in total, was recovered in the supernatant. These results differed significantly from those obtained in the case of smooth-surface membranes. RS was apparently more susceptible than the smooth-surface membranes to proteolytic digestion. The fact that a significant amount of lipid substances was recovered in supernatant suggests that RS might be digested with a different mechanism. Since a phospholipid is the major constituent of the membranes, the membranes of RS might not be digested only from the outside, as the smooth-surface membranes were attacked, but the membrane itself might disintegrate. Thereupon, it would be necessary to investigate further whether this strange behavior of RS against trypsin digestion is really due to the true nature of RS. Because RS was prepared with DOC from the rough-surface membranes, therefore, the nature of the membrane component of rough-surface membranes might be changed during the preparation. In order to study the effect of DOC on the membranes, the smooth-surface membranes were treated with DOC in exactly the same way as the rough-surface membranes were treated. The smooth-surface membranes thus treated were digested with trypsin and the results were compared with those obtained in the untreated smooth-surface membranes. The limit of solubilization of protein was about 40% of the total protein, which was higher than that of the untreated smooth-surface membranes, and the solubilization levels of protein without incubation, and with incubation in the absence of added protease were also higher. Therefore, it seemed likely that the smooth-surface membranes were altered and made more fragile by the DOC treatment. Judging from these experimental results,

234

the high susceptability of RS to proteolytic digestion may not be due to the true nature of the membranes of the rough-surface membranes. It appears to be unreasonable to emphasize that the membranes of the rough-surface membranes are more susceptible than the smooth-surface membranes.

Melanosomes isolated from Harding–Passey mouse melanoma were melanized *in vitro* by incubation with dopa. The reaction velocity of tyrosinase in melanosomes that had been thus melanized *in vitro* was significantly decreased. There was an inverse linear relationship between the concentration of dopa in the incubation mixture and the reaction velocity of tyrosinase in melanosomes. Using soluble purified tyrosinase, the relationship between tyrosinase activity and melanin formation was investigated. As shown in Fig. 6, a linear relationship between the concentration of dopa in the incubation mixture and the reaction velocity of soluble tyrosinase was shown. As tyrosinase in the melanosomes showed, soluble tyrosinase activity in the reaction mixture decreased with increasing melanin formation. The question arises whether the decrease in reaction velocity can be explained by the type of reaction inactivation or the type of product inhibition. Nelson and Dawson[13] have observed that plant orthodiphenolase or catecholase activity undergoes early and progressive inactivation with oxidation of the substrate. This reaction inactivation does not appear to arise from products formed during the oxidation of catechol but occurs at the time when catechol is oxidized. The tyrosinase inactivation observed after incubation in dopa also occurred only at the time when dopa was oxidized but not with products, such as melanin. Thus, the reduction in the reaction velocity of tyrosinase after incubation in dopa seems to result from a blocking of the active centers on the enzyme. Recently, using soluble tyrosinase and ^{14}C-phenol, Wood and Ingraham[14] obtained evidence to support the same view. On the other hand, Fig. 7 shows that synthesized dopa melanin in the reaction mixture precipitates rapidly and appears to be excluded from the reaction system. If the dopa melanin synthesized combined with active centers and blocked tyrosinase activity, tyrosinase should precipitate with melanin. The model experiment using ^{131}I-albumin clearly suggested that, as dopa oxidation takes place, albumin, which is not involved in the enzyme reaction, combined with melanin including the intermediates and precipitated.

From the foregoing, it does not seem unreasonable, at the present time, to assume that as melanin formation proceeds tyrosinase molecules combine with melanin including intermediates such as dopa quinone or indole-5,6-quinone, and the active centers of tyrosinase are blocked permanently and precipitate with melanin. If the active tyrosinase in the reaction system decreased in such a way, its inactivation mechanism can possibly be termed one type of product inhibition.

V. SUMMARY

Smooth-surface membranes (RS) were prepared with deoxycholate treatment from the rough-surface membranes of Harding–Passey mouse melanoma. The chemical composition of RS was compared with those of the smooth- and rough-surface membranes. The phospholipid content of RS was much higher than that of rough-surface membranes and lower that of smooth-surface membranes. RNA content, on the other hand, was 1.7 times as high as that of smooth-surface membranes. Tyrosinase activity of RS was one-fourth of that of the smooth-surface membranes. Digestion of RS by trypsin released protein with solubilizing phospholipids to a certain extent. The proteolysis was limited when about 70% of the protein had been solubilized. Tyrosinase activity, about 90% in all, had been solubilized. RS was apparently more susceptible than the smooth-surface membranes to proteolytic digestion. However, it was revealed that when the smooth-surface membranes were treated with DOC became more fragile against proteolytic digestion. Therefore the high susceptability of RS to proteolytic digestion may not be due to the true nature of the membranes of the rough-surface membranes.

In electon microscope studies, RS appeared as membrane-bound, closed vesicles with no recognizably attached ribosomes. There are a few membrane fragments with free margins. The residue of trypsin digestion was comprised of closed membranes in various sizes and limited by a unit membrane as they originally were, but their width was narrower than those of RS.

Soluble tyrosinase isolated from Harding–Passey mouse melanoma was incubated with dopa. The reaction velocity of soluble tyrosinase which had been thus reacted *in vitro* was significantly decreased. There was an inverse linear relationship between the concentration of dopa in the incubation mixture and the reaction velocity of tyrosinase. The tyrosinase inactivation observed after incubation in dopa occurred only at the time dopa was oxidized but not with product melanin. At the present time, it does not seem unreasonable to assume that as melanin formation proceeds, tyrosinase molecules combine with intermediates such as dopa quinone or indole-5,6-quinone, and the active centers of tyrosinase are blocked permanently and precipitate with melanin.

VI. ACKNOWLEDGMENTS

We would like to express our sincere thanks to Mr. H. Miyamoto and Miss S. Nakajima of the Tokyo Medical and Dental University for their assistance and cooperation throughout this study.

REFERENCES

1. Seiji, M., Fitzpatrick, T. B., Simpson, R. T., and Birbeck, M. S. C., Chemical composition and terminology of specialized organelles (melanosomes and melanin granules) in mammalian melanocytes. *Nature,* **197,** 1082 (1964).
2. Seiji, M. and Fitzpatrick, T. B., The reciprocal relationship between melanization and tyrosinase activity in melanosomes (melanin granules). *J. Biochem.,* **49,** 700 (1961).
3. Rothschild, J. A., Subfraction of rat liver microsomes. *Fed. Proc.,* **20,** 145 (1961).
4. Dallner, G., Orrenius, S., and Bergstrand, A., Isolation and properties of rough and smooth vesicles from rat liver. *J. Cell Biol.,* **16,** 426 (1963).
5. Shimao, K., Partial purification and kinetic studies of mammalian tyrosinase. *Biochem. Biophys. Acta,* **62,** 205 (1962).
6. Lowry, O. H., Rosebrough, N. U., Farr, A. L., and Randall, R. J., Protein measurement with the Folin phenol reagent. *J. Biol. Chem.,* **193,** 265 (1951).
7. Schneider, W. C., Phosphorous compounds in animal tissues. I. Extraction and estimation of desoxypentose nucleic acid and of pentose nucleic acid. *J. Biol. Chem.* **161,** 293 (1945).
8. Fiske, C. H. and Subbarow, Y., The colorimetric determination of phosphorus. *J. Biol. Chem.,* **66,** 375 (1925).
9. Ernster, L., Siekevitz, P., and Palade, G., Enzyme-structure relationship in the endoplasmic reticulum of rat liver: A morphological and biochemical study. *J. Cell Biol.,* **15,** 541 (1962).
10. Gianetto, R. and De Duve., Tissue fractionation studies: Comparative study of the binding of acid phosphatase, β-glucuronidase and cathepsin by rat liver particles. *Biochem. J.,* **59,** 433 (1955).
11. Nachlas, M. M. and Seligman, A. M., Evidence for specificity of esterase and lipase by use of 3 chromogenic substrates. *J. Biol. Chem.,* **181,** 343 (1949).
12. Seiji, M., Itakura, H., and Irimajiri, T., Tyrosinase in the membrane system of mouse melanoma. Presented at the Seventh International Pigment Cell Conference, Sept. 2–6, 1969, Seattle, Washington (in press).
13. Nelson, J. M. and Dawson, C. R., In *Advances in Enzymology and Related Subjects of Biochemistry* (F. F. Nord and C. H. Werkman, eds.), Interscience Publishers, New York, **4,** 99 (1944).
14. Wood, B. J. and Ingraham, L. L., Labelled tyrosinase from labelled substrate. *Nature,* **205,** 291 (1965).

DISCUSSION

DR. BLOIS: Dr. Seiji's experiment utilizing albumin in the *in vitro* melanizing system is a beautiful demonstration of the reciprocal effect in a defined system. One can conceive of two mechanisms being involved in this reciprocal effect: first, a direct chemical attack, perhaps by semiquinone intermediates, upon the active site of the tyrosinase molecule, thereby inactivating it chemically. The other possibility is the occlusion of the active site by a growing polymer which simply renders this site inaccessible to additional tyrosine or dopa. While Dr. Seiji's experiment does not distinguish between these two mechanisms, it is fully consistent with both and extends the validity of the reciprocal effect to an understandable system.

DR. QUEVEDO: Would you please comment on the relationship of your findings on tyrosinase in smooth and rough endoplasmic reticulum to those concerning the reciprocal relationship of tyrosinase activity and deposited melanin, in the light of the mechanics of melanosome assembly?

DR. SEIJI: This is a really difficult problem. I think nobody knows how the tyrosinase molecule moves through the channel of the membrane system. At the Golgi area, tyrosinase is somehow separated and condensed to form an internal network in the melanosome. Why does the tyrosinase not work, i.e., produce melanin in these processes? I think that, after tyrosinase is aligned or lined up so as to form an internal structure, melanin formation starts. This would be one possibility. The other possibility would be what Dr. Fitzpatrick's group is dealing with. That is, melanin could be formed even on the ribosome, but polymerization does not take place and move with proteins including tyrosinase through the membrane system to the Golgi area. Then, in the melanosome, they may be polymerized and become visible.

DR. CHAVIN:
1.   Would it not be possible to measure copper liberated as a function of tyrosinase destruction during the action of the proteolytic enzymes upon the membranes? This would provide some measure of tyrosinase inactivation during treatment and may provide some rationale for the limits of enzyme release during treatment.
2.   It may be possible that tyrosinase is bound to the melanosomal protein lattice by the tyrosine in the lattice. Therefore, until the "factors" or conditions are right, the enzyme would be inactive.
3.   In view of the discussion dealing with the "inactivation" of tyrosinase by the reaction product, it would appear reasonable that the outer region of the premelanosome would be melanized first, so that the melanosome would contain an impervious melanin shell. Under this circumstance the inner portion would not contain melanin. However, the inner regions of the melanosome are melanized. Thus, some additional factors or other concepts may need to be involved. Your comments regarding such mechanisms would be greatly appreciated.

DR SEIJI:
1.   It would be possible and valuable to study the active center of tyrosinase. I have not done this.
2.   This speculation is very interesting, but as a general rule, we must

know more about the chemical structure of the active center of tyrosinase. I think that your idea applies to affinity column chromatography.

3. So far, we have not seen any melanosomes in which melanization of the internal network was uneven. In other words, every tyrosinase molecule in the melanosome is thought to be able to produce equal amounts of melanin, and to be able to react with tyrosine equally.

DR. FITZPATRICK: It is quite amazing that in a squid ink sac very little tyrosinase is on the organelle, and 90% of the tyrosinase is in soluble form. When Dr. Seiji and I packed off for England in 1958, we arrived there with a paper on the reciprocal relationship of tyrosinase activity and melanization. The studies largely arose from some biological experiments in which it was shown that the black hair bulb contained *less* tyrosinase than the brown hair bulb. Harding–Passey melanoma, which is brown, contained more tyrosinase than B16, which is black. When we arrived, we showed the paper to the Oxford pharmacologist Blaschko. He said, "This is very nice but you've got to get a pure preparation." "How do you do that?" "Well, there is a new technique called sucrose-density gradient." This technique was used to isolate the melanosome as a particle and was the first demonstration of the melanosome as a disparate particle. Now, Dr. Seiji has come back full circle and has really pinned down more elegantly the true reciprocal relationship of tyrosinase melanization. I want to congratulate you on this superb study.

Ultrastructural and Autoradiographic Studies on Melanin Synthesis and Membrane System Using Cultured B16 Melanoma, Irradiated Melanoma, and Human Malignant Melanoma*

Tsutomu Kasuga, Takeshi Furuse, Ichi Takahashi, and Eiko Tsuchiya

Section of Pathology, Division of Physiology and Pathology, National Institute of Radiological Sciences, Chiba, Japan

I. INTRODUCTION

Neoplastic cells and/or tissues with a high turnover rate of cellular activity are well-suited for the study of the cellular metabolism,[1] as has been shown in many reports of experiments. Malignant melanomas have been used for analyzing the mechanism of melanin biosynthesis, and many excellent reports have appeared with relation to the methods and favorable progress of morphological and biochemical techniques. Namely, Seiji, et al. clearly demonstrated that there was tyrosinase activity not only in the melanosome fraction but also in the ribosome fraction by using the sucrose density-gradient technique.[2-4] Moyer suggested, from the standpoint of genetic variation, that unpigmented thin fibrillar melanosomes seem to be developed on the matrices consisting of ribosomal aggregations.[5,6] These observations have essentially disproved the hypothesis that melanosomes might be synthesized in relation to the Golgi apparatus[7-9] or mitochondria.[10,11] Therefore, we can agree with their interpretation of the reason protein biosynthesis is genetically controlled and of the relationship between nucleolar RNA and ribosomal RNA,[13] as is well known in protein synthesis of pancreas cells and plasma cells.

Using electron microscopic autoradiography, Greenberg and Kopac

* This work was supported by grants from the Science and Technology Agency and from the Japanese Ministry of Education

241

found that the incorporated of ^{14}C-dopa was evenly distributed through-out the tissue of the fish melanoma,[12] and Zelickson et al. demonstrated that the grains of ^{3}H-dopa occurred on the ribonucleoprotein particles (RNP) connected with the endoplasmic reticulum.[13] The reason the in-corporation of ^{14}C-dopa was revealed only in large-granule fraction melanosomes has not yet been identified, because of the fact that incorpo-ration of ^{3}H-dopa appeared in the microsome fraction or ribosomes.

On the other hand, the electron microscopic findings suggest that the membrane structures surrounding the melanosomes play the principal role in the process of melanization, in other words, maturation of melanosomes and degradation of melanosomes. The structural pattern of the membrane surrounding the melanosomes and melanin granules is not exactly same, that is, sometimes they are fibrillar and single-layered, and sometimes they are triple layered, and it has not been elucidated to what extent the activity of the enzyme at the limiting membrane plays a part in the maturation of melanosomes.

Recently, chemicals, radiation,[15] and so on have been used to study cellular kinetics. We have found that melanization has been promoted by X-irradiation, and melanosomes have greatly increased in the irradiated melanoma cells.[16] Therefore, we would like to report on the dynamics of membranes in association with melanogenesis and melanization, using human malignant melanomas and cultured B16-XI melanoma cells with X-irradiation, and on the intracellular structure that incorporates ^{3}H-DL-dopa (-2-3-T) and ^{3}H-L-tyrosine, using irradiated B16-XI, cloned cultured melanoma cells.

II. MATERIAL AND METHODS

A. *Materials for experiments*

Six cases of primary human malignant melanoma taken from various regions of the skin and mucosae were examined ultrastructurally. Of these one human malignant melanoma was also used for the examination of ^{3}H-dopa (DL-3,4-dihydroxyphenylalanine) incorporation and was established, further, as a continuous cultured cell strain (24 generations at present).

Melanotic B16-XI melanoma cell strain cloned from the tissue culture of transplantable B16 mouse melanoma obtained from Harvard University under the care of Dr. M. Seiji and Dr. K. Ishikawa was also required for the analysis of the intracellular site of radioactive incorporation. This cloned culture cell strain of B16 melanoma was established in our lab-oratory.[17]

B. *Histochemical procedures*

Tyrosinase activity was examined histochemically and was estimated by the method of Bloch[18] and Laidlaw and Blackberg.[19] Acid phosphatase activity was determined by Gomori's original method.[20] For the detection of lipid transformation of degraded melanin, Sudan III fat staining was applied to human melanoma tissues.

C. *Electron microscopic method*

For routine electron microscopy and ultrastructural autoradiography, the following procedure was used: the melanoma tissues were cut into pieces less than 1 cubic millimeter, and cultured cells were fixed in 2.5% glutaraldehyde buffered with 0.1 M sodium cacodylate, pH 7.4, for 1.5 hours at 4°C. After washing, they were refixed in 1% osmium tetroxide buffered with cacodylate buffer solution, pH 7.4, for 1 hour, to ensure good reservation of membrane detail. The materials fixed were then dehydrated in a graded series of ethanols and embedded in an Epon 812 and Araldite resin mixture. The specimens were sectioned with a Porter-Blum Ultramicrotome MT-1, and studied by a HU-11DS Hitachi electron microscope. The ultrathin sections were stained with saturated uranyl acetate solution for 1 hour and lead hydroxide solution for 5 minutes. Thick sections were also provided and these were stained with toluidine blue or Giemsa solutions for the examination of histology and cytology.[21]

D. *Electron microscopic autoradiography*

The following radioactive compounds were required for the incubation *in vitro*:

1. Tritiated DL-3,4-dihydroxyphenylalanine (dopa) (^{3}H-2-3); specific activity, 11 Ci/mM; radiochemical purity, 97%; produced by the Commissariat a l'Energie Atomique (CEA) in France.

2. Tritiated L-tyrosine (-3,5-^{3}H); specific activity, 36 Ci/mM; radiochemical purity, 97%, produced by the Radiochemical Centre in England.

These radioactive compounds were dissolved in F-10 culture medium containing tyrosine in $1 \times 10^{-5} M$/liter.[22] For studying the fresh human melanoma, the melanoma tissues cut into very small pieces were incubated in F-10 culture medium with 0.5 μCi/ml of tritiated dopa, for 6 hours. For study of the intracellular site of radioactive incorporation in culture cells, the B16-XI cells were seeded in glass petri dishes and were labeled in fresh F-10 medium with 10 μCi/ml of tritiated dopa for 6 hours, or with 10 μCi/ml of tritiated tyrosine for 6 hours when cultured cells seeded in dishes had fully grown in the dishes. Next they were rinsed several times in sterile physiological buffered saline solution; they were then collected by rubber spatula and mixed with 2.5% glutaraldehyde fixative, pH 7.41 at 4°C for 1 hour. One percent osmium tetroxide fixative, pH 7.4, at 4°C for 30 minutes was also used for staining. After that, the electron microscopic procedure described in *C* above was applied. Low-speed centrifugation at 1,500 rpm was used for collecting cells after each step. Finally, cells were embedded in epoxy resin and were cut by ultramicrotome. After being stained with saturated uranyl acetate solution, specimens were covered with an emulsion of NR-H2 obtained from Sakura Photo Film Company and exposed for 3 weeks and 5 weeks. Autoradiographic procedure was done according to the method of Uchida and Mizuhira.[23] Lead hydroxide staining was also done before they were treated with a last carbon-coating.

E. *External radiation method*

When the petri dishes were filled with monolayered cells, the cultured cells were irradiated with X-rays at a dose of 1,000 γ under the condition of rotation (200 kV, 20 mA, a half value layer of 0.3 mm copper and 1.0 mm aluminum). The X-ray machine was manufactured by the Shimazu Company. The cells were immediately incubated in the medium with radioactive compounds for 6 hours after irradiation and were then fixed in glutaraldehyde at 6, 24, and 72 hours after irradiation. These specimens were treated with electron microscopic and autoradiographic procedures and were studied.

III. OBSERVATIONS

A. *Ultrastructural study of melanoma cells*

1. General findings on human melanotic melanoma cells

At low-power, a melanoma cell consisted of a large, bizarre nucleus with prominent large nucleoli and cytoplasm rich in organelles, such as ribosomes, mitochondria, and melanosomes. The Golgi apparatus was located in a small area and was mild. Melanosomes and premelanosomes were dispersed throughout the entire cytoplasm with no relation to any mitochondria[24] or to the Golgi apparatus. The present study is based on this observation that the melanosomes and premelanosomes were distributed diffusely in the cytoplasm, and that the Golgi apparatus was located in a rather restricted area (Fig. 1).

2. Relationship between aggregated ribosomes and their limiting membrane in human and mouse neoplastic melanocytes

The human melanoma cells were rich in aggregated ribosomes (polysomes) which were sometimes free in the cytoplasmic matrix (Fig. 3) and sometimes appeared in connection with the endoplasmic reticulum (Fig. 8). The autoradiographical silver grains of tritiated dopa incorporated in the primary human malignant melanoma of the vagina did appear at the site of the aggregated ribosomes connected with the rough endoplasmic reticulum (Fig. 7). In spite of having few Golgi-derived membrane structures, human malignant melanoma cells contained a striking abundance of fibrillar filaments, tubular smooth membranes, and triple-layered

244

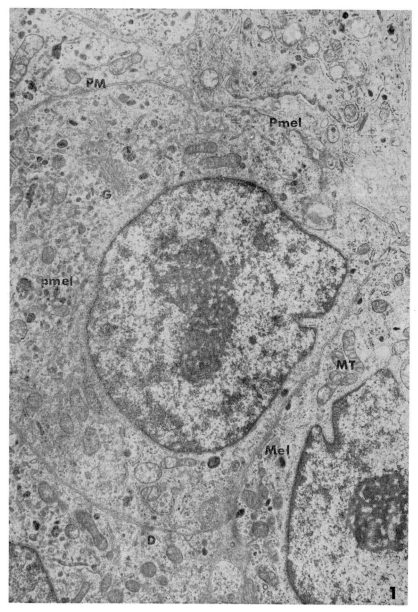

Figs. 1–8, 12–15. Human malignant melanoma originating in the vaginal wall; 65-year-old patient.

Fig. 1. Numerous premelanosomes (pmel) are dispersed throughout the cytoplasm, and there are abundant mitochondria (MT), large in size and with little rough endoplasmic reticula. Note the Golgi apparatus (G) restricted to a small area. Mel: melanosome; PM: plasma membrane; D: desmosome. Uranyl-lead staining. ×8,900.

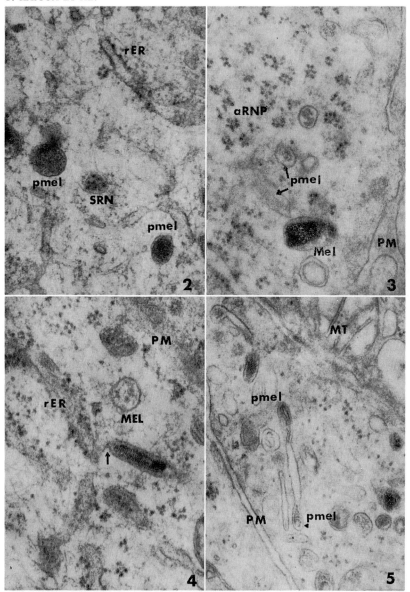

Fig. 2. Aggregated ribosomes surrounded by fine membrane (SRN) and premelanosomes (pmel) are present among the fibrillar structures. Note that the membrane of the SRN is thinner than that of the premelanosome being triple layered. Uranyl-lead staining. ×50,600.

Fig. 3. Aggregated ribosomes (aRNP), premelanosomes of two different stages (pmel), and melanin granule (Mel) near the plasma membrane (PM) are shown. Uranyl-lead staining. ×50,600.

Fig. 4. The triple-layered (unit) membrane of a melanosome (MEL) with melanin deposits is attached in part to the rough endoplasmic reticulum (rER). Uranyl-lead staining. ×50,600.

Fig. 5. Several types of premelanosomes (pmel) are investigated near the plasma membrane (PM). A segmental rod is present at each end of the tubular smooth membrane (indicated by triangles). MT: mitochondria. Uranyl-lead staining. ×50,600.

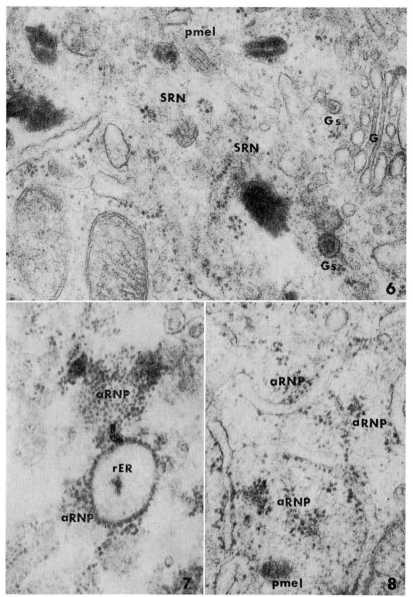

Fig. 6.   Two linearly arranged ribosomes are limited by fine filament (SRN); in another part, two SRN are seen. The Golgi apparatus (G), with a few Golgi saccules (Gs), is mild and small. A premelanosome surrounded by fine filamental sac (pmcl) is composed of relatively opaque ribosomes. Uranyl-lead staining. × 50,600.

Fig. 7.   Silver grains of tritiated dopa are visible on the aggregated ribosomes (aRNP) of the rough endoplasmic reticulum (rER). Uranyl-lead staining. No postfixation with osmium tetroxide. × 50,600.

Fig. 8.   Four aggregated ribosomes (aRNP) and a premelanosome (pmel) are present near the rough endoplasmic reticulum. Uranyl-lead staining. × 50,600.

247

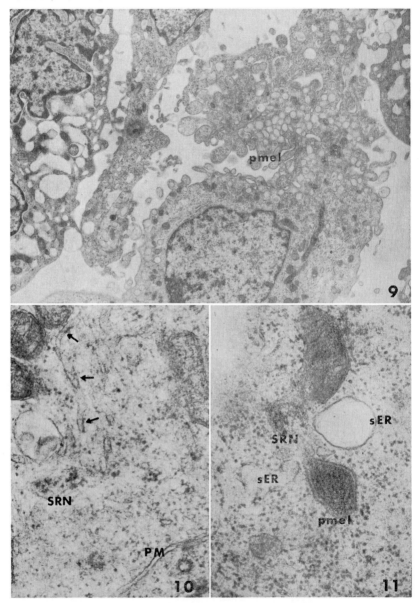

Fig. 9. B16-XI cultured melanoma cells 72 hours after X-irradiation of 1,000 R. The number of premelanosomes (pmel) in the viable cells is markedly increased. However, only slight dopa reaction appeared. Uranyl-lead staining. ×7,256.

Figs. 10, 11. B16-XI cultured melanoma cells 85 days post-irradiation.

Fig. 10. Fine fibrils and tubular smooth membranes (indicated by arrows) have increased in the irradiated cells. Ribosomes are surrounded by fibrillar structures (SRN). Uranyl-lead staining. ×50,600.

Fig. 11. The membranes of two structures of helical configuration (SRN and pmel) are of different thickness, and several vesicular membranes are found between them. The smooth endoplasmic reticulum (sER) is fibrillar in content. Uranyl-lead staining. ×50,600.

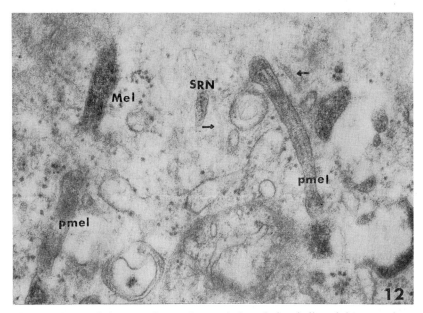

Fig. 12. Two tubular smooth membranes (microtubules, indicated by arrows), a membrane structure with linearly arranged ribosomes (SRN), two longitudinal premelanosomes (pmel), and a melanosome (Mel) are visible. Uranyl-lead staining. ×50,600.

membranes (Figs. 2–8). Although there were not many Golgi saccules (Fig. 6), numerous membrane structures containing periodic arrangement and longitudinal segmental arrays were scattered throughout the cytoplasm, especially near the plasma membrane (Fig. 5), and they were clearly separated from the Golgi apparatus. Aggregated ribosomes either free of or connected to the endoplasmic reticulum were delicately demarcated by a fine fibrillar membrane or a fine tubular smooth membrane. Their photographic pattern resembled that of the irregularly arranged fine filaments or microtubules in the cytoplasm. The thickness of those fine membranes was definitely different from that of the Golgi apparatus membrane. These structures surrounded by fine membrane (SRN) were classified as an early stage of melanosome (ultrastructural precursor; see Figs. 2,6,12). The appearance of a premelanosome connected with rough endoplasmic reticulum (rER) was commonly encountered, and the membranes surrounding melanosomes were thicker than that of rough endoplasmic reticulum (Fig. 4). The findings described above suggest that the melanosome precursor (SRN) will be completed when the aggregated ribosomes become engulfed by smooth membrane with the enzyme in melanin synthesis.

In the B16-XI cultured melanoma cells, cloned from the transplantable B16 mouse melanoma, numerous particles of an ovoid shape with longi-

249

tudinal periodic rods of low electron opacity appeared collectively in the cytoplasm at 3 days after X-irradiation at 1,000 R. Dopa reaction by means of histochemistry did not increase much. These particles were consistent with the premelanosomes classified by Seiji et al.[3] There was no peculiar change in the nucleus or other cytoplasmic organelles (Fig. 9). In comparison with human melanoma cells, B16-XI melanoma cells had fewer fibrillar components in the cytoplasm, but, after irradiation, there were notable increases in fibrillar filaments and tubular smooth membrane (indicated by arrow), and aggregated ribosomes surrounded by those membranes (SRN), similar to the pattern in the human melanoma cells, were also found (Fig. 10). An early premelanosome and a premelanosome were present in relation to vesicular smooth membranes (sER; see Fig. 11).

3. The role of the membrane in melanin degradation

Almost all of the premelanosomes seen in a human malignant melanoma showed lanky tubular shape. Numerous tubular smooth membranes (arrow) were scattered among the organelles (Fig. 12). We observed the disturbing pattern that a particle consisting of moderate electron opacity with three linearly arranged ribosomes and a membrane was phagocytized

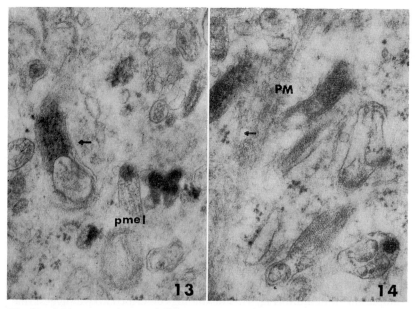

Fig. 13.   Melanosomes in several different stages are shown. Note a premelanosome consisting of four linearly arranged ribosomes and a fine membrane, which is engulfed by another membrane (arrow). Uranyl-lead staining. ×50,600.

Fig. 14.   Tubular smooth membrane (arrow), and four melanosome membranes which phagocytize the other membranes are shown. PM: plasma membrane. Uranyl-lead staining. ×50,600.

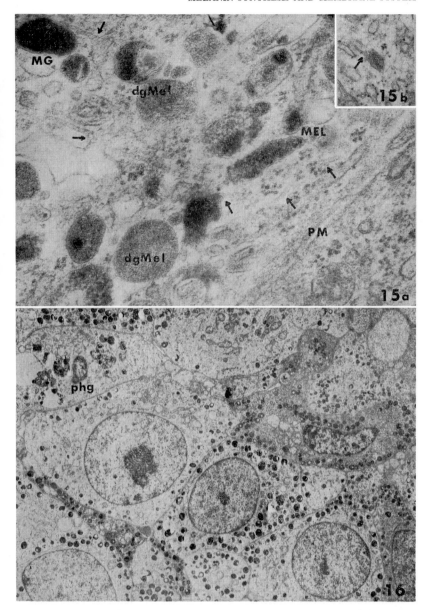

Fig. 15. Tubular smooth membranes are continuously connected with degradated melanosomes (dgMel) and other membranes (Fig. 15a) and are also connected with a premelanosome (Fig. 15b, indicated by arrows). Uranyl-lead staining. ×50,600.

Fig. 16. Human malignant melanoma from the sole of the right foot; 57-year-old female. Almost all of the melanosomes dispersed in the cytoplasm are and they can be identified as lipid-melanin by means of Sudan III staining. A few compound melanosomes are present. phg: phagosome. Uranyl-lead staining. ×3,375.

251

by another smooth membrane (Fig. 13). And it was a common finding that the unit membrane of the melanosomes also included other cytoplasmic organelles (Fig. 14). These patterns were generally observed near the plasma membrane. The transitional patterns from melanosome to degraded melanin are illustrated in Fig. 15; the tubular smooth membranes were directly connected with cystic smooth membrane, the unit membrane of degradated melanin granules, rough endoplasmic reticulm, and so on. These findings were also observed near the plasma membrane and were seen in many human melanomas. In another case of human melanoma, almost all melanosomes and melanin granules were changed to individual lipid-containing granules. In this case, the nature of pigment granules was examined by Sudan III staining for lipid, and they were identified to be lipid-containing melanin, because of the appearance of light microscopical fine lipid granules, evenly distributed in the entire cytoplasm of each melanoma cell, and also because of topographical correlation under electron microscopy. A few phagosomes were present in the melanoma cells as shown at the upper left corner of Fig. 16.

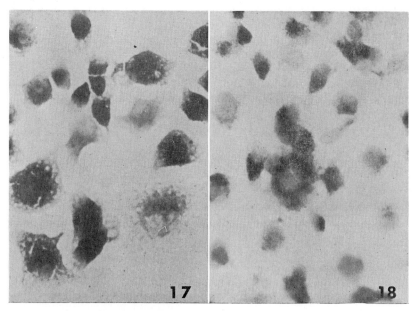

Figs. 17, 18. B16-XI cultured melanoma cells, nonirradiated.
Fig. 17.   Dopa reaction: cells in each of the different stages reveal noticeable tyrosinase activity and show similar levels of tyrosinase activity. Dopa reaction has decreased in a melanocyte of a later stage (lower right corner). ×99.
Fig. 18.   Acid phosphatase reaction: slight positive reaction revealed in melanocytes of early stage. This reaction is increased step by step in the cell maturation, and finally reaction reached in maximum in melanocyte of late stage. ×99.

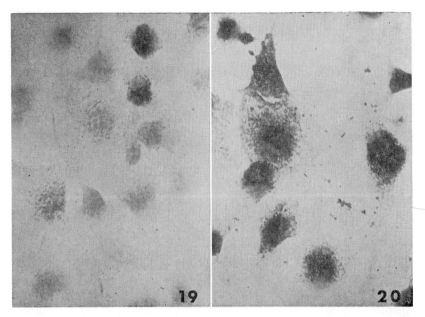

Figs. 19, 20.   B16-XI cultured melanoma cells 175 days post-irradiation with 1,000 rads:

Fig. 19.   Dopa reaction is decreased in the most mature cells, although two relatively young cells contain marked tyrosinase activity. The dendritic processes of the irradiated cells are more prominent than that of non-irradiated cells. ×99.

Fig. 20.   Acid phosphatase activity of the irradiated cells is more accelerated than that of non-irradiated cells. Note numerous dust-like granules (melanin) intra- and extra-cellularly. ×99.

B.   *Histochemistry of B16-XI melanoma cells in vitro*

Strikingly positive dopa reaction appeared throughout the entire cyto-plasm with no relation to the cytoplasmic location, and slight dopa re-action appeared beside the nucleus only in the mature melanocytes of epithelioid form (Fig. 17). In almost all of the irradiated melanoma cells, slightly positive dopa reaction was evenly dispersed in the cytoplasm, but marked reaction was found only in a few cells (Fig. 19).

In postmitotic melanocytes, acid phosphatase reaction was slightly positive and was located beside the nucleus only. This reaction was moder-ately positive in the intermediate melanocyte and was strongly positive in the entire cytoplasm of the mature melanocyte in the late stage (Fig. 18). In the irradiated melanocytes, acid phosphatase activity was diffusely prominent, and large amounts of melanin granules were found not only in the cytoplasm but also in the culture medium, in spite of low tyrosinase activity (Fig. 20).

253

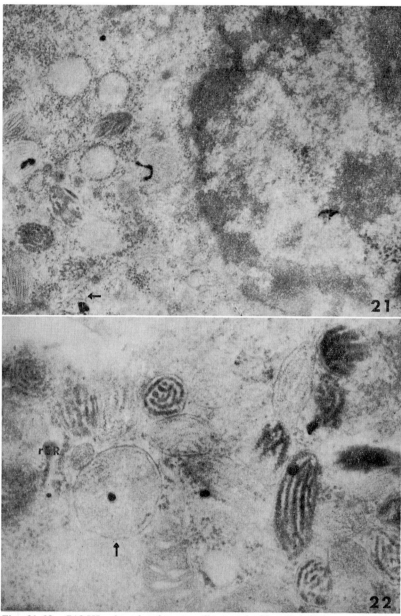

Figs. 21–29. B16-XI cultured melanoma cells treated with radioactive.
Fig. 21. Silver-grains of ³H-dopa are seen on the heterochromatin, premelanosomes, fibrillar melanosome (arrow) and ribosome attached to the rough endoplasmic reticulum. 175 days post-irradiation. Incubated for 30 hours, and Exposed for 5 weeks. ×27,000.
Fig. 22. Grains of ³H-dopa are present on the melanosomes and in the smooth endoplasmic reticulum with certain structure. Grain is also noted on the ribosome of the rough endoplasmic reticulum. Incubated for 6 hours, and exposed for 3 weeks. ×50,600. Uranyl-lead staining.

C. *Ultrastructural autoradiography*

In general, silver grains of tritiated dopa and tritiated tyrosine were present on the ribosomes in connection with the rough endoplasmic reticulm, pre-

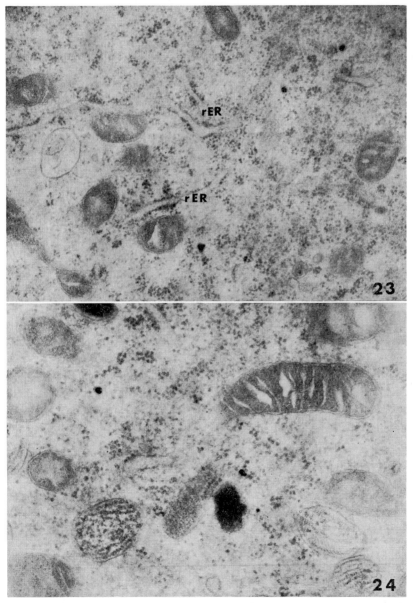

Figs. 23 (×33,750), 24 (×50,600). Grains of ^{3}H-dopa are observed on the only ribosomes. Incubated for 6 hours, and exposed for 3 weeks.

255

melanosomes, melanosomes, abortive premelanosomes, and mitochondria (Figs. 21,22). Grains were sometimes found at the heterochromatin of the nucleus (Fig. 21). Most grains, in general, were seen in relation to the free ribosomes (Figs. 23,24). Although a single grain was manifested on each organelle, the appearance of many concentrated grains was seen only on the areas of the aggregated ribosomes and on the helical configuration (Figs. 25,26). As has been described above, fine fibrillar filaments and tubular smooth membranes were well developed in the B16-XI cultured cells irradiated. Our purpose in the present study was to analyze the developmental pattern of melanosomes in the early stage, the precursor of melanosome, with tyrosinase activity by means of electron microscopic autoradiography, and we have described how the ultrastructural precursor of melanosome has been established with both the aggregated ribosomes and fine smooth membrane. Therefore, we attempted to examine next whether or not tritiated dopa is incorporated in the ultrastructural precursors of melanosomes, and we found that the fine grains were located

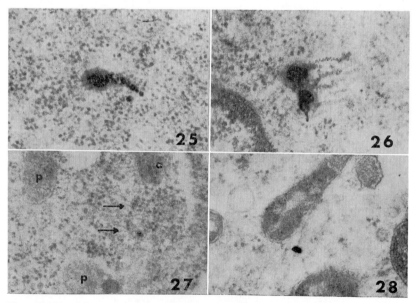

Figs. 25, 26. Many grains are appeared collectively on the ribosomes and fibrillar melanosome with helical configuration. ×50,600. Incubated for 6 hours, and exposed for 5 weeks. Uranyl-lead staining.

Fig. 27. The grain is noted on the architecture consisting of linear arranged ribosomes and fine limiting filament (arrows). An electron opaque with ribosomes(a) and two premelanosomes(p) is represented. ×50,600. Incubated for 6 hours, and exposed for 3 weeks. Uranyl-lead staining.

Fig. 28. Grain is on the fine fibrillar membrane. Incubated for 6 hours, and exposed for 3 weeks. ×50,600. Uranyl-lead staining.

directly on the ribosomal aggregations with some architecture limited by fine fibrillar membrane (Fig. 27), and on the fibrillar membrane (Fig. 28).

The intracellular site of tritiated tyrosine incorporation was also examined. Although tritiated tyrosine seemed to be nonspecifically distri-

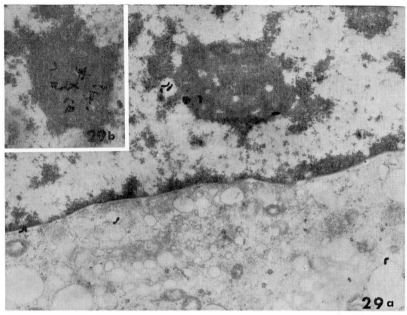

Figs. 29 a, b. Ultraautoradiography of ³H-tyrosine: Grains mainly occur in connection with the rough endoplasmic reticulum and nucleus (a), collectively in nucleolus (b). × 16,900. Uranyl-lead staining. Incubated for 6 hours, and exposed for 5 weeks.

Table 1. Intracellular site of tritiated DL-dopa and L-tyrosine incorporation in B16-XI cultured melanoma cells.

	Tyrosine-T nonirrad.	DOPA-T nonirrad.	DOPA-T 24 hours post-irrad.
CYTOPLASM	%	%	%
Free RNP	4.9	47.8	35.8
rER-RNP	33.2	8.7	7.5
sER	3.1	0.0	11.3
sER-aRNP	0.0	1.2	7.5
Premelanosome	1.2	15.5	22.6
Melanosome	8.3	11.8	7.5
Mitochondria	3.1	5.6	5.7
NUCLEUS			
Heterochr.	11.7	4.3	1.8
Euchr.	12.9	5.0	0.0
Nucleolus	21.5	0.0	0.0

buted in most of the cellular structures on the ultrastructural autoradiographs, we discovered by means of the grain-counting method that many grains did occur on the nucleolus and the ribosomes connected with the rough endoplasmic reticulum (Figs. 29a, b; Table 1).

D. *The intracellular site of tritiated dopa incorporation by means of grain counting*
The ratio of radioactivity associated with intracytoplasmic organelles of the B16-XI cultured cells, both irradiated and nonirradiated, is exhibited in Table 1. The total count of grains was an average of 325 in each series. As shown in the table, the radioactivity of dopa mostly appeared in the free ribosomes and melanosomes, in both irradiated and nonirradiated groups, though radioactivity did not occur in the nucleoli. The remarkable finding in the irradiated group was that 18.8% of the grains existed on the fine smooth membranes and ribosomes limited by fine fibrillar membrane, and 22.6% were in the unpigmented melanosomes of helical configuration.

In the case of human malignant melanoma cells treated with H^3-dopa *in vitro*, the incorporation ratio of radioactive compound in the intracellular organelles was 22.2% in free ribosomes, 25.0% in ribosomes attached to rough endoplasmic reticulum, 22.1% in premelanosomes, 11.1% in melanosomes, 2.7% in mitochondria, 16.6% in nuclei, and 0.0% in nucleoli. However, the number of grains counted was not enough to merit exhibiting in the table.

IV. DISCUSSION

A. *The intracellular site of DL-dopa incorporation and morphological precursors of melanosomes*
The difference between the sites where tyrosinase utilized for melanin synthesis reacts upon ^{3}H-tyrosine and upon ^{3}H-dopa was clearly demonstrated by means of ultra-autoradiography. Although tritiated tyrosine was incorporated in all kinds of organelles in the entire cell, it was mainly incorporated, as shown in Table 1, into the ribosomes relating to the endoplasmic reticulum (32.2%) and nucleus (46.1%). One half out of 46.1% (nucleus) did occur in the nucleolus (Figs. 29a, b). This experimental fact indicates that radioactive L-tyrosine is incorporated in the metabolic pathway of various protein biosyntheses. Tritiated DL-dopa plays a more specific role in the metabolic pathway of melanin synthesis than does tritiated L-tyrosine.[12] Subcellular localization of ^{14}C-dopa has been reported, using biochemical analysis; ^{14}C-dopa incorporation appeared only in the melanosome of Harding–Passey and B16 transplantable mouse melanomas.[4] With ultrastructural autoradiography, the silver grains of ^{14}C-dopa occurred mainly in the melanosomes (68.5%) of Harding–Passey

transplantable mouse melanoma.[14] In the present experiment, the intracellular site of tritiated dopa incorporated into the B16-XI melanoma cells *in vitro* was confirmed, by means of utilization of the electron microscopic autoradiographic method of Uchida and Mizuhira,[23] to be the ribosomes (56.5%) and melanosomes (27.3%). Zelickson et al. reported that the primary site of incorporation of tritiated dopa was on or in the ribosomes attached to the endoplasmic reticulum (66%) of Cloudman S-91 transplantable mouse melanoma.[13] The morphology of melanogenesis of human malignant melanomas and B16-XI cultured melanoma cells has demonstrated that the primary structure that is the earliest stage of melanosome development is brought to completion with the development of polysomes with tyrosinase activity and the evolution of fine membrane limiting it from other components of the cytoplasmic matrix. We call this construction a precursor of melanosome, electron microscopically, and it has been demonstrated that tritiated dopa has been incorporated into the precursor (1.2%). The fine membranes usually were single layered, and not triple layered, and it has been indicated that they might be derived from either intracytoplasmic fibrillar filaments or microtubules similar to the spindle fibers. As shown in Table 1, in the experimental cells, both irradiated and nonirradiated, tyrosinase activity by means of tritiated dopa incorporation increased step by step from precursors, to ribosomes with limiting membrane, to mature melanosomes, until melanosomes finally lose their interior structure. The most noteworthy observation in the cultured melanoma cells stimulated by X-irradiation was that tritiated dopa was incorporated in the smooth endoplasmic reticulum (11.3%) and in smooth endoplasmic reticulum with aggregated ribosomes (7.5%). This might be explained in relation to the phagocytizing activity of melanosomal membrane and the speed of polymerization.

B.   *The nature and role of the melanosomal membrane*

Novikoff et al. reported in favor of "the hypothesis that the granules arise from GERL," because of the fact that acid phosphatase and tyrosinase activity were confirmed in the Golgi-associated system of smooth endoplasmic reticulum.[24] In our experiment, the melanosomal membranes sometimes included viral particles, other melanosomes with unit membranes, and/or other cytoplasmic organelles, as also described previously by Novikoff et al.[24] Therefore, we think Novikoff's description clearly demonstrates the process of melanization, maturation, and degradation of melanin and of melanosomes.

The nature of the fine limiting membrane demarcating polysomes and the membrane surrounding melanin in various stages was demonstrated to be tubular and smooth, and is thought to consist of microtubules similar to spindle fibers. These microtubules were directly connected with smooth

259

endoplasmic reticula and membranes of the melanosomes and degradated melanin (Fig. 15). Therefore, it has been demonstrated that the histogenesis of lysosomal membrane and microtubular membrane is the same. It is interesting that acid phosphatase activity is lower in amelanotic cells with many abortive forms of melanosomes than in melanotic cells with compound melanosomes. Acid phosphatase activity is necessary for melanin formation in the human melanocyte and mouse melanoma,[4, 24] and it also occurred within the premelanosome.[26] Although tyrosinase activity was present in the premelanosomes in some unmelanized (amelanotic) melanomas, no melanin was formed.[25] On the basis of biochemical experimentation we can now say that the unit membrane related to melanin formation consists of lysosomal derivatives and has several kinds of enzymes for polymerization and/or maturation of the pigment compounds.[24, 26]

C. *The relationship between X-irradiation and melanin formation*
When an increase of melanin granules and premelanosomes following low-dose irradiation was found ultrastructurally, acid phosphatase activity increased,[16] despite the decrease of tyrosinase activity. The irradiated melanocytes have been maintained in a continuous culture system for 175 days; however they have shown no recovery of tyrosinase activity (Fig. 19) but increased acid phosphatase activity (Fig. 20) on histochemistry. Hu et al. mentioned that pigmentation of pigment cells was unaffected by irradiation with X-rays and ultraviolet rays,[15, 30] and acid phosphatase activity appeared in ultraviolet-irradiated melanocytes.[26]

D. *Stability of tritiated DL-dopa*
Only L-dopa is the precursor of melanin,[18] and it is incorporated in the ribosomes and melanosomes of melanocytes,[2-4, 13, 14] and in the adrenal medulla.[29] It has been confirmed that hydrogen atoms labeled in the aliphatic chain of DL-3,4-dihydroxyphenylalanine are extremely stable both physically and chemically,[27, 28] and that tritium labeled in the benzene ring of dopa is also reliably stable in hydrogen-3 and carbon-14 double-labeling experiments *in vivo*.[28, 29] In these experiments, considerable amounts of the radioactivity of those radioisotopes labeled in the benzene ring were included in the melanoma and adrenal medulla, although more than 95% of the total radioactivity was excreted by the second day after intraperitoneal injection of radioactive dopa.[28] Therefore, it is unnecessary to fear that nonspecific incorporation will occur in the intracytoplasmic matrix.

V. SUMMARY

Six cases of human malignant melanoma taken from the primary region were examined electron microscopically and histochemically. B16-XI melanotic cultured cells cloned from transplantable B16 mouse melanoma and cultured human melanoma cells (HMV) were used to study the intracellular site of tritiated DL-dopa-incorporation in comparison with L-tyrosine-^{3}H by means of ultra-autoradiography. For analyzing the role of the membrane system in relation to melanogenesis, X-irradiation was applied to the B16 cultured cells.

The ultrastructural precursor of melanosomes was established in close relation to the ribosomes with tyrosinase activity and tubular smooth membrane (microtubule). It was demonstrated that membranes of microtubules and melanosomes were quite similar, and that melanosomal membranes sometimes revealed activity of various enzymes. Acid phosphatase as well as tyrosinase are important enzymes in the processes of maturation and degradation of melanosome. The difference between the site where tyrosinase reacts on dopa and on tyrosine was clearly demonstrated by means of the grain-counting method on electron microscopic autoradiography. Stability of tritiated DL-3,4-dihydroxyphenylalanine was also discussed.

VI. ACKNOWLEDGMENTS

It is a pleasure to acknowledge the electron microscopic aids of Mrs. K. Shimazaki, the establishment of B16 cultured melanoma cell strain by Mrs. S. Ishii and Dr. T. Terasima, the opportunity for and assistance in this work given by Dr. T. Kawamura and Dr. M. Seiji, and the contribution of Dr. H. Sugano, who kindly supplied human malignant melanomas.

REFERENCES

1. Demopoulos, H. B., Kasuga, T., Channing, A. A., and Bagdoyan, H., Comparison of ultrastructure of B-16 and S-91 mouse melanomas, and correlation with growth patterns. *Lab. Invest.*, **14**, 108 (1965).
2. Seiji, M., Shimao, K., Fitzpatrick, T. B., and Birbeck, M. S. C., The site of biosynthesis of mammalian tyrosinase. *J. Invest. Derm.*, **37**, 359 (1961).
3. Seiji, M., Shimao, K., Birbeck, M. S. C., and Fitzpatrick, T. B., Subcellular localization of melanin biosynthesis. *Ann. N. Y. Acad. Sci.*, **100**, 497 (1963).
4. Seiji, M. and Iwashita, S., Intracellular localization of tyrosinase and site of melanin formation in melanocyte. *J. Invest. Derm.*, **45**, 305 (1965).
5. Moyer, F. H., Genetic effects of melanosome fine structure and ontogeny in normal and malignant cells. *Ann. N. Y. Acad. Sci.*, **100**, 584 (1963).
6. Moyer, F. H., Genetic variations in the fine structure and ontogeny of mouse melanin granules. *Amer. Zool.*, **6**, 43 (1966).
7. Dalton, A. J. and Felix, M. D., Phase contrast and electron micrography of the Cloudman S-91 mouse melanoma. In *Pigment Cell Growth* (M. Gordon, ed.), p. 267, Academic Press, New York (1953).

261

8. Wellings, S. R. and Siegel, B. V., Role of Golgi apparatus in the formation of melanin granules in human malignant melanoma. *J. Ultrastruct. Res.*, **3**, 147 (1959).

9. Wellings, S. R. and Siegel, B. V., Electron microscopy of human malignant melanoma. *J. Nat. Cancer Inst.*, **24**, 437 (1960).

10. Woods, M. W., du Buy, H. G., and Hesselbach, M. L., Cytological studies on the nature of the cytoplasmic particles in the Cloudman S-91 mouse melanoma, the derived algire S-91 A partially amelanotic melanoma and Harding-Passey mouse melanoma. *J. Nat. Cancer Inst.*, **9**, 311 (1949).

11. du Buy, H. G., Woods, M. W., Burk, D., and Lackey, M., Enzymatic activity of isolated amelanotic and melanotic granules of mouse melanomas and a suggested relationship to mitochondria. *J. Nat. Cancer Inst.*, **9**, 325 (1949).

12. Greenberg, S. S. and Kopac, M. J., Studies of gene action and melanogenic enzyme activity in melanomatous fishes. *Ann. N. Y. Acad. Sci.*, **100**, 887 (1963).

13. Zelickson, A. S., Hirsch, H. M., and Hartman, J. F., Melanogenesis: An autoradiographic study at the ultrastructural level. *J. Invest. Derm.*, **43**, 327 (1964).

14. Nakai, T. and Shubik, P., Electronmicroscopic radioautography: The melanosome as a site of melanogenesis in neoplastic melanocytes. *J. Invest. Derm.*, **43**, 267 (1964).

15. Silver, S. E. and Hu, F., The effects of ultraviolet light and X-irradiation on mammalian pigment cells *in vitro*. *J. Invest. Derm.*, **51**, 25 (1968).

16. Kasuga, T. and Furuse, T., Electron microscopic study of two different radiosensitive culture cell strains by means of X-irradiation. *J. Jap. Cancer Ther.*, **5**, 9 (1970).

17. Furuse, T., Kasuga, T., Takahashi, I., and Tsuchiya, E., Establishment of two sublines from original B-16-W mouse melanoma and its cellular kinetics. *Proceedings of the Japan Cancer Association 28th Annual Meeting*, Kanazawa (1969) (forthcoming).

18. Bloch, B., Das Pigment. In *Jadassohn's Handbuch der Haut und Geschlechts-krankheiten* **1** (1), 434, Springer-Verlag, Berlin (1927).

19. Laidlaw, G. F. and Blackberg, S. N., Melanoma studies. II. A simple technique for the dopa reaction. *Amer. J. Path.*, **8**, 491 (1932).

20. Gomori, G., Acid phosphatase. In *Microscopic Histochemistry: Principles and Practice* (G. Gomori, ed.), p. 189, University of Chicago Press, Chicago.

21. Cardno, S. S. and Steiner, J. W., Improvement of staining technics for thin sections of Epoxy-embedded tissue. *Amer. J. Clin. Path.*, **43**, 1 (1965).

22. Ham, R. G., An improved nutrient solution for diploid chinese hamster and human cell lines. *Exp. Cell Res.*, **29**, 515 (1963).

23. Uchida, K. and Mizuhira, B., Electron microscope autoradiography with special reference to the problem of resolution. *Arch. Histol. Jap.* (in press) (1970).

24. Novikoff, A. B., Albala, A., and Biempica, L., Ultrastructural and cytochemical observation on B-16 and Harding-Passey mouse melanomas. *J. Histochem. Cytochem.*, **16**, 299 (1968).

25. Mishima, Y. and Loud, A. V., The ultrastructure of unmelanized pigment cells in induced melanogenesis. *Ann. N. Y. Acad. Sci.*, **100**, 607 (1963).

26. Wolff, K. and Schreiner, E., Melanosomal acid phosphatase. *Proceedings of the Seventh International Pigment Cell Conference*, Sept. 2–6, 1969, Seattle, Washington.

27. Waterfield, W. R., Spanner, J. A., and Stanford, F. G., Tritium exchange from compounds in dilute aqueous solutions. *Nature*, **218**, 472 (1968).

28. Hempel, K., Investigation on the structure of melanin in malignant melanoma with [3]H- and ^{14}C-labeled at different positions. In *Structure and Control of the Melanocyte* (G. Della Porta and O. Mühlbock, eds.), p. 162, Springer-Verlag, Berlin.

29. Evans, E. A., *Tritium and Its Compound*, p. 306, Butterworth, London.

30. Kitano, Y. and Hu, F., The effects of ultraviolet light on mammalian pigment cells *in vitro. J. Invest. Derm.*, **52,** 25 (1969).

DISCUSSION

DR. QUEVEDO: Would you please repeat the evidence for conversion of melanosomes into lipid granules in some melanoma cells.

DR. KASUGA: The melanoma with numerous lipid-like granules was stained with Sudan III; then it became positive for it. And Sudan III-positive granules of a fine size were distributed throughout the cytoplasm.

DR. HORI: What is the reason you used tritiated tyrosine instead of tritiated dopa? Tyrosine can be metabolized to other proteins as well as to melanoprotein.

Dr. Kasuga: Tritiated tyrosine is ineffective in locating the site of melanogenesis. That is, it is nonspecific. We used tritiated dopa for our purpose. But there are still some problems; for example, it might be naturally oxidized.

DR. FITZPATRICK: I would like to clarify something in my mind. As I visualize it, there are two different quick mechanisms for formation of early melanosomes here. In his situation he visualizes polysomes making the vesicle and in the situation in which Dr. Miyamoto and Dr. Seiji and I visualized dilatation of the smooth membranes and pinching off, is this correct?

DR. KASUGA: I think so. But the ultra-autoradiographic patterns of the tritiated dopa incorporation in ribosomes limited by fine fibrillar membrane were not so frequent in the nonirradiated melanoma cells. When the melanoma cells are incubated with the tritiated dopa at 24 hours after irradiation of the cells with X-rays, you will find similar findings in the ultra-autoradiographic pictures. I think, in regard to the formation of early melanosomes, the primary site of tyrosinase activity will be in the ribosomes; however, I wonder whether or not the incorporation of tritiated dopa in ribosomes is an artifact.

DR. FITZPATRICK: When you spin out ribosomes you get a lot of tyrosinase activity. You apparently activated enough of it so that you can detect it. We were unable to detect it, as you could see in our preparations with dopa and with tyrosine with catalytic amounts of dopa.

Dr. Chavin: Perhaps the problem may be resolved by the use of combined approaches to avoid nonspecific oxidase and/or protein synthesis effects. We incubate tissue with the substrate (dopa, tyrosine) in the presence of inhibitors of protein synthesis and compare such findings with those of the tissue incubated in the same reaction mixture plus diethyldithiocarbamate or with DDC+ substrate. This type of approach provides a controlled consideration of tyrosinase localization and enzyme activity. Lack of such controlled consideration may yield misleading information.

Dr. fitzpatrick: I think in view of your results and also in view of some of the discrepancies here that the future solution of the problem is going to come from the development of a specific radioimmunoassay technique for tyrosinase in which you will be able to localize tyrosinase in electron micrographs. Then you will really be able to see.

The Origin of Melanosomes*

Kiyoshi Toda and Thomas B. Fitzpatrick

Department of Dermatology, Harvard Medical School, and
the Massachusetts General Hospital, Boston, Massachusetts, U.S.A.

I. INTRODUCTION

A concept was provided to explain the sequence in the development of an
organelle into a mature melanosome.[1] It was shown that tyrosinase, the
enzyme responsible for melanin formation, is synthesized by ribosomes and
is transferred into melanosomes,[1] where melanin formation takes place;
the melanosome is gradually transformed into a fully melanized melanosome.[1] It was believed that a melanosome originates as a small vesicle
about 0.05μ in diameter, which, either by fusion or by growth, becomes a
large vesicle in the Golgi area; these large vesicles, initially spherical, become elongated until they are approximately the size of a mature melanin
granule.[2] Recently, Maul[3] suggested that "premelanosomes" (which we
call Stage II; see Fig. 1) are connected to and develop within a tubular
smooth endoplasmic reticulum (ER) which is connected with the Golgi
apparatus during melanogenesis. Embryonic chick retinal pigment epithelium (ECRPE) is well known as a melanosome-synthesizing cell.
Several theories have been postulated concerning the formation of melanosomes in the ECRPE. Moyer's theory[4] was that granulogenesis appeared
to be initiated by the formation of the enzymes necessary for melanin
synthesis within the dilated cisterna of the ER, but Breathnach and

* This work was supported by U.S. Public Health Service Research Grant No. 5 RO1
CA 05010-11 from the National Cancer Institute.

Stage I: Large spherical membrane-limited vesicle that shows either a filament with periodicity by electron microscopy or tyrosinase activity by electron microscope histochemistry.

Stage II: The melanosome becomes ellipsoid and shows numerous filaments having distinct periodicity.

Stage III: The internal structure that was characteristic of Stage II is now partially obscured by electron-dense material.

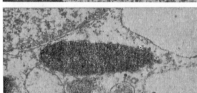

Stage IV: It is still ellipsoid and by this time is totally electron dense.

Fig. 1.

Wyllie's observation[5] suggested that the melanosome originates in the Golgi vesicle, and that the formation of the melanosome in human fetus pigment epithelium is the same as it is in the melanocyte. Our data,[6] however, revealed that the Stage I melanosome balloons out from the smooth-surface membrane in the Golgi area (see the classification of melanosome stages in Fig. 1). The activity of tyrosinase in the ECRPE changes with the development of the eye and reaches the highest level on the tenth day of development, thereafter decreasing until the eighteenth day, when it is almost unmeasurable.[7] Seiji[2] believed that, after the tenth day, the active center of tyrosinase was blocked by layers of melanoprotein that accumulated on the melanosome. Our data,[6] however, suggested that tyrosinase activity in the melanosome during the late phase of development (after the tenth day) is blocked by hydrolytic enzymes. Several theories have been postulated concerning the site of melanization. The "nucleus" theory[8] and the "mitochondria" theory[9] have been replaced by two newer theories suggesting that the Golgi apparatus may be associated with melanization. Zelickson et al.,[10] using electron microscope radioautography, observed that ^{14}C-dopa, injected intraperitoneally into mice, was incorporated into the rough-surface membrane. Using a biochemical procedure, however, Seiji et al.[2] observed that ^{14}C-dopa was incorporated into only the melanosome fraction. Novikoff[11] also proposed that the cisternae in the Golgi area may be the site of melanization.

The purpose of our experiment is to reinvestigate (a) the site of melanization by electron microscope histochemistry and electron microscope radioautography, and (b) the effect of glutaraldehyde fixation on tyrosinase activity.

II. MATERIALS AND METHOD

A. *Effect of glutaraldehyde fixation on tyrosinase*

Harding–Passey mouse melanoma was dissected and homogenized, and small-granule and large-granule fractions were prepared according to Seiji's method.[1] Each fraction was fixed with 2.5% glutaraldehyde in 0.1 M phosphate buffer for 30, 60, 120, and 180 minutes, and spun down at 20,000 $\times$ g for 5 minutes, washed with 0.1 M phosphate buffer; this procedure was repeated three times, and then the tyrosinase activity was determined by a manometric method.

B. *Electron microscope histochemistry*

1. Dopa reaction

Embryonic chick retinal pigment epithelium and a human hair bulb were each incubated in a solution of 0.05% dopa in 0.1 M phosphate buffer, 0.25 M sucrose, pH 6.8, for three hours without glutaraldehyde prefixation. After incubation, each preparation was fixed with 1% osmium tetroxide, 0.1 M phosphate buffer, pH 7.4, and embedded in Epon 812. Specimens were examined with the Siemens Elmiskop I.

2. Acid Phosphatase

The 16-day-old ECRPE and the human hair bulb were each fixed with 2.5% glutaraldehyde in 0.1 M sodium-cacodylate–hydrochloric acid buffer, pH 7.4. After 30 minutes of glutaraldehyde fixation, the tissues were washed in cacodylate–HCl buffer and incubated in a freshly prepared acid-phosphatase medium that contained 10 ml of 0.1 M Tris-maleate buffer, pH 5.0; 10 ml of distilled water; 10 ml of 1.25% sodium glycerophosphate; and 20 ml of 0.2% lead citrate, for 60 minutes at 37°C. After incubation, the tissues were fixed with glutaraldehyde and then with osmium tetroxide and were embedded in Epon 812.

C. *Electron microscope radioautography*

The large-granule fraction from the 10-day-old ECRPE was incubated at 37°C for one hour in the following two solutions (a) 3,5-^{3}H-L-tyrosine and

L-dopa in a 10:1 ratio (0.1 M phosphate buffer in 0.25 M sucrose solution; specific activity of 3,5-^{3}H-L-tyrosine$=40$ μc) and (b) the previous solution with the addition of 0.05% diethyldithiocarbamate (DDC).

The 10-day-old ECRPE was cultured in Eagle's culture medium with ^{3}H-dopa (specific activity$=50$ μc/0.95 mg dopa/ml), and the control specimen was cultured in Eagle's Culture Medium with L-dopa (0.95 mg dopa/ml) for one hour and fixed with 1% osmium tetroxide in 0.1 M phosphate buffer.

For cultured ECRPE and the large-granule fraction, electron microscope radioautography was carried out according to Caro and van Tubergen[12] with Ilford L-4 nuclear emulsion. After 4 weeks' exposure, the radioautographs were developed.

III. RESULTS

A. *Effect of glutaraldehyde fixation on tyrosinase*
Tyrosinase activities in the large-granule fraction and small-granule fraction were strongly inhibited by glutaraldehyde fixation (Table 1), and it is very interesting that the tyrosinase activity in the large-granule fraction was inhibited much more than was tyrosinase activity in the small-granule fraction.

Table 1.   Glutaraldehyde inhibition of tyrosinase.

	15 minutes	30 minutes	60 minutes	120 minutes	180 minutes
Small-granule fraction			50–60%	70–80%	85–90%
Large-granule fraction	25–35%	55–60%	80–85%	95–97%	100%

B. *Electron microscope histochemistry*
1.   Dopa reaction
Reaction products were observed in the smooth ER in the Golgi area, and in Stage I, II, III, and IV melanosomes (Figs. 2, 3).
2.   Acid Phosphatase
Reaction products were observed in melanosomes present in both ECRPE and in the hair bulb keratinocytes (Figs. 4, 5).

C. *Electron microscope radioautography*
Our data showed that Stage I melanosomes had ^{3}H-tyrosine uptake (Fig. 6), and that Stage IV melanosomes also have tyrosinase activity (Fig. 7) —in contrast with previously reported data.[1, 10, 11, 15] The control specimen did not show any ^{3}H-tyrosine incorporation.

In the radioautographs of ECRPE culture, ^{3}H-dopa incorporation was observed in Stage I, II, III, and IV melanosomes (Fig. 8).

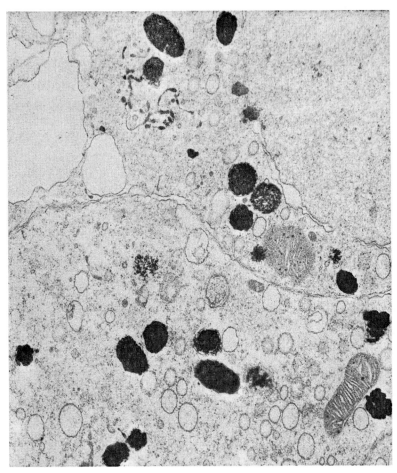

Fig. 2. Dopa reaction of ECRPE; smooth ER and Stage I and IV melanosomes show reaction products.

IV. DISCUSSION

Our data show that glutaraldehyde fixation very strongly inhibits tyrosinase activity, and that tyrosinase in different locations has a different reaction to glutaraldehyde; that is, when tyrosinase is bound to the membrane in the small-granule fraction, it is more stable than when it is bound to the membrane in the large-granule fraction. Our electron micrographs and electron radioautographs without glutaraldehyde prefixation reveal the same localization of tyrosinase in the melanocyte that Seiji[1] reported with a biochemical technique, but quite different from the

269

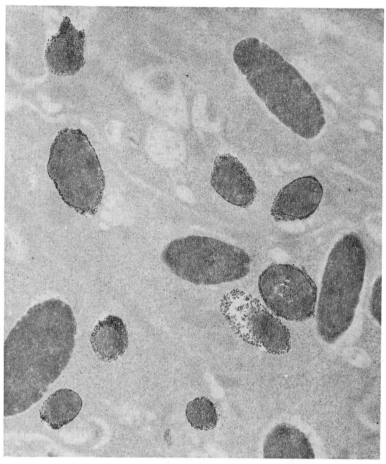

Fig. 3. Dopa reaction of hair bulb; Stage IV melanosomes show reaction products.

localization that Novikoff et al.[11] and Zelickson et al.[10] reported. When the specimen was fixed with glutaraldehyde, the electron micrograph did not reveal any dopa-positive Stage IV melanosomes but, contrary to expectation, it did reveal many Stage II melanosomes that were not dopa positive (Fig. 9). Therefore, we believe that, in the preparation of specimens for electron microscopic histochemistry of tyrosinase, glutaraldehyde prefixation should not be used. The data of Zelickson et al.[10] and of Novikoff et al.[11] could not show the localization of tyrosinase on the organelle, as can be done with the biochemical technique. Because their technique revealed localization on the smooth membrane and vesicles in the Golgi area, the tyrosinase must have been affected by the glutaraldehyde fixation. Hori's[13] observations of ³H-dopa uptake in ECRPE organ

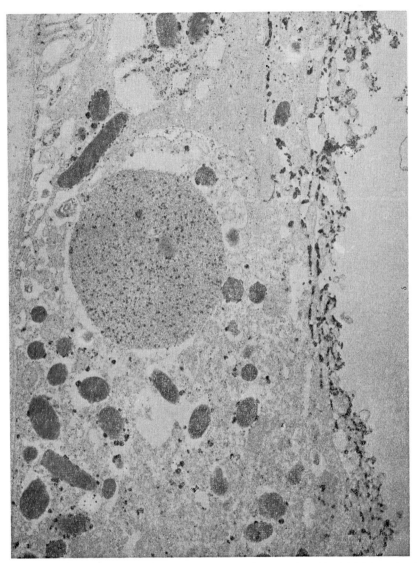

Fig. 4. Electron micrograph of acid phosphatase; in the 16-day-old ECRPE, reaction products are observed in smooth ER and around melanosomes.

culture revealed that all stages of melanosomes and ribosomes showed incorporation of ^{3}H-dopa. Model[14] used amphibian neural-crest tissue culture, and observed ^{3}H-dopa incorporation into melanosomes in all stages of development. Our data, based on ^{3}H-dopa uptake in organ culture of ECRPE, also revealed that ^{3}H-dopa uptake is localized in Stage I, II, III, and IV melanosomes.

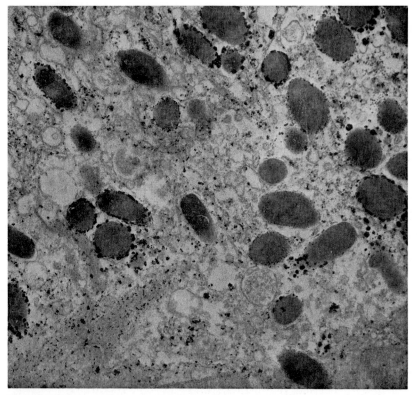

Fig. 5. Electron micrograph of acid phosphatase in the keratinocyte of hair; reaction products are observed around melanosomes.

Inasmuch as the time factors used by Model,[14] Hori,[13] and ourselves are basically comparable, the differences in the findings reported must be attributed to various other factors in the organ culture—among them, e.g., toxicity of the dopa. The studies of Model and ourselves suggest that the site of melanization is a melanosome.

Seiji's data[2] and ours[6] suggest that tyrosinase is attached to the smooth-surface membrane in the Golgi area of the melanocyte.

We found that dopa reaction products are precipitated on the outer membrane of melanosomes in all four stages. Silver grains in ECRPE organ culture are localized also on the outer membrane of melanosomes in all four stages, but, in Stage III and IV melanosomes only, the silver grains are sometimes localized in the internal structure of the melanosome. Based on these findings, we believe that tyrosinase in the melanosome is located on the outer membrane.

Novikoff et al.[11] and Seiji and Kikuchi[16] demonstrated acid phosphatase

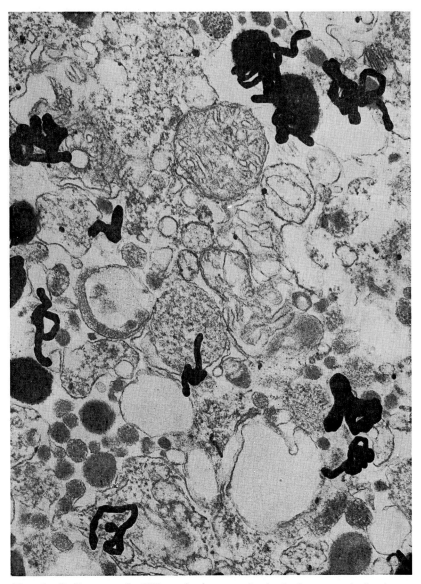

Fig. 6. Radioautograph of ³H-tyrosine incorporation in the large-granule fraction; a silver grain is observed in the Stage I melanosome.

activity in melanosomes of mouse melanoma. We also showed acid phosphatase activity in the melanosomes present in ECRPE.[6] Hori et al. observed acid phosphatase activity in the melanosome complex in human skin.[17] Our data in this study reveal that a single melanosome in the

273

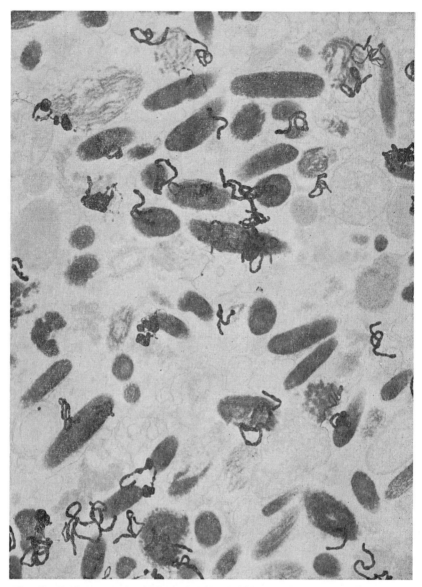

Fig. 7. Radioautograph of ³H-tyrosine incorporation in the large-granule fraction; silver grains are observed on Stage II, III, and IV melanosomes.

keratinocyte of Japanese hair has acid phosphatase activity. It is believed that the process of degradation of melanosomes occurs in either melanocytes or keratinocytes.[6, 11, 16, 17]

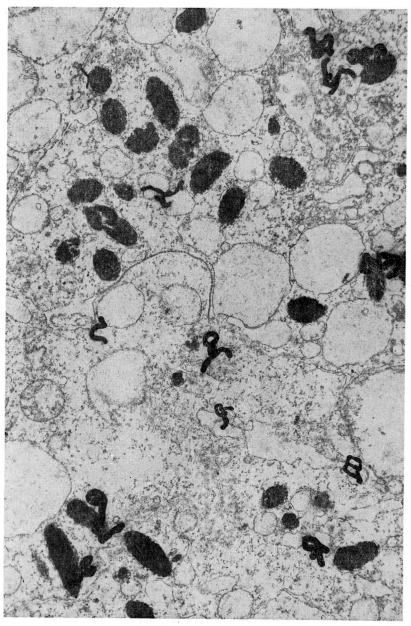

Fig. 8. Radioautograph of ^{3}H-tyrosine incorporation into the ECRPE organ culture; silver grains are observed on Stage I, II, III, and IV melanosomes.

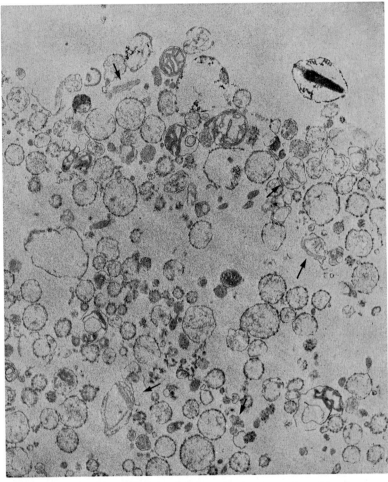

Fig. 9. Dopa reaction with glutaraldehyde prefixation; many Stage II melanosomes do not show reaction products (arrow).

Acid phosphatase activity in the different phases of ECRPE may play a role in the degradation of melanosomes and may be the control mechanism of tyrosinase activity in the ECRPE.[6] Our findings are summarized in Fig. 10: tyrosinase is synthesized in the ribosome; it then transfers to the Golgi area and attaches to the membrane; the membrane balloons out from the smooth-surface ER in the Golgi area, and the Stage I melanosome is formed, in which melanization begins. In the Stage II melanosome, the fine structure with periodicity is not the supporting structure of tyrosinase.[6]

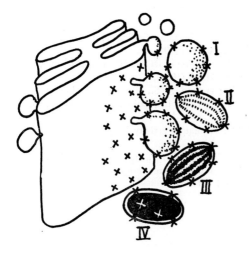

Fig. 10.   The formation of the melanosome (X=tyrosinase).

REFERENCES

1. Seiji, M., Shimao, K., Birbeck, M. S. C., and Fitzpatrick, T. B., Subcellular localization of melanin biosynthesis. *Ann. N. Y. Acad. Sci.*, **100,** 497 (1963).
2. Seiji, M., Subcellular particles and melanin formation in melanocytes. In *Advances in Biology of Skin*, vol. 8, *The Pigmentary System* (W. Montagna and F. Hu, eds.), p. 189, Pergamon Press, Oxford (1967).
3. Maul, G. G., Golgi-melanosome relationship in human melanoma *in vitro*. *J. Ultrastruct. Res.*, **26,** 163 (1969).
4. Moyer, F. H., Electron microscope observation on the origin, development, and genetic control of melanin granules in the mouse eye. In *The Structure of the Eye* (G. K. Smelser, ed.), p. 469, Academic Press, New York (1966).
5. Breathnach, A. S. and Wyllie, L., Ultrastructure of retinal pigment epithelium of the human fetus. *J. Ultrastruct. Res.*, **16,** 584 (1966).
6. Toda, K. and Fitzpatrick, T. B., Ultrastructural and biochemical studies of the formation of melanosomes in the embryonic chick retinal pigment epithelium. *Proceedings of the Seventh International Pigment Cell Conference,* Seattle, Washington, Sept. 2–6, 1969 (in press).
7. Miyamoto, M. and Fitzpatrick, T. B., On the nature of the pigment in the retinal pigment epithelium. *Science,* **126,** 449 (1957).
8. Lerche, W. and Wulle, K. G., Über die Genese der Melaningranula in der embryonalen menschelichen Retina. *Z. Zellforsch. Mikroskop. Anat.*, **76,** 452 (1967).
9. Woods, N. W. and Hunter, J. C., In *Pigment Cell Biology* (M. Gordon, ed.), p. 455, Academic Press, New York (1959).
10. Zelickson, A. S., Hirsch, H. M., and Hartman, J. F., Melanogenesis: An autoradiographic study at the ultrastructure level. *J. Invest. Derm.*, **43,** 327 (1964).
11. Novikoff, A. B., Albola, A., and Biempica, L., Ultrastructural and cytochemical observation on B-16 and Harding–Passey mouse melanomas. The origin of premelanosomes and compound melanosomes. *J. Histochem. Cytochem.*, **16,** 299 (1968).

12. Caro, L. G. and van Tubergen, R. P., High resolution autoradiography., *J. Cell Biol.*, **15,** 173 (1962).
13. Hori, Y., Radiographic electronmicroscopic study of chick embryonal retinal pigmented epithelium. *J. Invest. Derm.*, **54,** 88 (1969).
14. Model, P. G., The ultrastructural localization of DOPA-H^3 in amphibian melanoblasts grown *in vitro. J. Invest. Derm.*, **54,** 93 (1969).
15. Hunter, J. A., Mottaz, J. H., and Zelickson, A. S., Ultrastructure localization of dopa reaction product in skin following U. V. irradiation. *J. Invest. Derm.*, **52,** 369 (1969).
16. Seiji, M. and Kikuchi, A. Acid phosphatase activity in melanosomes. *J. Invest. Derm.*, **52,** 212 (1969).
17. Hori, Y., Toda, K., Pathak, M. A., Clark, W. H., Jr., and Fitzpatrick, T. B., A fine-structure study of the human epidermal melanosome complex and its acid phosphatase activity. *J. Ultrastruct. Res.*, **25,** 109 (1968).

Intracytoplasmic Activities in Malignant Melanoma: Viral, Melanogenic and Anti-Melanogenic*

Yutaka Mishima, Makoto Takahashi,** and Michael Cooper

Department of Dermatology, Wakayama Medical University, Wakayama, Japan, and
Wayne State University School of Medicine, Detroit, and Veterans Administration Hospital, Allen Park, Michigan, U.S.A.

I. INTRODUCTION

Despite the recent clarification of RNP particles as the site of tyrosinase-mediated melanization,[1] there appear to be subcellular control mechanisms of pigment cells which are independent of the level of activity of tyrosinase. Thus, although tyrosinase and premelanosomes have been shown to be present in albino melanocytes,[2] amelanotic melanomas,[3,4] and xanthic fish integument,[5] melanization fails to occur. In addition, recent studies have revealed the presence of virus-like particles in the ergastoplasm of Fortner's spontaneous melanomas of Syrian (golden) hamster,[6] S91 mouse melanoma,[7] B16 and Harding–Passey mouse melanomas.[8] However, the significance of these virus-like particles in melanoma carcinogenesis remains undetermined. For this reason the absence or presence of such particles in the spectrum of various malignant melanomas has become a subject of investigation. Current findings have led us to examine further cells of Greene's melanoma, which has not previously been reported to contain virus particles. In addition, we investigated whether these virus-like particles in hamsters were characteristic

* This investigation was supported in part by U.S. Public Health Service Research Grants No. CA-08891-02 from the National Cancer Institute.
** Present Address: Department of Dermatology, Sapporo Medical University, Sapporo, Japan.

279

only of hamster melanoma cells or whether they could be found elsewhere in hamster tissues.

It is the purpose of this paper to describe our recent findings on intrinsic melanogenic regulating factors, tyrosinase competitive,[9] and substrate limiting melanogenic inhibitors,[10] substrate availability,[11] nonpremelanosomal tyrosinase,[3, 4, 12, 13] and viral activities[14] in melanoma cells.

II. MATERIALS AND METHODS

A. *Tyrosinase-competitive inhibitor(s)*

For tyrosinase inhibitor(s) studies,[9, 15] Fortner's amelanotic melanoma type 3 and melanotic malignant melanoma type 1[16] were obtained from Joseph G. Fortner and transplanted subcutaneously in our laboratory. Non-necrotic melanomas were homogenized in 0.25 M sucrose-0.02 M phosphate buffer using a Teflon head-hand homogenizer, centrifuged at $700 \times g$ to remove cell debris and then centrifuged at $12,000 \times g$ to obtain a soluble fraction. Temperatures were maintained at 2°C during this preparation. These fractions were dialyzed against distilled water overnight. The diffusates were flash-evaporated and then tested for their inhibitory effect on mushroom tyrosinase.

Studies of tyrosinase enzyme activity were carried out in 1 cm cells in a Zeiss PMQ II spectrophotometer by the oxidation of L-3,4 dihydroxyphenylalanine (dopa) by measuring the initial rate of increase in optical density at 475 mμ. The final reaction mixture contained 0.5 μM/ml dopa and 35 μM/ml phosphate buffer, pH 6.8. Ion exchange chromatography was carried out on Dowex 50 resin as described elsewhere.[17]

B. *Substrate limiting melanogenic inhibitor*

Fortner's melanotic melanoma type 1 and B16 mouse melanoma (the darkly pigmented type, obtained in 1965 from the Jackson Laboratory, Bar Harbor, Maine) were homogenized and centrifuged at $105,000 \times g$ for 1 hour as previously described.[10] Human amelanotic (Y-1226-V) and melanotic melanomas (Y-1226-I) were homogenized with 5 ml sucrose per gram and assayed without further procedures. The radio-assay used here is a modification of the method of Kim and Tchen.[18] Various dilutions of the soluble fraction were incubated at 30°C for 16 hours with 0.27 μc, DL-tyrosine-2-^{14}C, 2.8×10^{-5} M (unless otherwise specified), in 0.1 M phosphate buffer, pH 6.8, along with 92 μg/ml chloramphenicol. When pyridoxal phosphate was added to the radio-assay incubation the concentration was 1 μg/ml. When α-methyl dopa was added the concentration was 1×10^{-5} M.

C. *Tyrosine availability in malignant melanomas*

Fortner's melanotic type 1 and amelanotic type 3 were homogenized in

0.25 M sucrose containing sodium phosphate buffer at 0.02 M, pH 6.8, and centrifuged at $700 \times g$. The supernatants were incubated at $37°C$ with and without 1×10^{-3} M sodium diethyldithiocarbamate. "Tyrosine" was estimated by the nitrosonaphthol method[19] at various intervals.

D. Non-premelanosomal tyrosinase and in vitro melanization

Fortner's melanotic melanoma type 1, amelanotic type 3, Greene's melanotic and amelanotic melanomas, originally obtained from H. S. N. Greene,[20] B16 mouse melanoma and human melanotic and amelanotic melanomas were used in this study. Homogenation, fraction and density gradient centrifugation for large granule subfractions were carried out as described elsewhere.[1] The soluble fraction was obtained by centrifuging at $105,000 \times g$ for 1 hour. Tyrosinase assay was carried out by either the spectrophotometric method with dopa as substrate,[11] or a modification[12] of the fluorometric method.[21, 22]

Electron microscopic examination to localize the intracytoplasmic sites of *in vitro* melanization was carried out using tissue fixed in 6.25% glutaraldehyde in cacodylate buffer or phosphate-buffered 4% paraformaldehyde in phosphate buffer after Peter followed with dopa incubation and post-osmification as described previously.[4]

E. Virus-like particles in malignant melanomas

Greene's melanotic and amelanotic melanomas received directly from Greene, prior to our transplantation, were used in this study. Syrian (golden) hamsters bred in our colony were used for the examination of normal hamster tissue. Excised tissues were fixed with 6.25% glutaraldehyde in cacodylate buffer followed by post-osmification as described elsewhere.[14]

III. RESULTS AND DISCUSSION

A. Tyrosinase-competitive inhibitors

The addition of melanoma diffusate equivalent to 0.04 g, 0.08 g, and 0.10 g original melanoma to partially purified mushroom tyrosinase results in similar inhibition curves (Fig. 1) for both melanotic and amelanotic melanomas. Figure 2 shows the elution pattern of amelanotic melanoma diffusates read at 280 mμ after ion exchange chromatography. Tubes 9–11, the only large peak in the chromatogram of these diffusates, contain the inhibitory material, and no other peak shows inhibitory activity, as evidenced by pooling the remaining eluates and testing for inhibitor activity against mushroom tyrosinase. Chromatography of melanotic melanoma diffusates shows the inhibitor activity to be in the first large peak in the chromatograph (Fig. 3), tubes 9 through 13. Again

281

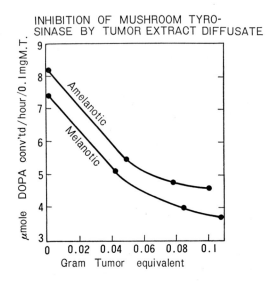

Fig. 1.   Inhibition of mushroom tyrosinase by diffusates of amelanotic and melanotic Fortner's melanomas.

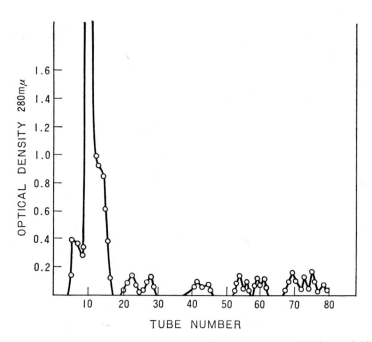

Fig. 2.   Elution pattern of Fortner's amelanotic melanoma $12,000 \times g$ soluble fraction after chromatography on Dowex 50 ion exchange resin.

282

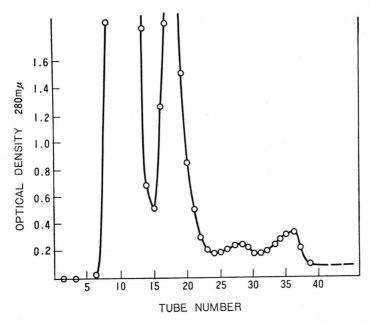

Fig. 3. Elution pattern of Fortner's melanotic melanoma 12,000 × g soluble fraction after chromatography on Dowex 50 ion exchange resin.

no activity was seen in the remaining peaks. Kinetic studies have revealed these inhibitors to be competitive in their action. This inhibition does not appear to be due to glutathione or sulfur-containing amino acids, since the addition of copper does not decrease the inhibitory activity and since acid hydrolysis and paper chromatography of the inhibitory peak from ion exchange chromatography do not reveal the Rf values of SH-containing amino acid. The above inhibitors obtained from amelanotic as well as melanotic melanoma of Fortner may be small peptides, as suggested by their insolubility in organic solvent, dialyzability, heat resistance, absorption maximum at 280 mμ, and paper chromatography data.[17] Concerning the defect in melanogenesis in amelanotic melanoma, further studies indicate that removing inhibitor from amelanotic fraction by dialysis fails to reveal tyrosinase activity. Thus the melanogenic defect in tyrosinase-negative Fortner's amelanotic melanoma is not due to masking the tyrosinase by this inhibitory activity.

B. *Substrate limiting melanogenic inhibitor*

When various dilution of 1 g/10 ml preparation of the soluble fraction of Fortner's melanotic hamster melanoma are incubated for 16 hours with 2.8×10^{-5} M ^{14}C-tyrosine, the total amount of ^{14}C-melanin formed in the TCA precipitate is equal to preparation concentration only in more

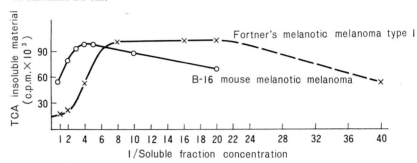

Fig. 4. Radio-assay of total melanin production at low substrate concentration for various dilutions of Fortner's and B16 melanoma 105,000 × g soluble fraction.

Table 1. Effect of addition of substrate or enzyme after incubation.

	Tyrosinase radio-assay of Fortner's melanoma soluble fraction of 1:1 and 1:20 dilutions					
Incubation	1:1	1:20	1:1+S	1:20+S	1:1+E	1:20+E
CPM of TCA	14,600	85,100	47,100	85,700	14,100	81,400
Precipitate (duplicates)	15,400	84,200	45,300	93,300	14,800	79,600

S, substrate added; E, soluble fraction added.

diluted preparations (Fig. 4). As the enzyme preparation becomes highly concentrated, the amount of melanin synthesized becomes inversely proportional to preparation concentration. Similar results are obtained with B16 mouse melanoma. The addition to the 1:1 (undiluted) incubation of an original amount of soluble fraction after the completion of 16 hours and further incubation for 4 hours fails to increase the amount of melanin formed, while the addition of an original amount of tyrosine-^{14}C increases the amount of melanin formed (Table 1). Thus all available tyrosine was already converted to TCA-insoluble material, and although active tyrosinase is present, a majority of the original tyrosine has become unavailable for melanin synthesis.

Table 2. Hypothetical interpretation of the molecular interaction occurring in the reaction mixture.

		1gm Melanoma/5ml	
A	Tyrosinase	Inhibitor	Tyrosine
	1000	80	100
	80 Tyrosine+80 Inhibitor→80 Tyrosine-Inhibitor		
	1000 Tyrosinase+20 Tyrosine→20 Melanin		
		1gm Melanoma/50ml	
B	Tyrosinase	Inhibitor	Tyrosine
	100	8	100
	8 Tyrosine+8 Inhibitor→8 Tyrosine-Inhibitor		
	100 Tyrosinase+92 Tyrosine→92 Melanin		

Table 3.   Radio assay with low substrate of total melanin production of 1:1 and 1:10 dilutions of human melanomas.

Human melanoma	1:1	1:10
Melanotic	7,000 cpm	44,000 cpm
Amelanotic	15,000	94,000

Table 4.   Radio-assay of total melanin production of Fortner's melanotic melanoma at various substrate concentrations with and without pyridoxal phosphate and α-methyl dopa.

0.27 μc DL-tyrosine	Control		Pyridoxal phosphate		DL-α-methyl dopa	
	1:1	1:10	1:1	1:10	1:1	1:10
2.8×10^{-5} M tyrosine CPM 1:10	8,200	27,800	6,700	24,600	6,100	24,000
CPM 1:1		3.3		3.7		3.9
4×10^{-4} M tyrosine CPM 1:10	48,500	28,200	49,300	26,100	40,000	19,400
CPM 1:1		0.58		0.53		0.48
8×10^{-4} M tyrosine CPM 1:10	34,600	16,700	39,000	15,100	37,400	18,000
CPM 1:10		0.48		0.39		0.48

We have interpreted this phenomenon as indicating the presence of a substance which reacts with tyrosine relatively rapidly compared to tyrosinase. We have termed this "substrate-limiting melanogenic inhibitor." Thus (Table 2 A), if a 1 gm/5 ml melanoma preparation contains 1,000 molecules of tyrosinase and 80 molecules of substrate limiting inhibitor, and if we add 100 tyrosine molecules, the inhibitor rapidly combines with the tyrosine, leaving only 20 tyrosine molecules for melanin synthesis. If we dilute this preparation to 1 g/5 ml (Table 2 B), we have only 100 tyrosinase molecules and 8 inhibitor molecules; 100 tyrosine molecules are again added. The substrate limiting inhibitor again combines with the tyrosine, but now it leaves 92 molecules of tyrosine available for melanin synthesis. We have also seen this phenomenon in human melanotic and amelanotic melanomas (Table 3), in which a diluted preparation forms more ^{14}C-melanin than a concentrated preparation. At higher substrate concentrations (Table 4) we do not see this phenomenon, as there is sufficient substrate remaining after combination with inhibitor to prevent substrate exhaustion during the incubation period. It should be noted here that all incubations were with 0.27 μc DL-tyrosine-^{14}C, and that the 1:1 preparation produces more ^{14}C-melanin when incubated with 4×10^{-4} M tyrosine than with 2.8×10^{-5} M tyrosine. Thus, a higher substrate concentration with lower specific activity can produce a larger amount of labeled melanin. In order to see if this inhibition might be due to tyrosine decarboxylase or transaminase, the coenzyme pyridoxal phosphate and decarboxylase inhibitor α-methyl dopa were added to the incubations. With the addition of pyridoxal phosphate, only a slight decrease in melanin formation occurred at low substrate concentrations and at higher substrate concentration there was a slight increase in melanin formation in the 1:1 preparation. Thus pyridoxal phosphate does not appear to influence significantly this activity. Furthermore α-methyl dopa fails to decrease the inhibitory activity. The independence of this activity from added ATP indicates that it is not due to *de novo* protein synthesis.

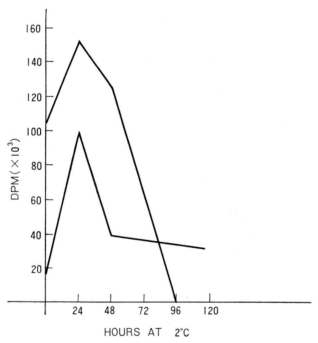

Fig. 5. Change in total melanin production before and after storage of Fortner's melanoma 105,000 × g soluble fraction at 2°C.

Storage of Fortner's melanotic melanoma soluble fraction at 2°C cause a marked change in the total amount of melanin produced when assayed under low substrate conditions (Fig. 5). For a 1g/5ml preparation there is an increase of approximately 400% after 24 hours' storage. However, when this preparation is diluted to 1:10, the increase after storage is only 50%. Before and after 24 hours' storage the 1:10 dilution forms more melanin than the undiluted soluble fraction. Only after 3 days' storage does the production of melanin vary directly with soluble fraction concentration.

Some transplantable tumors such as the Fortner's melanoma carried in our laboratory may lose substrate-limiting melanogenic inhibitor activity. However, the darkly pigmented B16 melanoma still retains this activity, and at present appears to be the best material for these investigations.

C. Tyrosine availability in malignant melanomas

We next studied the availability of tyrosine as tyrosinase substrate in amelanotic and melanotic melanomas. Incubation of the homogenate of Fortner's melanotic melanoma at 37°C with 1×10^{-3} M diethyldithiocarbamate (DDC) results in a distinct increase in nitrosonaphthol-positive material (Fig. 6). When DDC is excluded from the incubation, no increase is seen, and it therefore appears that this material is tyrosine, which is utilized as substrate by the intrinsic tyrosinase present. Incubation of Fortner's tyrosinase-negative amelanotic melanoma shows a similar release (Fig. 7) which is unaffected by the presence or absence of DDC. It thus appears that these amelanotic and melanotic melanomas contain a mechanism, presumably peptidases, which is able to release tyrosine, at least in vitro.

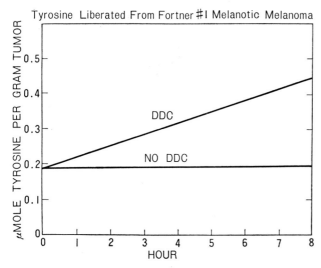

Fig. 6. Nitrosonaphthol-positive "tyrosine" before and after incubation of Fortner's melanotic melanoma $700 \times g$ supernatant at 37°C with and without DDC.

D. Non-premelanosomal tyrosinase and in vitro melanization

Although in the B16 mouse melanoma tyrosinase is primarily localized in the premelanosome fraction,[1] we have found a major portion of the

287

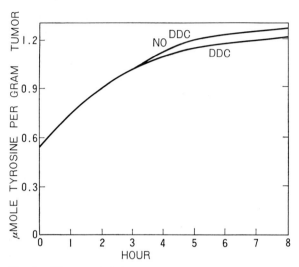

Fig. 7. Nitrosonaphthol-positive "tyrosine" before and after incubation of Fortner's amelanotic melanoma $700 \times g$ supernatant at 37°C with and without DDC.

tyrosinase activity of Fortner's melanotic and tyrosinase-positive amelanotic melanomas, as well as human melanotic melanoma, in the 105,000 $\times g$ soluble fraction. Studies of the subcellular tyrosinase distribution, measured spectrophotometrically using dopa as substrate, of Fortner's melanotic melanoma, reveal a large portion of the tyrosinase activity in the soluble fraction[11]. Although melanization *in vivo* has been observed only within premelanosomes and melanosomes, this does not preclude the existence throughout the cytoplasm of an inhibited or unactivated tyrosinase which may be temporarily incapable of catalyzing the sequence of reactions from tyrosine or dopa to melanin, or even of an active form of tyrosinase, if it is deprived of available substrate. In order to find out whether it is the Fortner's and human melanomas or B16 melanomas which are unusual regarding the presence of cytoplasmic tyrosinase, we have investigated this activity in Greene's melanotic and amelanotic melanomas. Using the fluorometric assay technique, it was found that the melanotic melanomas averaged 10 times more total tyrosinase activity, as recovered in the 3 major fractions, than the amelanotic tumors (Fig. 8). The average total activity for 3 melanotic melanomas was 2.4 μM/hr/g wet weight, compared with 0.24 μM/hr/g for the amelanotic melanomas. In all cases the melanotic melanoma soluble fraction contained more than 50% of the tyrosinase activity measured (Fig. 9). Each of the particulate fractions exhibited less than 30% of the measurable activity. Although a much lower tyrosinase activity was observed for amelanotic melanoma (Fig. 8), the subcellular distribution of the enzyme followed a pattern

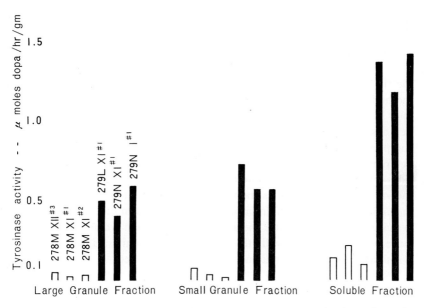

Fig. 8.   Specific activity of tyrosinase, measured fluorometrically, in Greene's amelanotic (white bar) and melanotic (black bar) melanoma fractions.

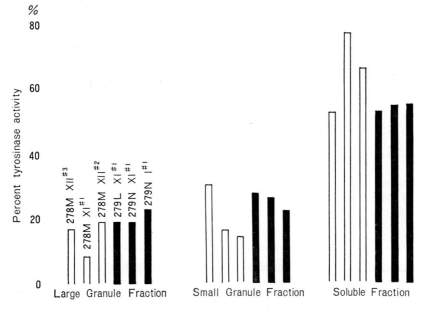

Fig. 9.   Percent tyrosinase activity distributed among the fractions of Greene's amelanotic (white bar) and melanotic (black bar) melanomas.

similar to that of the melanotic melanomas, that is, most of the activity was in the soluble fraction (Fig. 9).

The predominance of tyrosinase in the soluble fraction of Greene's melanotic and amelanotic melanoma, as shown by this fluorometric study, is in agreement with data obtained by oxygen uptake,[13] spectrophotometric method with dopa as substrate,[11] and ^{14}C-tyrosine-conversion methods[3, 4] in studies using Fortner's melanotic and amelanotic melanoma and human melanotic melanoma. Such soluble enzyme could not be demonstrated to represent leakage from the membrane of some subcellular components. Attempts to cause such leakage by freezing-thawing or detergent action were unsuccessful. Further tyrosinase assays of the subfractions of the large-granule fraction of Fortner's melanomas[3] have shown distinct differences between these amelanotic and melanotic melanomas (Fig. 10). Subfractions 3 and 5 show almost complete lack of activity in the amelanotic melanomas, while these fractions contain a large amount of activity in the melanotic melanomas. The absence of tyrosinase in these subfractions remains to be clarified.

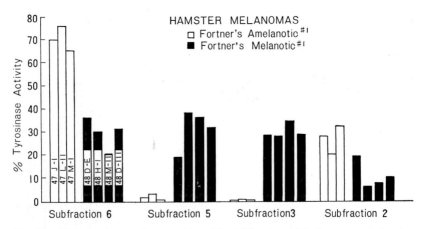

Fig. 10. Comparative tyrosinase activity of the subfractions of the large-granule fraction of Fortner's tyrosinase-positive amelanotic (white bars) and melanotic (black bars) melanomas.

In agreement with the biochemical assays which showed a predominance of tyrosinase in the soluble fraction of the mammalian melanomas, electron microscopic dopa reactions also reveal that newly synthesized dopa melanin appeared, not only within premelanosomes and Golgi-associated-endoplasmic reticulum, but also diffusely in the cytoplasm of Greene's, Fortner's, and human melanoma (Fig. 11). This reaction is not uniform, but varies from cell to cell.[23] In contrast, electron microscopy of B16 murine melanoma, in agreement with the biochemical assays, shows newly synthesized dopa melanin almost exclusively in the premelanosomes and in the Golgi-associated smooth membranes. This present study using Greene's melanotic and amelanotic melanomas, in confirming preceding ones on Fortner's melanotic and amelanotic as well as human melanoma, implies that the predominance of tyrosinase in the particulate fraction of B16 murine melanoma is not representative of the general mammalian pattern.

In 1966, M. Seiji showed that the incorporation of ^{14}C-dopa is distinctly different *in vitro* and *in vivo* for smooth membrane and premelanosomes. *In vivo* the uptake of ^{14}C-dopa is limited to the premelanosomes, whereas

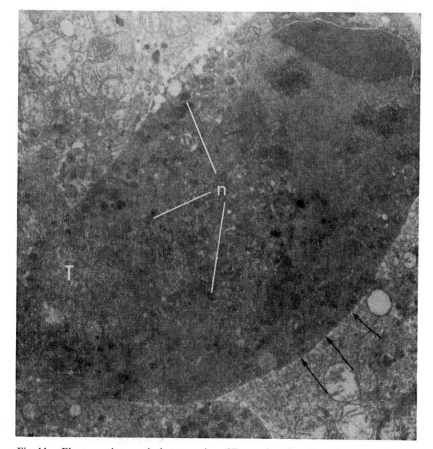

Fig. 11. Electron microscopic dopa reaction of Fortner's melanotic malignant melanoma cells revealing the deposition of newly formed dopa-melanin directly in the cytoplasm (arrows) as well as in premelanosomes (n). T: mitochondria. × 16,000.

291

in vitro the incorporation occurs both in smooth membranes and in pre-melanosomes, although at a slower rate in smooth membranes.[24] Although the radio-assay or electron microscopic radioautograpy of ^{14}C-dopa accumulation in various cell components can represent either non-tyrosinase-related uptake or the incorporation of dopa to form melanin by the action of tyrosinase, the electron microscopic dopa reaction reveals only the actual occurrence of melanization by the deposition of a newly formed melanin. After incubation of Fortner's melanotic melanoma in dopa, the vesicles and cisternae of the smooth endoplasmic reticulum in the vicinity of the Golgi apparatus often contain newly formed dopa melanin (Fig. 12). Similar findings are seen in Greene's melanotic melanoma.

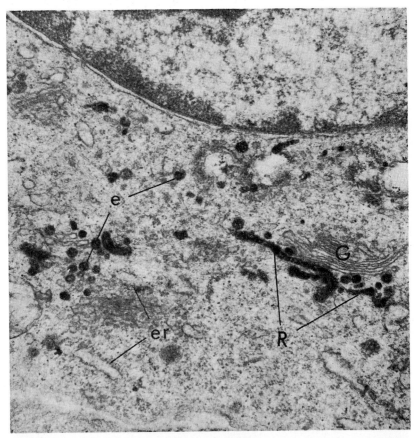

Fig. 12. Electron microscopic dopa reaction of Fortner's melanotic melanoma showing the deposition of newly formed dopa-melanin in the vesicles (R) and outer cisternae (R) of the Golgi apparatus (G) in addition to stage I–III melanosomes (e). This cell exhibits less distinct diffuse cytoplasmic melanization compared to Fig. 11. er: ergastoplasm. ×34,000.

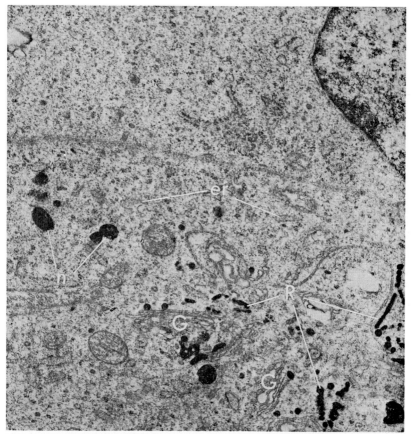

Fig. 13. Electron microscopic dopa reaction of B16 mouse melanoma, exhibiting the presence of newly formed dopa-melanin in the vesicles (R) and outer cisternae (R) of the Golgi apparatus (G) in addition to stage I–III melanosomes (n) appearing as strands or fused chains with varying electron density. × 18,900.

Dopa incubation of B16 melanoma results in varyingly increased electron density in the vesicles and outer cisternae of the Golgi apparatus, in addition to stage I–III melanosomes (Fig. 13). In agreement with the biochemical studies of B16 melanoma, we have not observed increased electron density diffusely in the cytoplasm of these cells. The above findings indicate that not only premelanosomes but even Golgi vesicles and cisternae may contain tyrosinase which is capable of forming melanin *in vitro*.

E. *Virus-like particles in malignant melanomas*

Electron microscopic examination of the cells of Greene's melanotic and amelanotic melanomas reveals distinctive spherical particles with a diameter of 90–110 mμ, consisting of an outer limiting membrane and a

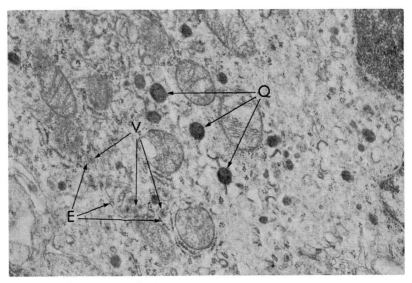

Fig. 14. General view of the subcellular organization of a Greene's melanotic malignant melanoma cell. V: virus-like particles; E: ergastoplasm; Q: melanosomes. ×18,800.

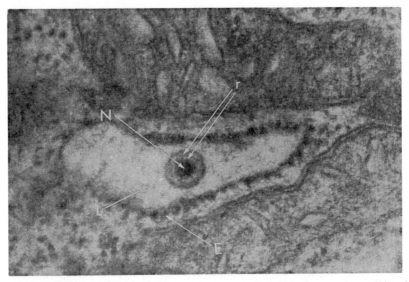

Fig. 15. Closer view of a virus-like particle demonstrating a limiting membrane (L) and centrally located electron-dense nucleoid structure (N) with spoke-like rays (r) within the sacs of the ergastoplasm (E). Greene's melanotic melanoma. ×87,750.

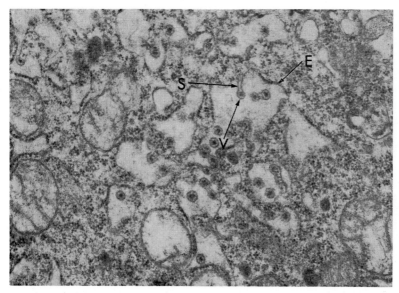

Fig. 16. Bulging stalk (S) of virus-like particle (V) from the ergastoplasm (E). Greene's melanotic melanoma. ×25,500.

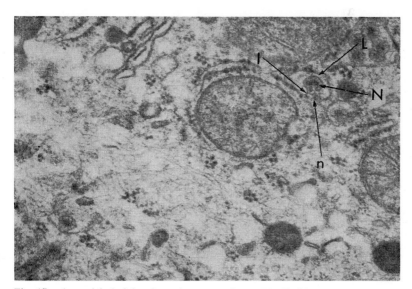

Fig. 17. A particle bulging from the ergastoplasm. The limiting membrane (L) and nucleoid structure (N) are in direct continuation with the membranes (I) and ribosomes (n) of the ergastoplasm. Greene's melanotic melanoma. ×40,100.

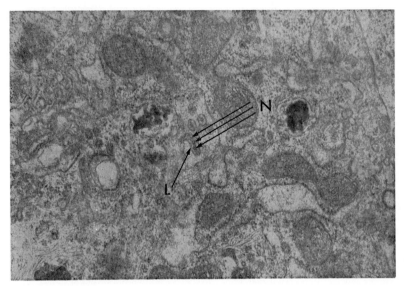

Fig. 18. Multiple nucleoid structures (N) in strands are enveloped within the bulging membrane (L). Greene's melanotic melanoma. ×26,100.

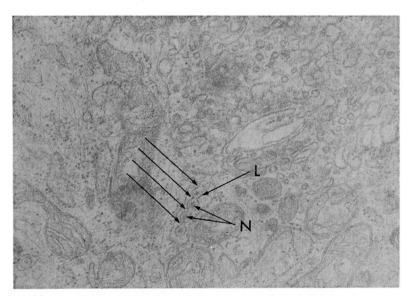

Fig. 19. Nucleoid structures (N) in strands are enveloped within continuous outer membranes (L) which show a pinched narrow sac (arrow) around each structure. Greene's melanotic melanoma. ×35,400.

296

spherical, centrally located, electron-dense nucleoid structure with a diameter of 40–50 mμ (Fig. 14). The nucleoid structure does not appear to have an enclosing membrane, although it shows some substructure and possesses thin spoke-like rays which may have contact with the outer limiting membrane (Fig. 15). Several variations of the particles within the rough-surfaced endoplasmic reticulum can be seen, suggesting developmental processes. Poorly organized nucleoid structures are seen within the mildly bulged ergastoplasm, extruding into the ergastoplasm sacs (Fig. 16). These immature central nucleoid structures closely resemble aggregates of ribosomes and are in continuity with the regularly aligned ribosomes on the surface of the surrounding ergastoplasm (Fig. 17). In Fig. 18, a number of these nucleoid structures can be seen, regularly spaced, in strands within the bulged, protruding membrane. Figure 19 shows the protruding membrane surrounding a strand of nucleoid structures narrowed and constricted between nucleoids.

In some areas, rows of separated, typical mature virus-like particles, consisting of limiting membranes, distinct nucleoid structures and spoke-like rays are seen within sacs of the ergastoplasm (Fig. 20). No virus-like particles have been found in the liver, kidney, or skin of non-melanoma-bearing control hamsters. According to Bernhard's morphologic typing,[25] the virus-like particles of Fortner's melanoma correspond most closely in structure to the C-type virus particles found in avian leucosis,[26] murine

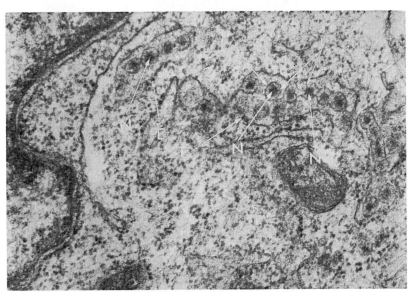

Fig. 20. The completed pinching off of the narrow limiting membrane (L) results in mature typical virus-like particles with distinct nucleoid structures (N) and with spoke-like rays (r) in the endoplasmic reticulum (E). Greene's melanotic melanoma. ×45,300.

Formation of Virus–like Particles in the Ergastoplasm

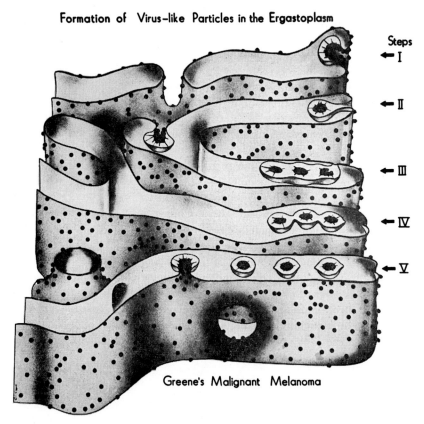

Greene's Malignant Melanoma

Fig. 21.   Proposed sequence of virus-like particle formation.

leukemia,[27] and lymphoid leukemia of the New-Zealand Black strain mice,[28] although the described C-type particles lack the spoke-like rays found here. The particles of S91 mouse melanoma have been reported to belong to the classification of A-type virus particles.

The particles of S91 mouse melanoma have been found to be formed in the rough-surfaced endoplasmic reticulum.[7] A number of authors have also indicated that type-A virus particles appear to originate from the membrane of the rough-surfaced endoplasmic reticulum.[29-33] The virus-like particles observed in our studies of Greene's spontaneous melanotic and amelanotic hamster melanoma consistently appear within the sacs of the rough-surfaced endoplasmic reticulum in various stages of structural maturation.

On the basis of the correlation of the bulging of the ergastoplasm at various stages, and the process of maturation of the interior of the particles,

298

we would like to propose the following model for the sequence of virus formation in hamster melanoma cells (Fig. 21). In addition to the illustrated sequence, steps I to V, it may be noted that the ribosomes on the surface of the ergastoplasm appear to be continuous with the aggregates of ribosome-like particles centrally located within the limiting membrane, suggesting that the nucleoid structure of the particles arises from the ergastoplasmic ribosomes and that the limiting membrane of the particles arises from the bulging membrane of endoplasmic reticulum.

The above findings, together with the recent report that malignant melanoma of Greene can be induced by the supernatant cell-free fraction of homogenized tumors,[34] indicate the necessity of further investigation on the biological significance of these particles in malignant pigment-cell growth.

IV. SUMMARY

The cytoplasmic control factors of melanogenesis other than ribosomal tyrosinase synthesis and premelanosome formation by the Golgi apparatus have been investigated. In addition, the occurrence and formation of virus-like particles in the cytoplasm of malignant melanoma cells are discussed.

REFERENCES

1. Seiji, M., Shimao, K., Birbeck, M. S. C., and Fitzpatrick, Subcellular localization of melanin biosynthesis. *Ann. N. Y. Acad. Sci.*, **100,** 497 (1963).
2. Mishima, Y. and Loud, A. V., The ultrastructure of unmelanized pigment cells in induced melanogenesis. *Ann. N. Y. Acad. Sci.*, **100,** 607 (1963).
3. Mishima, Y. and Cooper, M., Subcellular tyrosinase activity in amelanotic and melanotic Fortner's hamster melanomas. *J. Cell Biol*, **31,** 193A (1966).
4. Mishima, Y., Macromolecular characterization in neoplastic and dysfunctional human melanocytes. In *Structure and Control of the Melanocyte* (G. Della Porta and O. Mühlbock, eds.), pp. 135–55, Springer-Verlag, Berlin (1966).
5. Loud, A. V. and Mishima, Y., The induction of melanization in goldfish scale with ACTH, *in vitro*: Cellular and subcellular changes. *J. Cell Biol.*, **18,** 181 (1963).
6. Ito, R. and Mishima, Y., Particles in the cisternae of the endoplasmic reticulum of Fortner's amelanotic and melanotic melanomas. *J. Invest. Derm.*, **48,** 268 (1968).
7. Zelickson, A. S. and Hirsch, H. M., An electron microscope study of a mouse melanoma. *Acta Dermatovener.* (Stockholm), **44,** 399 (1964).
8. Novikoff, A. B., Albala, A., and Biempica, L., Ultrastructural and cytochemical observations on B-16 and Harding-Passey mouse melanomas. *J. Histochem. Cytochem.*, **16,** 299 (1968).
9. Satoh, G. J. Z. and Mishima, Y., Tyrosinase inhibitor in Fortner's amelanotic and melanotic malignant melanoma. *J. Invest. Derm.*, **48,** 301 (1967).
10. Cooper, M. and Mishima, Y., Substrate limiting melanogenic inhibitor in malignant melanomas. *Nature*, **216,** 189 (1967).

11. Satoh, G. J. Z. and Mishima, Y., Tyrosine release and tyrosinase activity in amelanotic and melanotic malignant melanoma. *J. Invest. Derm.*, **52**, 107 (1969).

12. Mishima, Y. and Adams, J., Non-premelanosomal and premelanosomal tyrosinase activity in Greene's amelanotic and melanotic malignant melanoma. *Cancer*, **23**, 952 (1969).

13. Mishima, Y. and Johnson, K. E. E., Subcellular tyrosinase activity in Fortner's malignant melanoma. *Brit. J. Derm.*, **80**, 293 (1968).

14. Takahashi, M. and Mishima, Y., A sequence of virus-like particle formation in the ergastoplasm of Greene's malignant melanoma cells. *Cancer*, **23**, 906 (1969).

15. Satoh, G. J. Z. and Mishima, Y., Subcellular enzyme-substrate relationships between amelanotic and melanotic malignant melanoma. *Fed. Proc.*, **25**, 294 (1966).

16. Fortner, J. G., Moht, A. G., and Schrodt, G. R., Transplantable tumors of the Syrian golden hamster. I. Tumors of the alimentary tract, endocrine glands, and melanomas. *Cancer Res.*, **21**, 161 (1961).

17. Satoh, G. J. Z. and Mishima, Y., Tyrosinase inhibitors in amelanotic and melanotic malignant melanoma. *Dermatologica*, **140**, 9 (1970).

18. Kim, K. and Tchen, T. T. Tyrosinase of the goldfish *Carassius auratus* L. I. Radioassay and properties of the enzyme. *Biochem. Biophys. Acta*, **59**, 569 (1962).

19. Udenfriend, S. and Cooper, J. R., The chemical estimation of tyrosine and tyramine. *J. Biol. Chem.*, **196**, 227 (1952).

20. Green, H. S. N., A spontaneous melanoma in the hamster with a propensity for amelanotic alteration and sarcomatous transplantation. *Cancer Res.*, **18**, 122 (1958).

21. Bertler, A., Carlsson, A., and Rosengren, E., A method for the fluorimetric determination of adrenaline and noradrenaline in tissues. *Acta Physiol. Scand.*, **44**, 273 (1958).

22. Adachi, K. and Halprin, K. M., A sensitive fluorometric assay method for mammalian tyrosinase. *Biochem. Biophys. Res. Commun.*, **26**, 241 (1967).

23. Mishima, Y. and Widlan, S., Enzymically active and inactive melanocyte population and ultraviolet irradiation: Combined dopa-premelanin reaction and electron microscopy. *J. Invest. Derm.*, **49**, 273 (1967).

24. Seiji, M., Subcellular tyrosinase activity and site of melanogenesis in melanocytes. In *Structure and Control of the Melanocyte* (G. Della Porta and O. Mühlbock, eds.), p. 123, Springer-Verlag, Berlin (1966).

25. Bernhard, W., The detection and study of tumor viruses with the electron microscope. *Cancer Res.*, **20**, 712 (1960).

26. Haguenan, F. and Beard, J. W., The avian sarcoma-leukosis complex: Its biology and ultrastructure. In *Tumors Induced by Viruses; Ultrastructure Studies*, pp. 113–50, Academic Press, New York (1962).

27. Dalton, A. J., Microcorphology of Murine tumor viruses and of affected cells.. *Fed. Proc.*, **21**, 936 (1962).

28. Yumoto, Y. and Dmochowski, L., Light and electron microscope studies of organs and tissues of the New Zealand Black (NZB/BL) strain mice with lymphoid-leukemia and autoimmune disease. *Cancer Res.*, **27**, 2083 (1967).

29. Adams, W. H. and Prince, A. M., An electron microscope study of the morphology and distribution of the intracytoplasmic "virus-like" particles of Ehrlich ascites tumor cells. *J. Biophys. Biochem. Cytol.*, **3**, 161 (1957).

30. Deharven, E. and Friend, C., Further electron microscope studies of a mouse leukemia induced by cell free filtrates. *J. Biophys. Biochem. Cytol.*, **7**, 747 (1960).

31. Friedlaender, M. and Moore, D. H., Occurrence of bodies within endoplasmic reticulum of Ehrlich ascites tumor cells. *Proc. Soc. Exp. Biol. Med.*, **92**, 828 (1956).

32. Parsons, D. F., Darden, E. B., Jr., Lindsley, D. L., and Pratt, G. T., Electron microscopy of plasma cell tumors of the mouse. I. MPC-1 and X5563 tumors. *J. Biophys. Biochem. Cytol.*, **9,** 353 (1961).

33. Pearson, H. F. and Baker, R. F., Persistent Theioer's virus in ependymoma tissue culture and the problem of virus-like bodies seen by electron microscopy. *J. Nat. Cancer Inst.*, **27,** 793 (1961).

34. Epstein, W. L., Fukuyama, K., and Benn, N., Transmission of a pigmented melanoma in golden hamsters by a cell-free ultrafiltrate. *Nature*, **219,** 979 (1968).

DISCUSSION

DR. QUEVEDO: Would you kindly comment on the nature of the nucleic acid composition of the virus-like particles and on potential infectivity?

DR. MISHIMA: Our experiment is not aimed in this direction; however, the structural continuity of the nucleoid structure of our virus-like particles with ribosomes of endoplasmic reticulum was clearly found under an electron microscope.

Melanin Production in Mammalian Cell Cultures*

Funan Hu

Oregon Regional Primate Research Center, Beaverton, Oregon, U.S.A.

I. INTRODUCTION

Study of the growth behavior of pigment cells in culture[1] indicates that melanin production in a pigment cell is regulated by:

A. The cell's genetic make up, which defines its limit and its potential to form melanin.

B. The cellular environment, which determines and modifies the expression of that potential.

The present communication is a review of what has been done in the past and a report of new data that lend support to these conclusions.

II. GENETIC CONTROL OF MELANIN PIGMENTATION

In a few attempts to investigate the mechanism of gene control of melanin pigmentation, the techniques of somatic cell hybridization have been used. Davidson et al.,[2,3] who hybridized pigmented Syrian hamster melanoma cells with mouse fibroblasts, suggested that genetic control of differentiation in hybrid cells involves a diffusible regulatory substance that functions negatively. Silagi[4] hybridized B16 melanoma cells with an

* Publication No. 440 of the Oregon Regional Primate Research Center, supported in part by grant CA 08499 of the National Cancer Institute, and grant FR 00163 of the National Institutes of Health, U.S.A.

azaguanine-resistant variant of L cells (mouse fibroblasts) and observed that the aptitude for progressive growth *in vivo* appears to be "dominant" in the hybrid that inherits from the C57BL melanoma parent the aptitude for proliferative growth *in vivo*, genes for H–2[b] histocompatibility antigens, and the ability to produce inosinic acid pyrophosphorylase. From the C3H parent, hybrids inherit genes for H–2[k] histocompatibility antigens and a possible suppressor of melanin production. The authors, therefore, concluded that the genetic control of differentiation in the hybrid cells is under a negative influence.

Attempts to convert amelanotic cells to melanotic ones or vice versa have not been consistently successful. Claunch et al.,[5] who tried to test the histone repressor theory,[6,7] were unable to convert the pigmented cells to nonpigmented cells by mouse liver histone. Glick and Salim[8] have reported the conversion of nonpigmented to pigmented cells with DNA prepared from the melanotic cells. Lipkin[9,10] recently reported success with RNA prepared from the pigmented cells, but Claunch et al.[5] were unsuccessful in similar attempts. These pigment-induction efforts, even when successful, are short-lived; none of the workers has been able to carry the converted cells as pigmented varieties on a long-term basis. Silagi's experiments on intraspecific somatic cells hybridization (B16 mouse melanoma cells and azaguanine-resistant variant of mouse fibroblasts) and tumor transplantation indicate that the capability for progressive growth *in vivo* and the ability to produce melanin in the pigment cells are controlled by separate genes, since the hybrid cells lose one characteristic and retain the other of each of the parent cells.[4]

III. GROWTH CHARACTERISTICS OF PIGMENT CELLS IN CULTURE

Two pigment cell lines (440B and P/51), sublines of line HFH-18, were used in this study.[1,11] Both have pigmented and nonpigmented cells. The cells are normally nonpigmented at the beginning of each subculture. Pigmented cells appear when confluent sheets are formed in about 7 to 10 days, and they increase in number when cells grow into multiple layers.[1]

Medium 199 supplemented with 10% fetal calf serum (GIBCO) and containing antibiotics (100 units of penicillin G, 20 μg of neomycin sulfate, and 50 μg of polymyxin B per ml of medium) was used in all cultures.

A. *Morphological variation of melanoma cells in culture*

Pigment cells vary greatly in size and shape. Different clones often show different morphologies (Figs. 1–10). They may be small, round, or ovoid, spindle, or epithelial-like, or they may be large polydendritic or epithe-

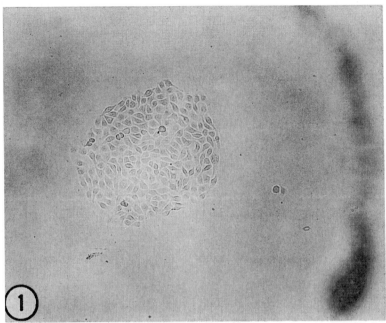

Figs. 1–3.   One clone of P/51 melanoma cells. ×85.
Fig. 1.   13 days. All nonpigmented cells.

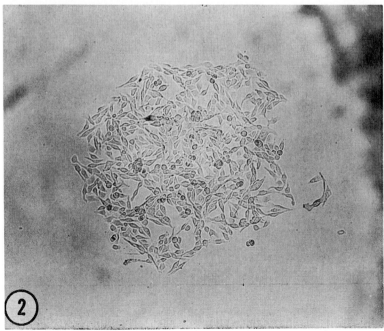

Fig. 2.   15 days. All except one or two cells are nonpigmented.

F. HU

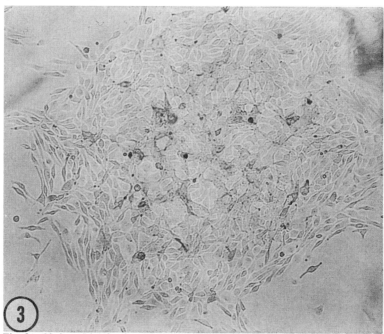

Fig. 3. 20 days. Lightly pigmented cells in the center, nonpigmented at the periphery.

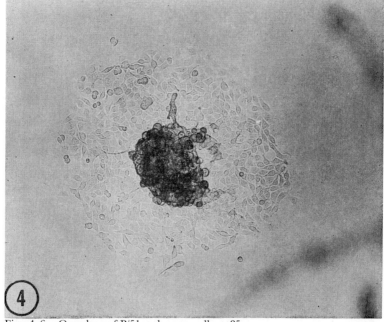

Figs. 4–6. One clone of P/51 melanoma cells. ×85.
Fig. 4. 13 days. A clone with a black center of many pigmented cells surrounded by nonpigmented ones.

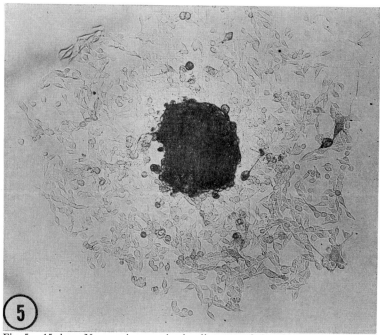

Fig. 5.   15 days. Note an increase in the diameter of the clone and a few scattered pigmented cells in addition to those in the black center.

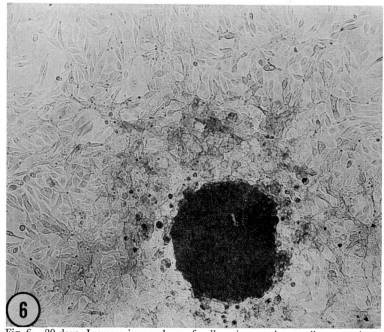

Fig. 6.   20 days. Increase in numbers of cells—pigmented as well as nonpigmented.

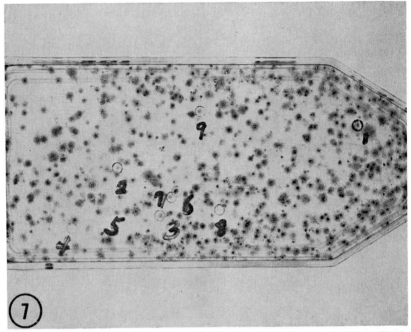

Fig. 7. P/51 melanoma cells in culture; 10,000 cells in 3 ml of medium in Falcon Plastic T-30 flask. Note the variation in size and in degree of pigmentation of these colonies. Almost all colonies have a pigmented center, surrounded by a nonpigmented periphery.

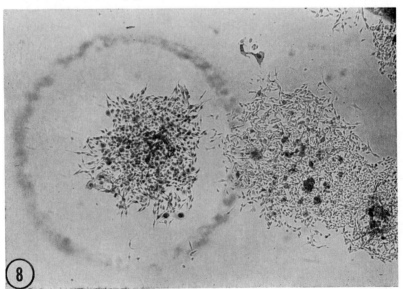

Figs. 8–11. A pigmented clone of P/51 melanoma cells.
Fig. 8. 15 days. ×30.

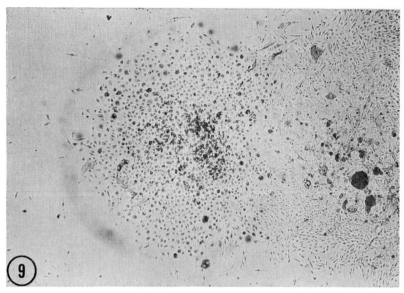

Fig. 9. 20 days. ×30.

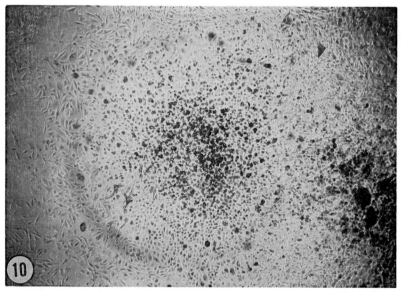

Fig. 10. 27 days. ×30.

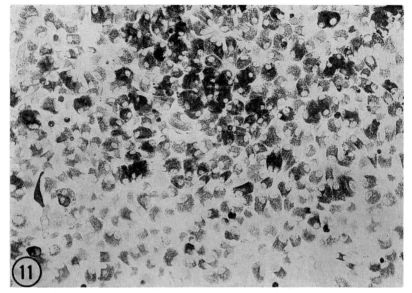

Fig. 11. Higher magnification of Fig. 10. ×120. Note epithelial-like cells with perinuclear melanin pigment.

lioid. In general, the cells with the highest aptitude for proliferation are small, round, ovoid, or spindle-shaped. As the cells mature, they increase in size and become polydendritic or epithelioid. The presence or absence of melanin pigment depends on the genetic background of the individual cells. As a rule, with the exception of the nonpigmented strain, the cells become more pigmented as they mature. The aging process appears to be accelerated by treatment with agents that either affect cell proliferation or are somewhat harmful to the cells. Increase of pigmentation by the appearance of large cells after treatment with antimitotic agents such as Colcemid or Velban[1] is probably the result of one or the other of these effects.

The morphology of cells is influenced also by the morphology of the company they keep. Co-cultivation of pigment cells with either fibroblasts or retinal epithelial cells favors the formation of a syncytium of dendritic cells or large epithelioid cells respectively. The orientation of the dendrites of pigment cells often follows that of the long axis of the fibroblasts.[1]

B. *Growth in mass cultures*

All cultures were initiated by an inoculation of 250,000 cells suspended in 3 ml of growth medium into Falcon Plastic T-30 flasks or on coverglasses in 60×15 mm plastic culture dishes. Cells multiplied and formed confluent sheets in about 7 to 10 days. At this time there were usually only a few scattered pigmented cells. At 10 to 14 days, when cells began to form multiple layers, small aggregates of pigmented cells appeared in centers where multiple layers were more evident. The number of pigmented cells increased as incubation was continued. Further incubation led to the formation of many colonies that coalesced to form plaques containing many pigmented cells; some of them became very large and epithelioid.

The same cultures, when subcultured at weekly intervals, showed a predominance of nonpigmented, small round, ovoid, or spindle-shaped cells.[1]

C. *Cloning*

For cloning, instead of the usual 250,000-cell inoculum, 1,000, 5,000, or 10,000 cells suspended in 3 ml of medium were used to initiate subcultures. At the end of 24 to 48 hours incubation single cells were marked for follow-up observation. Microscopic examinations were made every 1 to 3 days, and photomicrographs were taken as often as needed to record the development.

Our results indicate that almost all clones started as nonpigmented small cells that multiplied and formed aggregates in monolayers. The aggregates remained as monolayers up to certain sizes (which varied somewhat in each individual clone); cells then began to pile up in the center

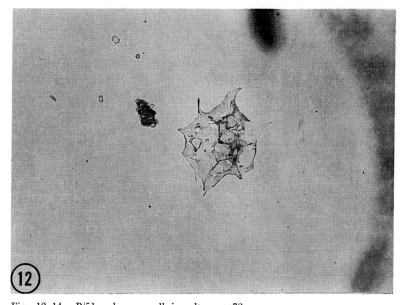

Figs. 12–14. P/51 melanoma cells in culture. ×72.
Fig. 12.   13 day. Two aggregates of pigmented cells in the field—one consists of small round cells, the other of large epithelial-like cells.

311

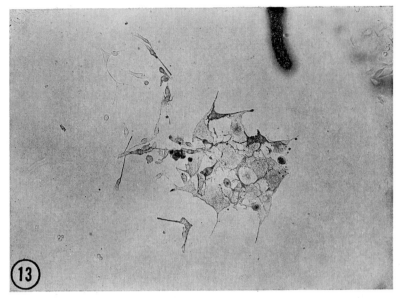

Fig. 13. 15 days. Note that most of the small cells have become spindlelike in shape and appear to have moved into the large cells' **area**; the large cells have increased in size but not in number.

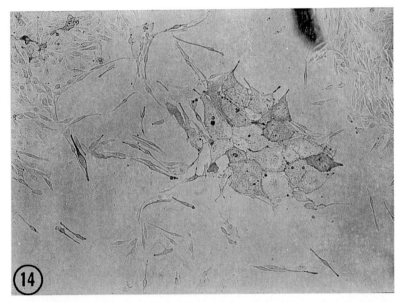

Fig. 14. 20 days. Note further increase in the size of the large cells but still without increase in number. The field is invaded from all directions by the proliferating (predominantly nonpigmented) small cells.

of each aggregate. The cells in these areas became pigmented and increased in number as incubation continued (Figs. 1–6).

Although clones have some variation in size when cells are piling up, as well as during the time when pigmented cells appear in them, all clones eventually become pigmented (Figs. 1–7). Almost all had a black center and a small rim of nonpigmented periphery made up of small, usually spindle-shaped cells. Only very rarely was a clone entirely black.

An occasional aggregate may consist of pigmented cells at the beginning; these may be small ovoid or spindle (Fig. 8), or large epithelioid (Fig. 12). Those aggregates, which are made of small cells, proliferate to form colonies (Figs. 8–11). The large epithelioid cells, however, increased only in size, not in number; and they did not develop into a colony (Figs. 12–14). They are regarded as aging cells that survive trypsinization and are carried over during subculture; usually they do not engage in DNA synthesis but continue to synthesize RNA and protein.[12]

IV. EFFECTS OF CHANGES IN CULTURAL ENVIRONMENT ON MELANIN PRODUCTION

More pigmented cells developed when they were grown on agar, as pellets or aggregates, with the use of a rotary shaker.[1] Sections of the aggregates

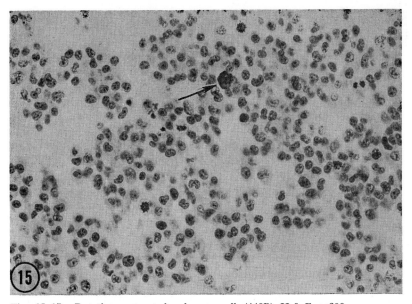

Figs. 15–17.   Rotation-aggregated melanoma cells (440B). H & E. ×300.
Fig. 15.   1 day. Note that except for one pigmented cell (arrow) all cells in the field are nonpigmented. There are at least 6 mitotic figures.

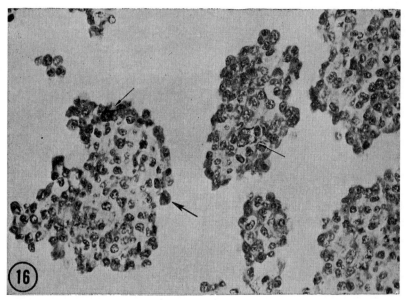

Fig. 16. 3 days. There are more pigmented cells in the aggregates (small arrows) and only one mitotic figure (large arrow).

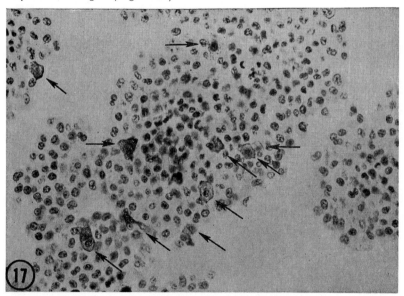

Fig. 17. 6 days. There are 10 pigmented cells in the field but no mitosis.

at different days of incubation showed that after 24 hours' incubation there were many mitotic figures but very few pigmented cells (Fig. 15). At 3 days the number of dividing cells decreased, and that of pigmented ones increased (Fig. 16); at 6 days there was no evidence of mitosis, but many cells were pigmented (Fig. 17).

When grown in a Rose multipurpose culture chamber under a dialysis membrane, many pigmented cells developed, but the same cells grown without the membrane were predominantly nonpigmented.[1,13] Similar results were obtained when the cells were grown under perforated cellophane. The pigmented cells were usually seen under the nonperforated part of the membrane.

Growing cells on agar, as pellets or as aggregates, helps to establish an increase in cell density and a reduction in growth rate. The membrane possibly produces these effects by: (1) preventing loss of intermediate metabolites that may be essential in the synthesis of melanin, (2) separating the macromolecular components of the medium from contact with the cells, or (3) increasing relative cell density by reducing the available free spaces around the cells.

V. EFFECTS OF AGENTS KNOWN TO AFFECT MELANIN PIGMENTATION

A. *Hyperpigmenting agents*

None of the pigmenting agents tested (copper sulfate, dopa, tyrosine), stimulated pigment formation.[14] ACTH and MSH at low concentrations (0.2 to 0.5 $\mu g/ml$) had no effect on pigment cells except occasionally to produce slightly more, larger, and darker pigmented cells when treatment was carried for 2 to 4 subcultures.[14] At high concentration (5 $\mu g/ml$) there were definite decreases in the total number, along with the appearance of very large pigmented and nonpigmented cells (Figs. 18–20). These are similar to the large cells observed in the Colcemid or Velban-treated cultures and cells exposed to cold temperature[1] or X-rays.[15]

B. *Depigmenting agents*

Among the depigmenting agents (ascorbic acid, melatonin, hydroquinone), only hydroquinone was found to have a selective effect on the pigmented cells. This agent, when incorporated in the culture medium at a concentration of 0.625 $\mu g/ml$, inhibited the appearance of pigmented cells but not of nonpigmented ones.[14]

Two chemical agents with reactive SH groups, ergothioneine and Cleland's reagent, were also tested in our culture system. Neither showed a depigmenting effect on the pigmented cells. Both were toxic to the cells at higher concentrations but they permitted normal growth at lower concentrations (0.0001 M/ml).

In primary cultures of adult guinea pig skin, α-MSH, acetylcholine, melatonin, and norepinephrine were also ineffective in changing pigmentation in the melanocytes.[16] Ultraviolet irradiation inhibited cell proliferation but failed to induce hyperpigmentation.[15,17]

VI. CO-CULTIVATION WITH HETEROGENOUS CELLS

Mouse melanoma cells, when grown with epithelial-like rhesus retinal

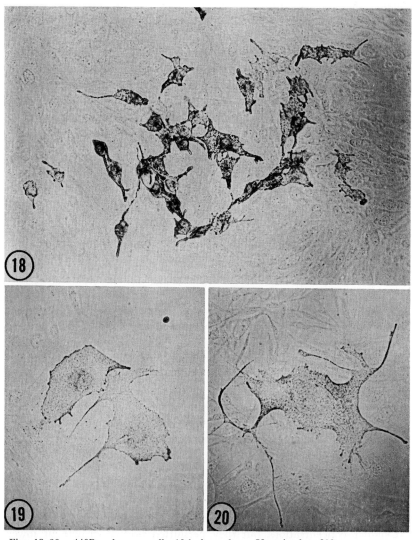

Figs. 18–20.   440B melanoma cells. 10th-day culture. Unstained. ×300.
Fig. 18.   Treated with ACTH 0.5 μg/ml. Growth like untreated controls.
Figs. 19 and 20.   Treated with ACTH 5 μg/ml. Note the large pigmented and non-pigmented cells.

cells, formed colonies among the retinal cells. Many of the colonies were pigmented. The mouse cells became pigmented more easily than they did without the monkey cells.[1] Similar results were obtained when mouse melanoma cells were grown with either monkey or human fibroblasts.

The presence of normal monkey or human cells, which ordinarily grow at a slower rate than mouse cells, tends to slow down the growth rate of the pigment cells. In all cases the addition of a different cell type to the culture of mouse melanoma cells, whether rhesus retinal cells, rhesus fibroblasts, or human fibroblasts, helps to accelerate the emergence of large pigmented cells.

Co-cultivation with heterogenous cells probably achieves its effect by increasing cell density, although, as mentioned earlier, the presence of cells from normal tissue origin also slows down the overall growth rate of melanoma cells. The morphology of pigmented cells appears to follow the pattern of the surrounding cells. They are epithelial-like when grown with the retinal cells (Fig. 21); grown with fibroblasts, they are polydendritic and their processes tend to align with the long axes of the nearby fibroblasts (Figs. 22, 23). Epidermal melanocytes have also been shown to change their shape and activity according to their immediate cellular environment.[18, 19]

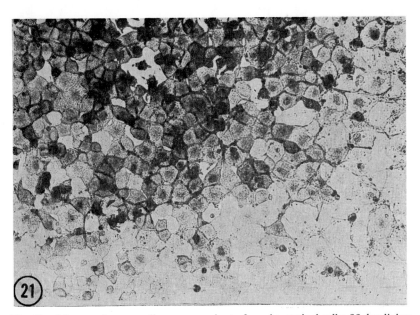

Fig. 21. Mouse melanoma cells grown on sheet of monkey retinal cells; 26-day living culture. Note sheets of pigmented epithelial-like cells, darker in the center and light at the periphery. ×120.

317

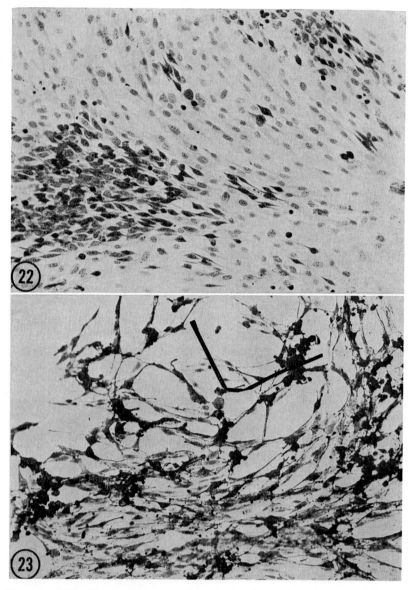

Figs. 22 and 23.   Mouse melanoma cells grown on sheets of monkey skin fibroblasts.
May-Grünwald-Giemsa. ×120.
Fig. 22.   13-day culture. Note the darkly stained melanoma cells lining up along the
long axes of the lightly stained monkey fibroblasts.
Fig. 23.   14-day culture. Note a syncytium formation of polydendritic melanoma cells.
A few lightly stained cells in the background are monkey fibroblasts.

VII. DISCUSSION

Analysis of the data so far shows clearly that, except for those that affect cell proliferation, none of the factors consistently influences pigment formation. Silagi has also commented that treatments that are successful in evoking pigment formation also drastically inhibit growth.[20] She suggested that these effects may result from the production of an enzyme inhibitor. Production of this enzyme inhibitor ceases or decreases under conditions of stationary growth, permitting the tyrosinase gene to re-express itself. Although cell proliferation inhibits pigment formation, it cannot be the only factor involved, since UVL, X-rays, and others that also inhibit cell proliferation are not effective in promoting pigmentation significantly *in vitro*.

Cell interaction seems to play a role in pigment production. Growing cells on agar, as pellets and aggregates, or under membranes, all of which help to establish cell density and increase contact between cells, appears to promote pigment cell development. Crowding and formation of multiple layers have similar effects. The appearance of pigmented cells in the multilayered central zone of the clone is noteworthy. It is possible that an interaction between cells is required for, or at least is favorable to, pigment formation. Metabolic cooperation by cell-to-cell transfer occurs in tissue culture.[21-25]

In the rotation-aggregated cells, an inverse relationship exists between mitosis and pigmentation. In early cultures, the emergence of many pigmented cells occurs as mitotic activity ceases. This concurs with the general idea that cells maintained in a nonmitotic and viable population of tissuelike density retain their phenotypic differentiation function. For example, steroid synthesis is maximal in the stationary phase culture of mouse testicular interstitial cells;[26] the rapidly dividing cells of chick embryonic chondrocytes in monolayer cultures lose their ability to synthesize chondroitin sulfate more rapidly than cells in pellets that divide slowly.[27] Chick retinal pigment cells[28] and mouse[1] and hamster[29] melanoma cells lose their melanotic phenotype during proliferation. These observations suggest that the commencement of specific synthesis correlates with a switch from replication to nonreplication. It is conceivable that slowing the growth rate enables the cells to channel their efforts from synthesis of protein for growth to a synthesis of specific products. Silagi[20,30] as well as Romsdahl and Hsu[31] have observed respectively in mouse and human melanoma cell cultures a longer doubling time of the pigmented clones as compared with their nonpigmented counterparts. Whittaker has suggested that the loss of melanotic phenotype from chick retinal pigment cells in monolayer culture results when the new messenger RNA produced by rapidly growing cells competes for translation with the existing tyro-

Table 1. Factors influencing melanin production in mammalian cell cultures.

Cell origin	Type of culture	Agents	Melanin production	Reference
B16 mouse melanoma	Permanent cell line	α-MSH, ACTH Tyrosine, dopa $CuSO_4$	No effect	Hu, 1966[14]
		Melatonin Ascorbic acid	No effect	
		Hydroquinone	Selective inhibition on pigmented cells	
B16 melanoma	Permanent cell line	Ergothioneine Cleland's reagent	No effect	Hu
		Vincristine (Velban)	Increase in size of pigmented and nonpigmented cells	Hu, 1969[1]
		Colcemid	Melanin production	Hu, 1969[1]; Silagi, 1969[20,30]
		1-β-D-Arabinofuranosyl cytosine Actinomycin D Puromycin Cycloheximide p-fluorophenylalanine FUdR, BUdR, IUdR Amethopterin Edeine Mitomycin	No effect	Silagi, 1969[20,30]
B16 melanoma	Permanent cell line	Mouse liver histone	Little or no repression of pigment formation	Claunch et al., 1967[5]
		RNA extracted from black tumors or from black cells	Little or no effect	
		Melanosomes from black tumor	Temporary pigmentation (phagocytosis)	

Cell type		Treatment/Condition	Effect	Reference
B16 melanoma	Permanent cell line	UVL	No increase	Kitano and Hu, 1969[7]; Silver and Hu, 1968[5]
		X-ray	No increase	Silver and Hu, 1968[5]
		Cold (4°C)	Increase in size of pigmented and nonpigmented cells	Hu, 1969[1]
		Growth on agar	Increase	Hu, 1969[1]; Silagi, 1969[20]
		Crowding	Increase	Hu, 1969[1]
		Rotation-aggregated cultures	Increase	Hu, 1969[1]
		Co-cultivation with heterogenous cells	Increase	Hu, 1969[1]
		Growth under dialysis membrane	Increase	Rose, unpublished data
		Hybridization with mouse fibroblasts	Complete lack of pigment	Silagi, 1967[4]
Hamster melanoma	Permanent cell line	Hybridization with mouse fibroblasts	Complete lack of pigment	Davidson et al., 1966, 1968[2,3]
		DNA from melanotic melanoma	3-fold increase in pigment producing cells	Glick and Salim, 1967[8]
		RNA from melanotic melanoma	Temporary increase in pigmented cells	Lipkin, 1969[9]
Guinea pig adult skin	Primary cell culture	α-MSH	No effect	Klaus and Snell, 1967[6]
		Acetylcholine	No effect	
		Melatonin	No effect	
		Norepinephrine	No effect	

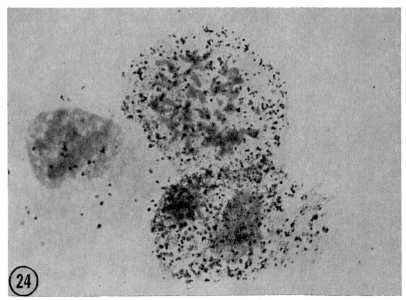

Fig. 24. 9-day melanoma cell culture. Three cells in the field: one unlabeled nonpigmented cell and two labeled pigmented cells, one of which is in mitosis. Melanin granules are smaller, lighter in color, and are not in exact focus. Autoradiography ^{3}H-dopa. $\times 1200$.

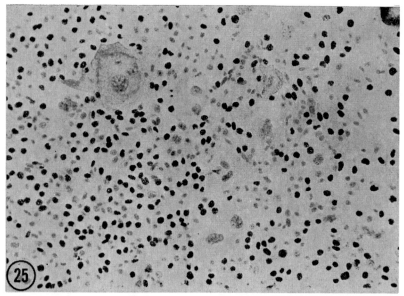

Fig. 25. 28-day melanoma cell culture. Note that the nuclei of most of the small cells are labeled. The large cells, pigmented and nonpigmented, are not labeled. Autoradiography ^{3}H-thymidine. $\times 120$.

sinase messenger RNA. When monolayer cultures become crowded, the cells begin synthesizing melanin more rapidly. Whittaker believes that some mechanism other than tyrosinase gene inactivation causes the decrease in tyrosinase synthesis in these cultures.[32]

On the other hand, proliferation is probably necessary for "differentiation" because without cell proliferation the much-needed cell density and interaction would not have been possible. Certain connective tissue cells perform their specific functions best when they are growing most actively.[33, 34] Coon and Cahn[35] have discovered culture conditions that permit proliferation without apparent loss of phenotypic traits. In cloning experiments, melanoma cells do not lose their melanizing function when they are still actively growing. Thus proliferation and synthesis of a specialized product are not mutually exclusive; in our cultures, pigmented cells divided as easily as nonpigmented ones. Other investigators have reported similar observations in pancreatic, plasma, and other cells.[36-39] With autoradiography we have shown that mitotic division occurs in cells that have incorporated dopa (Fig. 24) and that pigmented cells can engage actively in DNA synthesis and cell division. The nonpigmented small ovoid or spindle cells are usually more active in DNA synthesis than in melanin synthesis, whereas the pigmented large epithelioid cells are more active in melanin synthesis than in DNA synthesis (Figs. 24, 25).

VIII. SUMMARY AND CONCLUSIONS

Table 1 summarizes the various factors that have been utilized to influence melanin production in cell cultures. Except for those that affect cell proliferation, none of the factors listed consistently influence pigment formation. At present we have been able to alter the melanin-producing capacity in cells within, but not beyond, the limits of their genetic makeup.

REFERENCES

1. Hu, F., Proliferation and melanin production in melanoma cell culture. Presented at the Seventh International Pigment Cell Conference, Seattle, Washington, Sept. 2–6, 1969.
2. Davidson, R., Ephrussi, B., and Yamamoto, K., Regulation of melanin synthesis in mammalian cells, as studied by somatic hybridization. I. Evidence of negative control. *J. Cell Physiol.*, **72**, 115 (1968).
3. Davidson, R. and Yamamoto, K., Regulation of melanin synthesis in mammalian cells, as studied by somatic hybridization. II. The level of regulation of 3, 4-dihydroxyphenylalanine oxidase. *Proc. Nat. Acad. Sci. U.S.A.*, **60**, 894 (1968).
4. Silagi, S., Hybridization of a malignant melanoma cell line with L-cells *in vitro. Cancer Res.*, **27**, 1953 (1967).
5. Claunch, C., Oikawa, A., Tchen, T. T., and Hu, F., Biochemical studies on certain

"pigmented" and "nonpigmented" strains of melanoma cells. In *Advances in Biology of Skin*, vol. 8, *The Pigmentary System* (W. Montagna and F. Hu, eds.), p. 479, Pergamon Press, Oxford (1967).

6. Littau, V. C., Burdick, C. J., Allfrey, V. G., and Mirsky, A. E., The role of histones in the maintenance of chromatin structure. *Proc. Nat. Acad. Sci. U.S.A.*, **54**, 1204 (1965).

7. Huang, R. C. and Bonner, J., Histone-bound RNA, a component of native nucleohistone. *Proc. Nat. Acad. Sci. U.S.A.*, **54**, 960 (1965).

8. Glick, J. and Salim, A., DNA-induced pigment production in a hamster cell line. *J. Cell Biol.*, **33**, 209 (1967).

9. Lipkin, G., RNA-mediated pigment induction in amelanotic malignant melanocytes. *Clin. Res.*, **17**, 275 (1969) (abstract).

10. Lipkin, G., Induction of pigment in hamster amelanotic malignant melanocytes by RNA from a melanotic strain. Presented at the Seventh International Pigment Cell Conference, Seattle, Washington, Sept. 2–6, 1969.

11. Hu, F. and Lesney, P. F., The isolation and cytology of two pigment cell strains from B16 mouse melanomas. *Cancer Res.*, **24**, 1634 (1964).

12. Kitano, Y. and Hu, F., DNA, RNA, and protein synthesis of pigment cells in culture. *J. Invest. Derm.* (in press) (1970).

13. Rose, G. G., personal communication.

14. Hu, F., The influence of certain hormones and chemicals in mammalian pigment cells. *J. Invest. Derm.*, **46**, 117 (1966).

15. Silver, S. E. and Hu, F., The effects of ultraviolet light and x-irradiation on mammalian pigment cells *in vitro*. *J. Invest. Derm.*, **51**, 25 (1968).

16. Klaus, S. and Snell, R. S., The response of mammalian epidermal melanocytes in culture to hormones. *J. Invest. Derm.*, **48**, 352 (1967).

17. Kitano, Y. and Hu, F., The effects of ultraviolet light on mammalian pigment cells *in vitro*. *J. Invest. Derm.*, **52**, 25 (1969).

18. Kitano, Y. and Hu, F., DNA, RNA, and protein synthesis of pigment cells in culture. *J. Invest. Derm.* (in press) (1970).

19. Kitano, Y. and Hu, F., Morphological changes in melanocytes of the great bushbaby *in vitro*. Presented at the Seventh International Pigment Cell Conference, Seattle, Washington, Sept. 2–6, 1969.

20. Silagi, S., Control of pigment production in mouse melanoma cells *in vitro*: Evocation and maintenance. *J. Cell Biol.*, **43**, 263 (1969).

21. Subak-Sharpe, H., Burk, R. R., and Pitts, J. D., Metabolic cooperation by cell to cell transfer between genetically different mammalian cells in tissue culture. *Heredity*, **21**, 342 (1966).

22. Stoker, M. P., Transfer of growth inhibition between normal and virus-transformed cells: Autoradiographic studies using marked cells. *J. Cell Sci.*, **2**, 293 (1967).

23. Stoker, M. P., Contact and short-range interactions affecting growth of animal cells in culture. In *Current Topics in Develop. Biology* (A. A. Moscona and A. Monroy, eds.), Academic Press, New York, **2**, 107 (1967).

24. Eagle, H. and Piez, K., The population-dependent requirement by cultured mammalian cells for metabolites which they can synthesize. *J. Exp. Med.*, **116**, 29 (1962).

25. Rein, A. and Rubin, H., Effects of local cell concentrations upon the growth of chick embryo cells in tissue culture. *Exp. Cell Res.*, **49**, 666 (1968).

26. Shin, S., Studies on interstitial cells in tissue culture: Steroid biosynthesis in monolayers of mouse testicular interstitial cells. *Endocrinology*, **81**, 440 (1967).

27. Stockdale, F. E., Abbott, J., Holtzer, S., and Holtzer, H., The loss of phenotypic

traits by differentiated cells. II. Behavior of chondrocytes and their progeny *in vitro. Develop. Biol.*, **7**, 293 (1963).

28. Whittaker, J. R., Changes in melanogenesis during the dedifferentiation of chick retinal pigment cells in cell culture. *Develop. Biol.*, **8**, 99 (1963).
29. Moore, G. E., *In vitro* cultures of a pigmented hamster melanoma cell line. *Exp. Cell Res.*, **36**, 422 (1964).
30. Silagi, S., Control of pigment production in synchronized melanoma cells *in vitro*. Presented at the Seventh International Pigment Cell Conference, Seattle, Washington, Sept. 2–6, 1969.
31. Romsdahl, M. and Hsu, T. C., Establishment and characterization of human malignant melanoma cell lines. Presented at the Seventh International Pigment Cell Conference, Seattle, Washington, Sept. 2–6, 1969.
32. Whittaker, J. R., Translational competition as a possible basis of modulation in retinal pigment cell cultures. *J. Exp. Zool.*, **169**, 143 (1968).
33. Davidson, E. H., Gene activity in differentiated cells. In *Retention of Functional Differentiation in Cultured Cells* (V. Defendi, ed.), *Wister Inst Symp. Monogr.*, **1**, 49 (1964).
34. Morris, C. C., Quantitative studies on the production of acid mucopolysaccharides. *Ann. N. Y. Acad. Sci.*, **86**, 878 (1960).
35. Coon, H. G. and Cahn, R. D., Differentiation *in vitro*: Effects of Sephadex fractions of chick embryo extract. *Science*, **153**, 1116 (1966).
36. Wessells, N. K. DNA synthesis, Mitosis, and differentiation in pancreatic acinar cells *in vitro. J. Cell Biol.*, **20**, 415 (1964).
37. Nossal, G. J. and Makela, O., Autoradiographic studies on the immune response. I. The kinetics of plasma cell proliferation. *J. Exp. Med.*, **115**, 209 (1962).
38. Davies, L. M., Priest, J. H., and Priest, R. E., Collagen synthesis by cells synchronously replicating DNA. *Science*, **159**, 91 (1968).
39. Herrmann, H., Marchok, A. C., and Baril, E. F., Growth rate and differentiated function of cells. *Nat. Cancer Inst. Monogr.*, **26**, 303 (1967).

Discussion

Dr. Quevedo: Dr. Whittaker has suggested that, when retinal melanocytes are not dividing, "growth protein" mRNA is translated instead of mRNA for tyrosinase owing to the much greater concentration of the former.

Dr. Bagnara: I don't know if I agree with Dr. Whittaker's hypothesis as explained by Dr. Quevedo, for Dr. Pehlmann showed us rather nicely at Seattle that MSH stimulates cell division of melanophores and at the same time melanin synthesis. In other words, pigmented melanophores divide rapidly and synthesize melanin.

Dr. Hu, does MSH stimulate those melanocytes grown in co-culture on human fibroblasts? Many of those melanocytes resemble those seen *in vivo.*

Dr. Hu: No, it does not. These large epithelial-like or dendritic cells are filled with pigment but no longer actively synthesize melanin. In this respect they are quite similar to the melanophores of the fishes and amphibians; they do not, however, respond to MSH like the melanophores do.

DR. OIKAWA: Concerning Dr. Quevedo's comment, I observed three types of pigmentation patterns of cultured melanoma lines. In the first type, pigmentation accelerated at the stationary stage of culture, for which Dr. Quevedo gave a beautiful explanation. In the second type, maximum melanin synthetic activity was observed at the exponentially growing stage. This type is consistent with Dr. Bagnara's comment. The third type is a mixture of the first two.

DR. HU: To me, each clone has its own growth rate and ability to form pigment. There are various reports about the doubling time of pigment cells. Nonpigmented cells always seem to grow faster and to have a shorter doubling time than pigmented cells. I think Dr. Whittaker's idea is certainly a good one. I don't know whether it applies to our cells or not. It is logical to assume that when cells are busily engaged in one activity some of the other activities have to slow down somewhat, unless the cells' energy is unlimited. But how much or even whether the cells are limited in their activity is not known. More than one factor must be involved in regulating these activities. As Dr. Oikawa just indicated, it probably depends on which cell you are dealing with.

DR. BLOIS: Dr. Hu has shown by means of her beautiful experimental system that light has no stimulating effect upon pigment synthesis in a pure melanocyte culture. Since the mixed cell culture is an equally well-defined system, I am wondering if Dr. Hu has looked for light stimulation of melanin synthesis in a culture containing melanocytes plus a second cell type. This approach could greatly increase our understanding of the mechanism of action of light-induced pigmentation.

DR. HU: We have no such data as yet, but we are thinking along the same line.

DR. FITZPATRICK: The human malignant melanoma, in contrast with some other human malignant tumors (e.g., thyroid carcinoma), retains its ability to differentiate and forms pigmented melanosomes in cells that are undergoing mitotic proliferation.

DR. HU: From these cells, we have developed two transplantable tumors, black and white. Their rate of proliferation is about the same, the black tumor growing slightly faster than the white one, for some unknown reason. But in sections the pigmented tumor is not completely black, because the proliferating cells are predominantly nonpigmented. This is comparable to our *in vitro* system, where in the same culture we have cells with or without pigment. The nonpigmented cells are primarily responsible for proliferation, whereas the black cells usually are older and more mature. But when we plate the cells out, we find clones of various degrees of pigmentation: completely melanotic, partially melanotic, and amelanotic. For the melanotic clones, naturally the cells that are proliferating are also pigmented. I believe that cells in a tumor derive from one or several clones, and therefore their pigment content, like that of the cultured cells, also varies. Similarly, malignant melanomas in man, like those of the mouse, can be melanotic or amelanotic.

Dr. Mishima: I am interested in your findings concerning the increased melanin synthesis of melanoma cells *in vitro* after cold storage or culturing in dialysis membrane. Could you give us your present thoughts about these findings of yours? Is there any possibility of relating these changes to the diminution of melanogenic inhibitors in these circumstances?

Dr. Hu: I have found that cold temperature, which is one of the ways of arresting mitosis, works more or less like Colcemid. Here again, the slowing of the growth rate seems to make the difference. Under the dialysis membrane, the cells are protected from the flooding of nutrient on top of it, so that anything diffusible coming from these cells is not carried away by the constant flushing of the media. The dialysis membrane probably prevents some loss of macromolecular elements, e.g., the intermediates that are essential for pigment formation. It would seem that, if the cells constantly lose the intermediates, melanin formation would never reach the end stage. So far, we have not been able to demonstrate an inhibitor in our pigmented or nonpigmented cells.

Dr. Seiji: Have you ever determined the acid phosphatase activity of cultured melanocytes? It could be quite interesting to know in relation to the change of tyrosinase activity during the culture.

Dr. Hu: We have not done any determinations. Under light microscopy, some cells in cultures, especially the older ones, do have vacuoles that usually indicate acid phosphatase activities. By electron microscopy, those large cells have been shown to have melanosome complexes that suggest a lysosomal type of structure. Hence, I would postulate some phosphatase activity.

Dr. Chavin: Dr. Myron Gordon has pointed out that the rapidly growing hybrid swordtail melanotic melanoma has areas of decreased cellular melanin content where growth is most rapid. This appears to coincide with the amelanotic melanoma (with considerable tyrosinase activity) which does not grow as rapidly and is smaller than the melanotic melanoma of the same species. Thus, the situation here is reversed. In addition, the amelanotic melanomas of older albino animals develop the ability to synthese melanin and become pigmented. Metastases of such pigmented cells in these albino fish are evident. It would appear, therefore, that a variety of controls are present, even in a given system, and one should not expect to ascribe pigment content to a single agent.

Human Malignant Melanomas of the Skin and Their Pre-existing Conditions
—*A Comparative Electron Microscopic Study*—

Takae Hirone, Tadashi Nagai, Tameaki Matsubara, and Ryoichi Fukushiro

Department of Dermatology, Kanazawa University School of Medicine, Kanazawa, Japan

I. INTRODUCTION

In recent years, it has been suggested by several investigators[1-4] that human malignant melanomas may arise from apparently normal skin, from Hutchinson's melanotic freckle, or from junctional nevus. Assuming this to be true, it is further supposed that these different kinds of melanomas may be different in ultrastructure and in biological potential. However, few attempts to prove this supposition have been made. It is the purpose of this study to demonstrate the ultrastructural features of various melanomas, with special reference to melanin-containing organelles. In addition, the ultrastructures of various melanomas and those of their pre-existing conditions will be compared.

The terminology of melanin-containing organelles is used in accordance with the proposal[5] made at the Sixth International Pigment Cell Conference held in Sofia, Bulgaria, in 1965.

II. MATERIALS AND METHODS

Materials used for this study were as follows: (1) two cases of melanoma *de novo*, consisting of melanotic and amelanotic parts, (2) two cases of melanotic melanoma arising in Hutchinson's melanotic freckle, (3) two cases of melanotic melanoma, presumably arising from junctional nevus, and (4) two cases of junctional nevus and 10 cases of compound nevus.

329

These cases were classified by their anamnestic, clinical, and histological data. In all cases of melanomas, except a case presumably arising from junctional nevus, the primary tumor was examined. In the exceptional case, only the metastatic cutaneous tumor was examined, since the primary tumor, on which we made a microscopic diagnosis, was excised at another hospital.

For electron microscopy, specimens were fixed for 2 hours in 2.5% glutaraldehyde buffered to pH 7.4 with cold (4°C) 0.1 M cacodylate buffer[6] and then postfixed for 2 hours in 2% osmium tetroxide buffered to pH 7.4 with cold veronal-acetate to which sucrose had been added.[7] After dehydration in graded concentrations of acetone, specimens were embedded in Epon 812.[8] Sections were cut with glass knives on a Porter-Blum MT-2 microtome, stained with uranyl acetate followed by lead citrate, and examined in a Hitachi HU-11Ds electron microscope.

III. RESULTS

Melanoma cell nests usually contain variable numbers of melanophages which are identified by the presence of melanosome complexes and by the absence of premelanosomes disposed individually in the cytoplasm. In this study our attention was focused on melanoma cells, and the melanophage was excluded from observations.

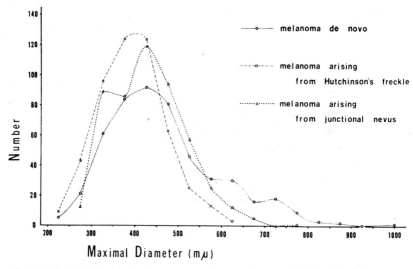

Fig. 1. Variation in size of melanosomes and premelanosomes in three kinds of melanomas. A total of five hundred melanosomes and premelanosomes in each kind were measured. Extraordinarily large melanosomes and premelanosomes are more numerous in melanoma *de novo* and melanoma arising from junctional nevus than in melanoma arising from Hutchinson's melanotic freckle. Normal range is 120 to 480 mμ.[21]

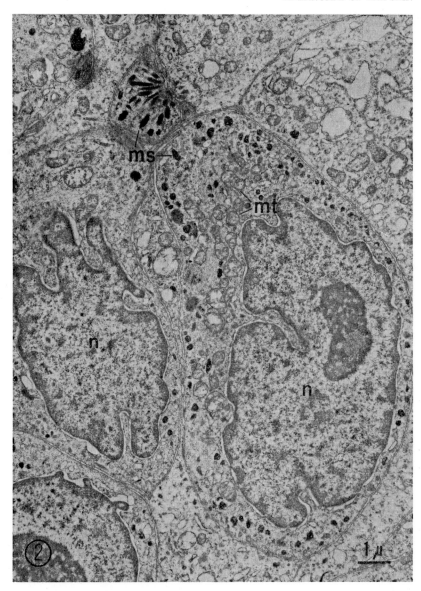

Fig. 2. Melanotic melanoma *de novo*. The following key is used in all figures: n: nucleus; mt: mitochondria; er: endoplasmic reticulum; g: Golgi apparatus; ms: melanosome; pms: premelanosome; v: vacuole. ×8,900.

A. *Melanoma de novo*

In melanotic melanomas of this kind (Figs. 2, 10), the cells usually have indented nuclei with ramified nucleoli, a large number of mito-

331

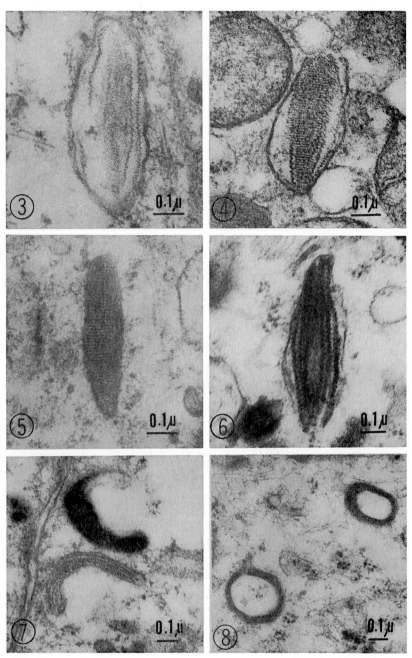

Figs. 3–8.   Various stages in the development of premelanosomes in melanotic melanoma *de novo*. Fig. 3: ×85,000; Figs. 4, 5: ×84,000; Fig. 6: ×72,000; Fig. 7: ×60,000;
Fig. 8: ×54,000.

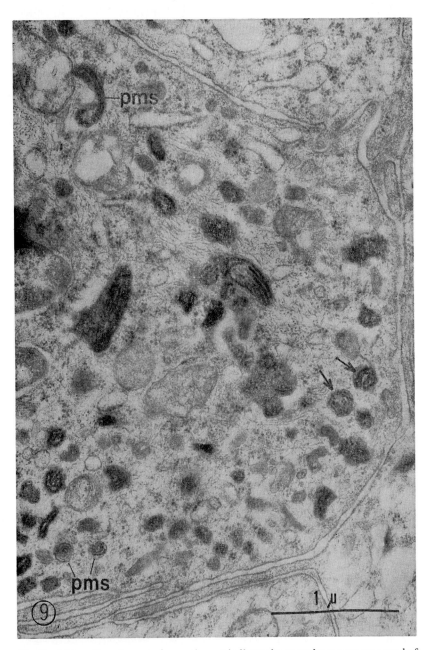

Fig. 9. Melanotic melanoma *de novo*. Arrows indicate the premelanosomes composed of interrupted denser membranes. ×35,000.

chondria, and numerous free ribosomes. The Golgi complex is developed in varying degrees, and a small amount of cytoplasmic filaments are found. Variable numbers of melanosomes and premelanosomes are contained in the cytoplasm. In some cells the melanosomes are predominant, and in other cells the premelanosomes predominate; these are usually disposed in individuals. Both melanosomes and premelanosomes are usually spindle-shaped or ovoid, varying from 210 to 1,000 mμ in maximal diameter (Fig. 1). Sometimes they show crescent- or ring-shaped profiles (Figs. 7–9). These profiles appear to be identical with those of red cell-shaped melanosomes found by Mishima[9] in the melanoma from Hutchinson's melanotic freckle. While the melanosomes are revealed as uniformly dense and structureless granules, the premelanosomes show distinct internal structures. Longitudinal sectioning of the early premelanosomes shows that they consist of several longitudinal fibrils which are loosely disposed in a smooth-surfaced vesicle (Figs. 3–6). It appears that the fibrils are helical structures with small granules, 35 to 45 Å in diameter, occupying the medial and lateral apogees along the helix at definitive intervals. In advanced premelanosomes the longitudinal fibrils are laterally associated and cross-linked together in two dimensions, forming distinctive membranes with cross-striation of a period of about 85 Å. In cross sections (Fig. 9), the membranes resulting from the cross-linkage of longitudinal fibrils are revealed as concentric or irregularly arranged lamellae. As melanin deposits, the membranes become thicker and denser. In cross sections, however, some of the premelanosomes are found to be composed

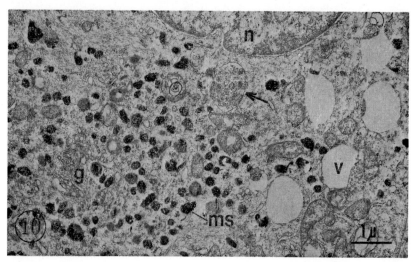

Fig. 10. Melanotic melanoma *de novo*. Arrow indicates an autophagic vacuole containing premelanosomes. ×10,900.

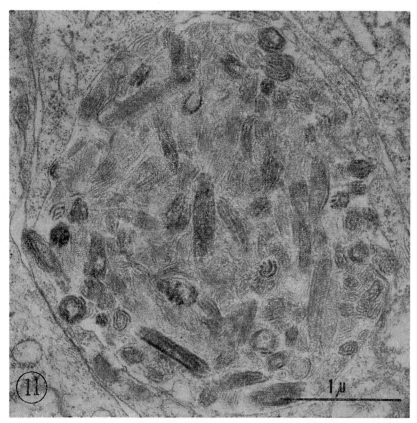

Fig. 11.   Autophagosome in a cell of melanotic melanoma *de novo*. The internal structure of premelanosomes is clearly visible. ×33,000.

of interrupted denser membranes, on which melanin is deposited (Fig. 9). This finding suggests that they are composed of fibrils with incomplete cross-linkage. In addition to the individual melanosomes, compound melanosomes are seen in a few cells (Figs. 10, 11). They are membrane-bound bodies filled with melanosomes which undergo degradation. The fact that the remnants of other cytoplasmic organelles are sometimes encountered within the compound bodies suggests that the compound bodies arise not by phagocytosis but by autophagy. Occasionally, it is found that in the autophagosomes the melanin-containing organelles undergo degradation into the elementary fibrils (Fig. 12).

The ultrastructure of amelanotic melanoma cells is similar to that of melanotic melanoma cells except for melanin-containing organelles (Figs. 13, 14). In the cytoplasm of the cells a small number of premelanosomes, composed of fibrils with either complete or incomplete cross-

335

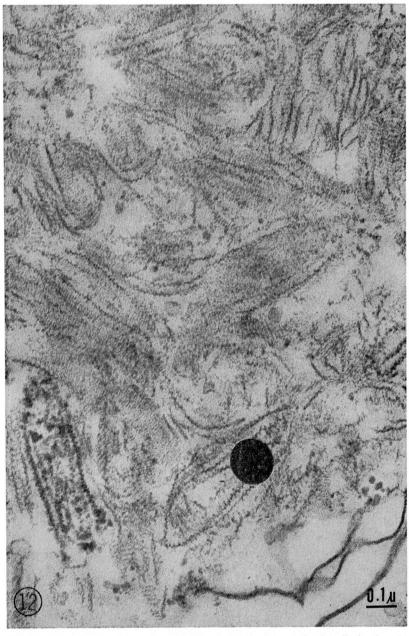

Fig. 12.   Degradation of premelanosomes in an autophagic vacuole of a cell of melanotic melanoma *de novo*. ×85,000.

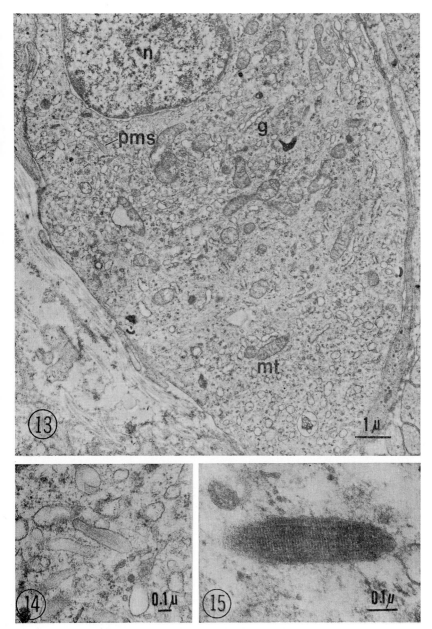

Fig. 13.   Amelanotic melanoma *de novo*. ×10,500.

Fig. 14.   Enlargement of the premelanosome shown in Fig. 13. ×37,500.

Fig. 15.   Premelanosome of moderate density, occasionally found in amelanotic melanoma *de novo*. ×85,000.

337

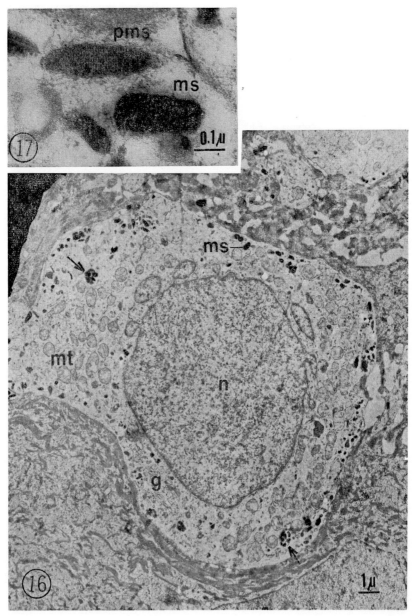

Fig. 16. Intra-epidermal melanocyte in Hutchinson's melanotic freckle. Arrows indicate presumable autophagosomes. ×6,600.

Fig. 17. Melanosomes and premelanosomes of intra-epidermal melanocyte in Hutchinson's melanotic freckle. ×85,000.

338

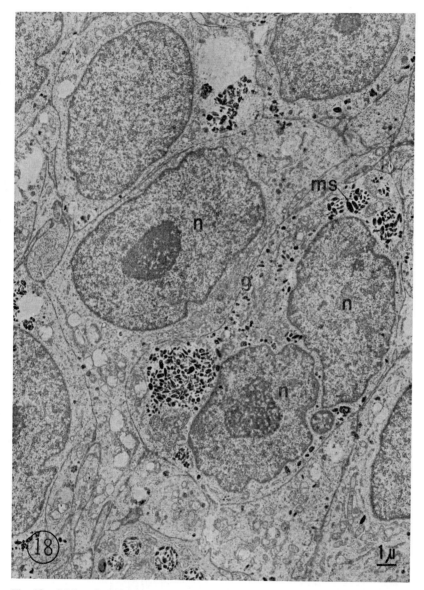

Fig. 18.   Melanotic melanoma arising from Hutchinson's melanotic freckle. ×5,400.

linkage, can be found. These premelanosomes are identical in shape and size to those of the melanotic melanoma cells. Although it is difficult to determine whether or not melanization occurs, some of the premelanosomes are moderately increased in density (Fig. 15). Autophagic vacuoles containing premelanosomes are infrequently encountered.

339

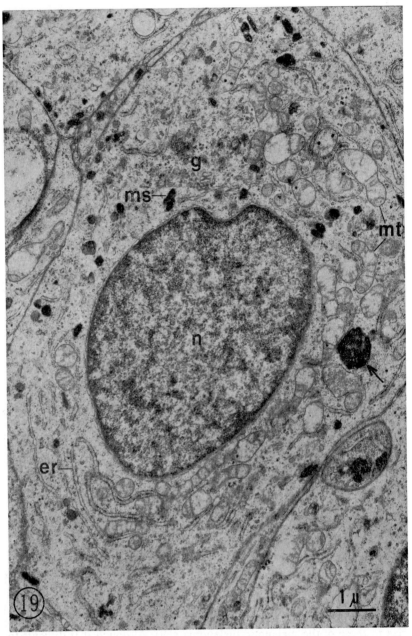

Fig. 19. Melanotic melanoma arising from Hutchinson's melanotic freckle. Arrow indicates an autophagosome containing melanosomes. ×12,900.

B. *Melanoma arising in Hutchinson's melanotic freckle*

In Hutchinson's melanotic freckle (Fig. 16), polymorphous melanocytes are seen in the basal cell layer and the prickle cell layer of the epidermis. They have features that are considered abnormal. These consist of large nuclei with indentation or protrusion and a large number of mitochondria. Such atypical melanocytes have variable numbers of melanin-containing organelles, most of them being fully melanized melanosomes. The melanosomes are usually disposed in individuals and occasionally in aggregates with or without a surrounding membrane. The premelanosomes are occasionally encountered. They are composed of membranes with cross-striation of a period of about 85 Å (Fig. 17).

In arising melanomas, the same features as the above are observed in the cells (Figs. 18, 19); nuclei show varying degrees of indentation, and mitochondria are numerous, some of them being vacuolated in varying

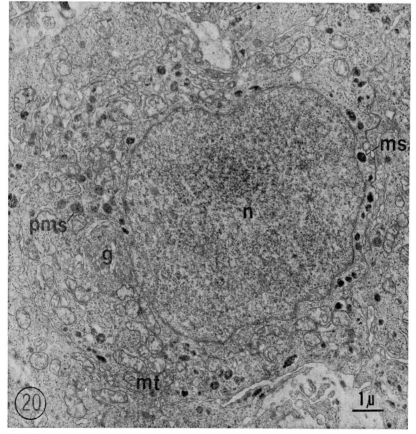

Fig. 20.   Melanotic melanoma presumably arising from junctional nevus. ×8,400.

341

degrees. The endoplasmic reticulum is well developed, consisting mostly of rough-surfaced, tubular and cisternal elements whose cavities appear slightly dilated. Free ribosomes are less numerous compared with those of the other kinds of melanoma cells. The Golgi apparatus is prominent. Variable numbers of melanosomes and premelanosomes are contained within the cytoplasm, and melanosomes are always predominant. They show spindle-shaped or ovoid profiles, ranging from 200 to 620 mμ in maximal diameter (Fig. 1). No premelanosomes of ring-shaped profile are found. The autophagosomes containing variable numbers of melanosomes are more frequently encountered. The internal structure of premelanosomes of the melanoma cells is identical with that of the intra-epidermal atypical melanocytes.

C.   *Melanoma presumably arising from junctional nevus*

In melanomas of this kind (Figs. 20, 22), there are abundant organelles in the cytoplasm of the cells. Mitochondria having the usual ultrastructure are numerous. The endoplasmic reticulum consists of randomly disposed, rough-surfaced, tubular and cisternal elements, and smooth-surfaced, vesicular elements. The cytoplasmic matrix contains numerous free ribosomes, most of them being disposed in clusters. The Golgi apparatus is prominent. Cytoplasmic filaments are infrequently encountered. The most striking feature is seen in the ultrastructures of melanosomes and premelanosomes. Two types of premelanosomes can be distinguished (Figs. 21, 22). One is characteristic of this kind of melanoma cell; it shows round or ovoid profiles, varying from 270 to 730 mμ in maximal diameter (Fig. 1) and is composed of granules which are 70 to 150 Å in

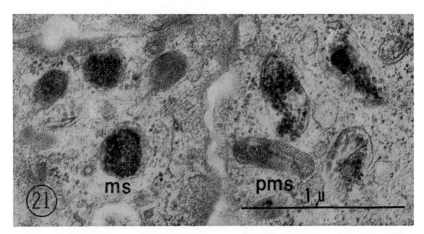

Fig. 21.   Melanosomes and premelanosomes in melanotic melanoma presumably arising from junctional nevus. ×45,500.

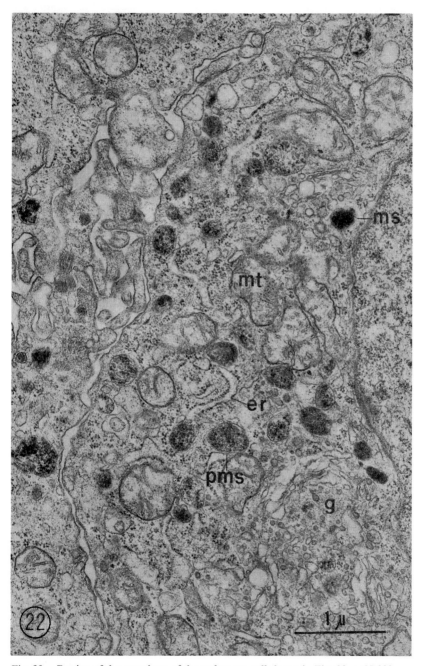

Fig. 22.   Portion of the cytoplasm of the melanoma cell shown in Fig. 20.   ×25,000.

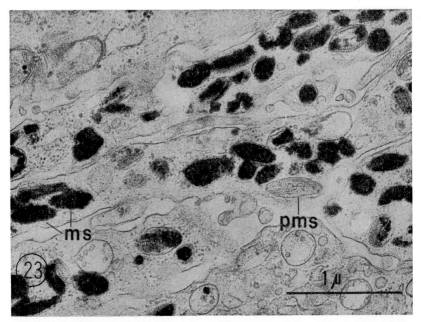

Fig. 23.   Melanosomes and premelanosomes of junctional nevus cells. ×32,000.

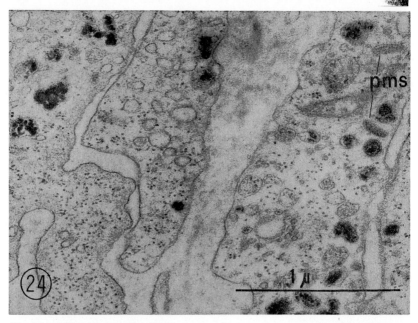

Fig. 24.   Melanosomes and premelanosomes of nevus cells beneath the epidermis in compound nevus. Premelanosomes of striated structure are seen in the upper right corner. The other premelanosomes show granular structure. ×45,000.

diameter. The other type of premelanosomes, similar to that of premelanosomes in melanoma *de novo*, is composed of fibrils with either complete or incomplete cross-linkage. In melanoma cells of this kind, premelanosomes with granular structure are always much more numerous than those with fibrillar structure. In both types of premelanosomes, melanization is usually incomplete and not uniform. Additional observations of junctional and compound nevi showed that the nevus cell nests of junctional nevus and junctional theques of compound nevus contained numerous premelanosomes composed of membranes with cross-striation of a definite period (Fig. 23), and that the nevus cell nests beneath the epidermis contained premelanosomes showing either fibrillar structure or granular structure (Fig. 24), although these premelanosomes are much smaller than those of the junctional nevus cells.

IV. DISCUSSION

A. *Ultrastructures of premelanosomes in various melanomas*
The ultrastructures of premelanosomes observed in this study were essentially similar to those found by others[9-16] in normal and neoplastic human melanocytes. They could be grouped as follows: (1) spindle-shaped premelanosomes of fibrillar structure with either complete or incomplete cross-linkage, and (2) round premelanosomes of granular structure. The frequently encountered cross-striated structure was interpreted as the longitudinal profile of the membrane resulting from the complete cross-linkage of the fibrils. These premelanosomes were not always uniformly and completely melanized.

B. *Comparison of the different kinds of melanomas*
Two points are worthy of mention here. One is on the subject of the melanin-containing organelles (Table 1). The cells of melanoma *de novo* contained premelanosomes of fibrillar structure with either complete or incomplete cross-linkage. These premelanosomes showed marked variation in shape, size, and melanization. Some of them showed unusual

Table 1. Comparison of three kinds of melanomas.

	Premelanosome	
Melanoma	Fibrillar	Granular
de novo	+*	—
From Hutchinson's melanotic freckle	+**	—
From junctional nevus	+	+
	(Very few)	(Numerous)

* Premelanosomes are abnormal in shape, size, and melanization.
** Premelanosomes are nearly normal in shape, size, and melanization.

crescent- or ring-shaped profiles; a considerable number of them were extraordinarily large; most of them were not completely melanized. In the melanoma arising from Hutchinson's melanotic freckle, there were premelanosomes of fibrillar structure with complete cross-linkage in the cells. These premelanosomes were nearly normal in size, shape, and melanization, and, moreover, the great majority were completely melanized. In the melanoma presumably arising from junctional nevus, premelanosomes of granular structure were predominant in the cells. The evidence strongly suggests that these three kinds of melanomas can be distinguished by the ultrastructures of synthesized premelanosomes.

The other interesting point concerns the relation of some organelles to biological potential. In the melanoma *de novo*, the presence of numerous clusters of free ribosomes, as well as the greater number of mitochondria in the cells, suggests that melanoma cells of this kind have a highly increased capacity to synthesize the proteins for themselves and a higher metabolic activity compared with normal epidermal melanocytes. In other words, our suggestion is that this kind of melanoma cell has a higher biological potential. The same situation as the above was seen in the cells of melanoma presumably arising from junctional nevus. On the other hand, the cells of melanoma arising from Hutchinson's melanotic freckle show a comparative scarcity of clusters of free ribosomes, suggesting that the cells of melanoma of this kind have a lower capacity to synthesize the proteins utilized by the cells themselves.

Table 2.   Cell types of three kinds of melanomas and their pre-existing conditions.

Melanoma	Pre-existing condition	Neoplasm
de novo	Normal epidermal melanocyte	Malignant epidermal melanocyte
From Hutchinson's melanotic freckle	Malignant epidermal melanocyte	Malignant epidermal melanocyte
From junctional nevus	Naevocyte	Malignant naevocyte

C.   *Comparison between various melanomas and their pre-existing conditions*

Electron microscopic studies of Hutchinson's melanotic freckles revealed that the intra-epidermal melanocytes in this condition showed obvious abnormal features, very similar to the cells of melanoma arising from this condition. The evidence indicates that the intra-epidermal melanocytes in Hutchinson's melanotic freckle are malignant melanocytes themselves and supports the view[17-19] that Hutchinson's melanotic freckle is a melanoma *in situ* (Table 2).

Other electron microscopic evidence indicated that the premelanosomes showed fibrillar structure in junctional nevi and junctional theques of compound nevus, and that many of them showed granular structure after the nevus cells had dropped into the dermis. This variation appears to be primarily related to the nature of the nevus cells, although it is difficult to explain exactly. The same variation was observed in the cells of melanoma arising from junctional nevus. Taking into consideration the fact that synthesis of melanin-containing organelles is under genetic control,[20] these findings suggest that the cells of melanoma of this kind may still have a nevoid character. Therefore, we are in agreement with Mishima's opinion[9] that cells of melanoma arising from junctional nevus should be termed "malignant naevocytes."

V. SUMMARY

Malignant melanomas which arise from normal skin, from Hutchinson's melanotic freckle, and from junctional nevus were studied by electron microscopy. From the results obtained, it was strongly suggested that these three kinds of melanomas can be distinguished by the ultrastructures of synthesized premelanosomes. A comparison between various melanomas and their pre-existing conditions led to the conclusion that Hutchinson's melanotic freckle should be interpreted as a melanoma *in situ,* and supports the view that cells of melanoma arising from junctional nevus should be termed "malignant naevocytes."

REFERENCES

1. Miescher, G., Über Klinik und Therapie der Melanome. *Arch. Derm. Syph.*, **200**, 215 (1955).
2. Lund, H. Z., Pigment cell neoplasm: Problems in the diagnosis of early lesions. In *Proceeding of the Fourth National Cancer Conference*, p. 619, Lippincott, Philadelphia (1961).
3. Trapl, J., Paleček, L., Ebel, J., and Kučera, M., Origin and development of skin melanoblastoma on the basis of 300 cases. *Acta Dermatovener.*, **44**, 377 (1964).
4. Jackson, R., Williamson, G. S., and Beattie, W. G., Lentigo maligna and malignant melanoma. *Canad. Med. Ass. J.*, **95**, 846 (1966).
5. Levene, A., McGovern, V. J., Mishima, Y., and Oettle, A. G., Terminology of vertebrate melanin-containing cells, their precursors, and related cells: A report of the nomenclature committee of the 6th International Pigment Cell Conference. In *Structure and Control of the Melanocyte* (G. Della Porta and O. Mühlbock, eds.), p. 1, Springer-Verlag, Berlin (1966).
6. Sabatini, D. D., Bensch, K., and Barrnett, R. J., Cytochemistry and electron microscopy: The preservation of cellular ultrastructure and enzymatic activity by aldehyde fixation. *J. Cell Biol.*, **17**, 19 (1963).
7. Caulfield, J. B., Effects of varying the vehicle for OsO₄ in tissue fixation. *J. Biophys. Biochem. Cytol.*, **3**, 827 (1957).
8. Luft, J. H., Improvements in epoxy resin embedding methods. *J. Biophys. Biochem. Cytol.*, **9**, 409 (1961).

9. Mishima, Y., Melanocytic and nevocytic malignant melanomas: Cellular and subcellular differentiation. *Cancer*, **20**, 632 (1967).
10. Breathnach, A. S., Electron microscopy of a small pigmented cutaneous lesion. *J. Invest. Derm.*, **42**, 21 (1964).
11. Gottlieb, B., Brown, A. L., Jr., and Winkelmann, R. K., Fine structure of the nevus cell. *Arch. Derm.*, **92**, 81 (1965).
12. Tröger, H. and Klingmüller, G., Die Melaningranula im Naevuszellnaevus. *Arch. Klin. Exp. Derm.*, **226**, 1 (1966).
13. Birbeck, M. S. C., Electron microscopy of melanocytes: The fine structure of hairbulb premelanosomes. *Ann. N. Y. Acad. Sci.*, **100**, 540 (1963).
14. Welling, S. R. and Siegel, B. V., Electron microscopic studies on the subcellular origin and ultrastructure of melanin granules in mammalian melanomas. *Ann. N. Y. Acad. Sci.*, **100**, 548 (1963).
15. Drochmans, P., The fine structure of melanin granules (the early, mature, and compound forms). In *Structure and Control of the Melanocyte* (G. Della Porta and O. Mühlbock, eds.), p. 90, Springer-Verlag, Berlin (1966).
16. Toshima, S., Moore, G. E., and Sanberg, A. A., Ultrastructure of human melanoma in cell culture: Electron microscopic studies. *Cancer*, **21**, 202 (1968).
17. Becker, S. W., Critical evaluation of the so-called "junction nevus." *J. Invest. Derm.*, **22**, 217 (1954).
18. Couperus, M. and Rucker, R. C. Histopathologic diagnosis of malignant melanoma. *A.M.A. Arch. Derm. Syph.*, **70**, 199 (1954).
19. Gartman, H., Besteht ein histologischer Unterschied zwischen der problastomatosen Melanose und dem "activated junctional nevus" (Allen)? *Hautarzt*, **13**, 507 (1962).
20. Moyer, F. H., Genetic variations in the fine structure and ontogeny of mouse melanin granules. *Amer. Zool.*, **6**, 43 (1966).
21. Hirone, T., Supplementary electron microscope studies on human skin, especially on epidermal melanocyte and cutaneous pigment. *Jap. J. Derm.*, **70**, 530 (1960).

DISCUSSION

DR. HORI: What is the ultrastructure of melanosomes in the deeply invasive part of the melanoma derived from Hutchinson's melanotic freckle?

DR HIRONE: I examined a couple of portions of melanoma cell nests arising from Hutchinson's freckle, located in the upper and middle parts of dermis. But there was no difference between them in the ultrastructures of melanosomes and premelanosomes.

DR. FITZPATRICK: Drs. Britton and Clark have shown the same kind of incomplete cross-linking.

DR. HIRONE: Yes, but there were very few. I would like to point out that the melanoma arising from junctional nevus contains mainly premelanosomes of granular structure.

DR. OHKUMA: How did you reach the conclusion that the phagosomes in the melanoma cells and in the cells of Hutchinson's melanotic freckle are autophagic and not phagocitized ones?

DR. HIRONE: An autophagosome is usually identified by the presence of remnants of cell organelles such as mitochondria, endoplasmic reticulum, etc., in a vacuole limited by a single unit membrane. As already shown, there were many aggregates of melanosomes in the cells of melanoma arising from Hutchinson's melanotic freckle, and some of the aggregates were contained within a vacuole, in which remnants of membranous structure were present. Therefore, I regarded them as autophagosomes. I think that, in a degenerate melanoma cell, melanosomes and premelanosomes may be segregated into the autophagic vacuole of the cell.

DR. MISHIMA: Until about ten years ago, when I distinguished melanoma into nevocytic and melanocytic melanoma, most peoples had been thinking that malignant melanoma was a uniform, single entity, despite observed variations in its prognosis. I am very happy to see distinct subcellular differences between melanoma developing from junctional nevus and that developing from Dubreuilh's precancerous melanosis. However, now we have another task. That is the clarification of so-called Pagetoid melanoma, or superficially spreading melanoma, which was recently identified.

DR. FITZPATRICK: In lentigo-maligna melanoma there are all gradations of malignant cells, from single neoplastic cells all the way to frankly malignant cells. In the superficial spreading melanoma (Pagetoid) there is a uniform cell population; almost all melanocytes are in the same stage of mitotic proliferation.

DR. MISHIMA: I agree with Dr. Fitzpatrick on most points. However, I might add that the distinct entity of nevocytic melanoma can be observed in the melanoma developing within giant pigmented nevus and compound nevus.

Pigmentary Disorders in Asiatics*

Taro Kawamura, Shigeo Ikeda, Yoshiaki Hori, Hiroko Obata, and Michihito Niimura

Department of Dermatology, Faculty of Medicine, University of Tokyo, Tokyo, Japan

I. INTRODUCTION

The level of skin pigmentation in Asiatics, intermediate between that of Caucasians and Negroes, has made pigmentary disorders easy to find in Asiatics. Thus, many pigmentary conditions first noticed in Asiatics have later been found to be not specific for Asiatics. The Mongolian spot, for instance, namely the accumulation of dermal melanocytes in the skin of the sacral region in babies and children, was discovered also in Caucasians through the early microscopic studies of the late Professor Adachi.[1] The nevus of Ota (Naevus fuscocaeruleus ophthalmomaxillaris Ota, 1939), first noticed by the late Professor Ota[2] was comprehensively studied by his pupil Tanino[3,4] in 1939 and 1940 and has later been referred to also as ocular and dermal melanocytosis by Fitzpatrick and his co-workers.[5] The nevus of Ota has recently been studied by Hidano et al.[6-8] in farther detail, and it has been found to occur also in Caucasians.[8] Data concerning normal melanocytes in Asiatics, Caucasians, and Negros do not necessarily differ. The populations of melanocytes in various parts of the skin in Japanese, estimated by Ikeda[9] and Hayashi[10] do not differ significantly from those estimated by Staricco and Pinkus[11] for Caucasians and Negroes.

* This work was supported by a grant for scientific research from the Japanese Ministry of Education.

The melanoblast in the embryonic skin has been found to appear first in the 10th fetal week in both the Negro[12] and the Japanese.[10]

Thus, the normal and abnormal pigmentation conditions in Japanese to be discussed in this paper are not necessarily specific for Asiatics.

II. NORMAL AND ABNORMAL PIGMENTATION OF THE PALMS AND SOLES

In the skin of the palms and soles, the epidermal ridges are classified into two different categories: the crista profunda intermedia beneath the crista superficialis, and the crista profunda limitans beneath the sulcus superficialis; the crista profunda intermedia is penetrated by sweat ducts (Fig. 1). Hayashi[10] found epidermal melanocytes in Japanese fetuses to appear first between the 13th and 14th fetal week. They are situated in this stage not only in the basal layer but also in higher levels of the epidermis (Fig. 2). This finding is particularly interesting if considered in the context of the findings of nerve elements in the skin. Studied by means of an optical microscope, they were reported to be present in the epidermis of fetal skin but to be lacking in the adult epidermis, though Emmi Hagen and her co-worker[13] recently demonstrated their presence in adult epidermis through a scrupulous electron microscope search. Between the 15th and 16th fetal weeks, Hayashi found the crista profunda intermedia to develop, while the crista profunda limitans was not recognized. In this stage, the epidermal melanocytes begin to arrange themselves on the crista profunda intermedia. In the 24- to 26-week-old fetus, epidermal melanocytes were revealed exclusively on the crista profunda intermedia (Fig. 1).

This pattern of distribution does not persist throughout life. In the palmo-plantar epidermis of adults, they were found by Ikeda[9] to be distributed almost evenly all over the junctional surface of the epidermis (Fig. 3). However, there seems to be a certain difference between the

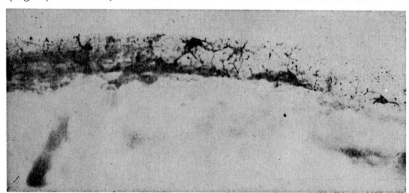

Fig. 2. The epidermis of a 14-week-old human fetus. Dopa stain.

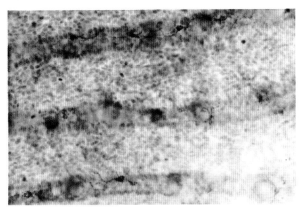

Fig. 1.   The epidermis of 28-week-old human fetus. Ammoniac silver stain. The arrow indicates sweat ducts penetrating the crista profunda intermedia. Black stained melanocytes are revealed exclusively on the crista profunda intermedia.

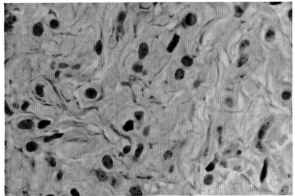

Fig. 7.   Nevus fibers (fibers naeviques) of Masson. Goldner's modification of Masson's trichrome stain technique.

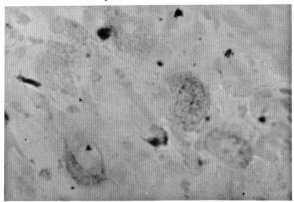

Fig. 8b.   Balloon cell nevus. Feyrter's Einschussfärberei with tartaric acid thionine.

353

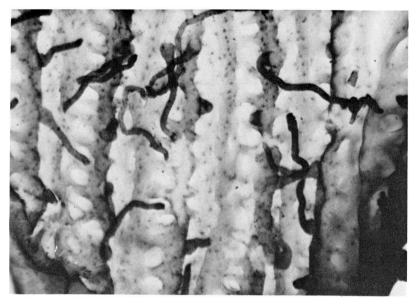

Fig. 3. The distribution of melanocytes in the epidermis of the palmar skin of an adult. Dopa stain.

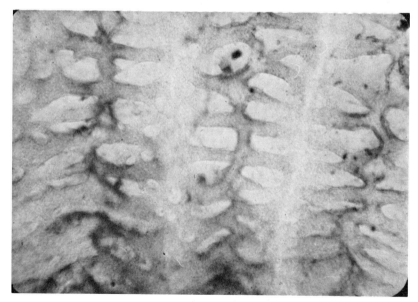

Fig. 4. The more prominent pigmentation in the crista profunda intermedia revealed in a pigmented patch on a palm of an Addison's disease patient.

function of melanocytes in the crista profunda intermedia and that in the crista profunda limitans. In the Japanese patients with Addison's disease, patchy pigmentation on the palms and soles is frequently encountered. On microscopic examination of a pigmented patch, the crista profunda intermedia is revealed to be more heavily pigmented than the crista profunda limitans (Fig. 4).[14] Since selection between the crista profunda intermedia and limitans may not be possible for MSH and/or ACTH circulating in the blood, the difference in the pigmentation between the two kinds of cristae might be due to a difference between the melanocytes on the different epidermal ridges in their ability to react to MSH and/or ACTH. A characteristic pattern of pigmentation has also been revealed in the palmo-plantar pigmented patches in the Peutz–Jeghers (Peutz–Touraine) syndrome. Here too, the pigmentation is localized in the crista profunda intermedia. This pattern of pigmentation, first reported by Mori[15] and repeatedly confirmed by Kitamura et al.[16] is one of the important diagnostic signs for the diagnosis of this syndrome.

In the nevocytic pigmented spots on the palms and soles, the pigmentation is more conspicuous on the crista profunda limitans[14] or is irregularly distributed.[9]

III. NEVOCELLULAR NEVUS AND ALLIED CONDITIONS

The nevocellular nevus and allied conditions have been intensively studied by Japanese dermatologists. When I began to study the nevocellular nevus twenty years ago, I preferred the term "pigmented cell nevus" to the term nevocellular nevus, because there was at that time no agreement concerning what was the nevus cell. Some[17] would define the nevus as being a skin lesion composed of nevus cells, without giving any definition to the nevus cell. Others defined nevus cells as those cells which are the main component of nevus tissue, without defining the nevus. I thought, therefore, that the first task to be performed was to find a way out of such an endless circle of definitions. I defined the compound nevus in 1956[18] as the lesion which fulfills all of the three following conditions:

1. A benign growth composed of cells. The area occupied by them expands itself, beginning from the epidermis and continuing to the deepest layer of the corium.

2. The cells change not only in their individual characteristics but also in their mode of arranging themselves, beginning with those in the epidermis and extending to those in the deepest corium, gradually and uninterrupted by any skip.

3. The cells in the higher layer contain melanin, which tapers off in quantity gradually as the deeper layer is reached, usually disappearing in the cells situated in the deepest layer of the corium.

Starting from this definition of the compound nevus, the junctional nevus was defined as being the upper half of the compound nevus. The dermal nevus was defined as being the deeper half of it. And the nevus cell was defined as being the specific constituent cell of the compound, junctional, or dermal nevus.

The pattern of distribution of cells originating from the neural crest is illustrated by the scheme on the left in Fig. 5. The differentiation between the melanocytes and Schwannian cells is clear-cut, leaving no intermediate cells between them. Tyrosinase activity, for instance, turns abruptly negative beneath the basement membrane, where tyrosinase-negative Schwannian cells are revealed, attached to fine nerve axons. As for the contiguity or continuity between the nerve fibers and melanocytes, opinions differ greatly. Matsumoto[19] suggested contiguity between them by staining the skin supravitally with methylene blue. Niebauer and Sekido,[20, 21] using a new staining technique with methylene blue developed in our department at the University of Tokyo,[22] have beautifully shown their continuity. Such findings as these may suggest the role of the autonomous nerve in controlling melanogenesis. These finding should perhaps be studied with an electron microscope in order to look for submicroscopic evidence which might contribute to our understanding of the functional relationship between the nerve fiber and the melanocyte.[13]

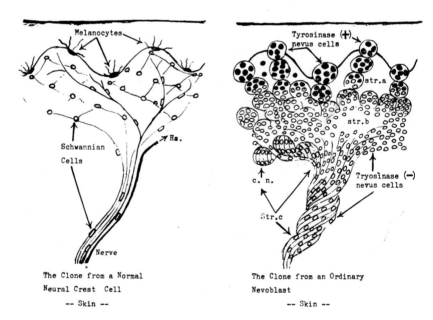

Fig. 5. The normal cell system of neural crest origin, i.e., a clone from a normal neural crest cell (left) and the nevus tissue, i.e., a clone from a nevoblast (right).

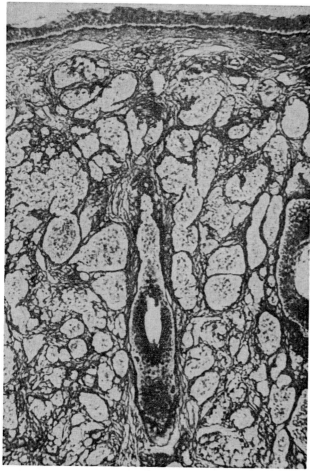

Fig. 6.   The shift of the pattern of reticulum fibers at various levels in the nevus tissue.

Coming back to the nevocellular nevus, as shown schematically in the right side of Fig. 5, tyrosinase activity, for instance, shifts gradually from the top to the bottom of the tissue. That is, nevocytes on the higher levels of the nevus tissue are tyrosinase active, as are the epidermal melanocytes. The tyrosinase activity in nevus cells tapers gradually as the deeper layers of the corium are reached. Namely, there is not such a clear-cut differentiation between melanocyte-like and non-melanocyte-like nevus cells as there is between melanocytes and Schwannian cells. As another instance of the gradual shift in characteristics of the nevus tissue, the pattern of the reticulum fiber should be mentioned. In the nevocellular nevus (Fig. 6), the pattern of the reticulum fibers gradually shifts from coarse meshes in the uppermost layer to fine meshes in the deepest corium, where each

357

individual nevus cell may sometimes be surrounded by reticulum fibers. A similar gradual shift in the reticulum pattern[23] was revealed in the so-called juvenile melanoma, which might be none other than a subtype of the nevocellular nevus.

The production of reticulum fibers in nevus tissue might suggest their resemblance to Schwannian cells, which are covered by a thin layer of reticulum fiber web not only in the normal condition (Plenck[24]–Laidlaw's web) but also in experimental and human neoplastic conditions.[25] Reticulum fibers were reported to be produced by Schwannian cells in certain *in vitro* conditions.[26] However, the cell that actually produces reticulum fiber has recently been proved to be the fibrocyte. Thus, whether or not the similarity between nevus cells and Schwannian cells in their relation to the reticulum fibers is actually due to the activity of those cells in producing them is open for further study.

Nevus cells have additional features in common with Schwannian cells. These features are rather conspicuous in the deeper layer of the corium. One of these is the nevus fibers of Masson (Fig. 7), i.e., the elongated, undulating, and slightly pink-colored cytoplasm of nevus cells covered with green-colored fine collagen sheaths, in specimens stained by the trichrome technique of Masson.[27-29] These findings are common to Schwannian cells and Schwannoma cells.[25] Such a conspicuous development of nevus fibers as illustrated in Fig. 7 is rather rare. However, less conspicuous nevus fibers are usually revealed by scrupulous microscopic examination. Another feature of the nevus cells suggesting their intimate relationship to the Schwannian cells is their contiguity to the nerve axon. This feature was first noticed by Soldan[30] by staining the nevus tissue for myelin and was more definitely shown by staining it with silver or gold.[31-34] According to light-microscopic examination, the nevus cells are in direct contact with nerve axons. The question whether or not mesoaxons mediate between the nevus cells and nerve axons is open for further study by electron microscopy.

Both epidermal melanocytes and Schwannian cells contain a sort of peculiar lipoid substance staining metachromatic with thionine.[35] This substance is also found in the pigmented and nonpigmented nevus cells,[36,37] and is most conspicuously revealed in the balloon cell nevus, as illustrated in Fig. 8. The existence of this substance in nevus cells might suggest their intimate relationship to both melanocytes and Schwannian cells. The rather recent finding reported by Ishibashi[38] that mycobacterium leprae having primary affinity for Schwannian cells also has conspicuous affinity also for nevus cells might suggest a certain community of both kinds of cells in their biological disposition. Ishibashi's findings were later confirmed by Klingmüller[39] with electron microscopy.

The gradual shift of the cell and tissue characteristics from the high-

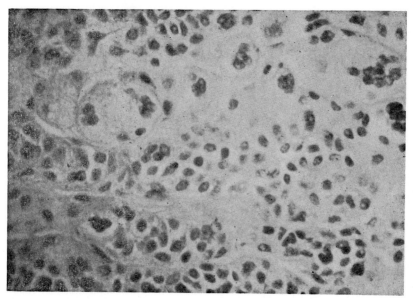

Fig. 8a.   Balloon cell nevus. H. E. stain.

level nevus cells which are tyrosinase active in common with melanocytes to the low-level nevus cells lead me to the following conclusion in 1956:[18] the variously shaped nevus cells arise neither from the already differentiated melanocytes nor from the already differentiated Schwannian cells. Rather, they arise from an abnormal cell originating in the neural crest, whose development is from the beginning misdirected. If one assumes that such an abnormal cell arises in the neural crest, it follows that thier daughter cells are only able to differentiate into various intermediate cells on the spectrum between the two polar-type cells, namely the melanocytes and the Schwannian cells, without being able to differentiate into either of the polar-type cells. By comparing Fig. 5 (right) with Fig. 5 (left), one might conclude that the nevus tissue, composed of variously shaped nevus cells, is an awkward imitation of the normal pattern of melanocytes and Schwannian cells. Although a few dermatologists in the United States and Europe had approved[40] or cited[41] our unitary view of the origin of nevus cells, familiarity with it was largely limited to Japan until Pinkus and Mishima[42] coined the term nevoblast in 1961 for the abnormal cell assumed by us to arise in the neural crest. Through use of the term nevoblast, students of dermatology all over the world are becoming acquainted with the unitary view of the origin of the nevus cell.[43]

Our unitary view is valid for the origin of typical and atypical blue nevi[18,44-46] (Fig. 9), the mélanose neurocutanée of Touraine[47], and phakomatosis Recklinghausen,[18,48] the differentiation and the distribu-

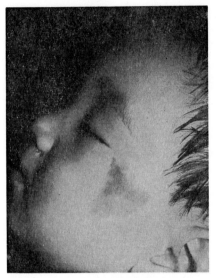

Fig. 9.   Atypical blue nevus type 2, composed of brown epidermal pigmentation without théque and blue dermal pigmentation composed of blue nevus cells. The darker areas in this picture are those where the brown epidermal patch covers the blue dermal patch.

tion patterns of the offspring of the nevoblast or phakomatoblast being different in each of the disorders.

IV.   MELANOEPITHELIOMA OF OTA, THE DEGRADATION OF MELANOSOMES IN EPIDERMAL MELANOCYTES

Pigmented epithelial tumors are frequently encountered on the skin of the Japanese. Among the basalomas seen in the Tokyo-Yokohama area, approximately 37% were found to be pigmented.[49] Until thirty years ago, confusion prevailed in Japan in the diagnosis of pigmented skin tumors. The first step toward elucidating the pathology of pigmented skin tumors was made by the late Professor Ota[50] in 1940, who classified pigmented skin and mucous membrane tumors into two groups, namely, nevocellular nevus and malignant melanoma, and melanoepithelioma. His classification may have drawn its inspiration from the view of Masson,[51, 52] who regarded the epidermis as being composed of Malpighian cells (keratinocytes) and dendritic cells (melanocytes and Langerhans cells). The demonstration by dopa-reaction experiments of the fact that melanoepitheliomas are composed of a "continuous phase" of epithelial cells and a "dispersed phase" of melanocytes was first carried out by my former associate Kamide,[53] and later by Mishima and Pinkus,[54] and by Ikeda.[9]

Since knowledge about melanoepitheliomas is widespread, only the

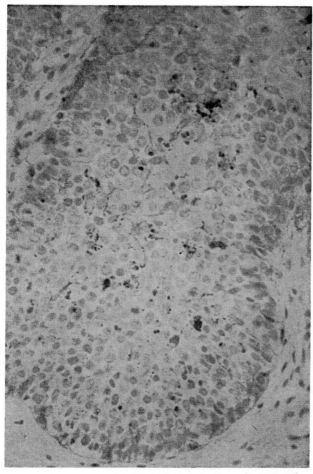

Fig. 10.   Bowen lesion in the vulva. H. E. stain. Dendritic melanocytes with coarse melanin "granules" are revealed among atypical epithelial cells without melanin.

degradation of melanosomes in them will be discussed here. Melano-epitheliomas are a suitable site for studying the degradation of melano-somes for two reasons: (1) the discernment of melanocytes from melano-phages is easy in melanoepitheliomas because the melanocytes are sur-rounded by epithelial cells; (2) the melanocytes, if not normal, at least are not neoplastic. The lysosomal degradation of melanomas was first demon-strated in melanophages of experimental melanoma.[55] Then Mishima demonstrated it in nevus cells and melanoma cells, that is, in cells that are primarily melanogenic.[56] In melanoepitheliomas, heavily pigmented melanocytes with large "pigment granules" are observed. The configura-tion of the latter is clearly visible without the aid of an oil immersion lens

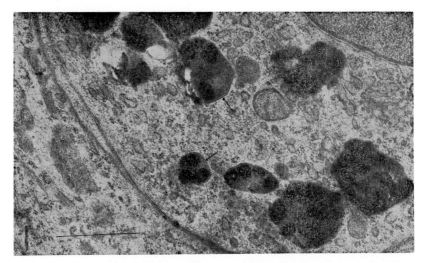

Fig. 11.   Bowen lesion in the vulva. Electron microscopic finding of a part of a melanocyte from the same lesion as shown in Fig. 10. The length of the vertical line represents 1 μ.

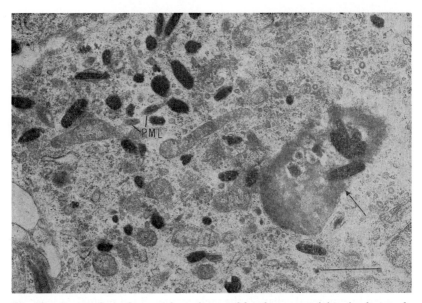

Fig. 12.   A part of a melanocyte in a pigmented basaloma, containing simultaneously
a lysosome (indicated by arrow), degrading melanosomes, and melanosomes with grid
patterns (PML) The length of the vertical line represents 1 μ.

(Fig. 10). This means that their dimensions are definitely larger than the
limit of the resolving power of the light microscope.[52] Since the dimensions of epidermal melanosomes were demonstrated by Mishima[56] to be

almost on the borderline of the resolving power of the light microscope, the light-microscopic findings mentioned suggest that the large pigment granules might not be melanosomes but "conglomerates" of them. An electron microscopic examination performed by Obata et al.[57,58] of a case of pigmented Bowen disease, illustrated in Fig. 11, demonstrated that the large pigment granules are vacuoles with surrounding membranes. They contain melanosomes and less electron-dense amorphous material filling the space between the latters. Fully melanized melanosomes in the vacuoles decrease in electron density to merge into the amorphous material. These findings might be enough to identify these vacuoles with lysosomes, though Obata has not run the histochemical stain for the hydrolytic enzyme. Another of her findings is illustrated in Fig. 12. In this melanocyte of a pigmented basaloma, striated, possibly tyrosinase-active melanosomes were revealed along with lysosomes degrading melanosomes.

One question I would like to leave open concerns the actual relationship between the "melanophages" and melanocytes. Since Miescher's beautiful experimental work aimed at producing them experimentally, melanophages have been believed to be histiocytes. However, I still question whether another way of origination of melanophages should strictly be excluded, because there is a striking similarity between melanocytes filled with lysosomes, and melanophages. Leaving the conclusion open, I would like to mention the works of Szabó,[60] Sekido,[22] and Furuya[61] which suggest the possibility that epidermal melanocytes drop off into the corium. Neoplastic melanocytes, namely, the cells of precancerous melanosis of Dubreuilh, and melanoma cells are known to have the tendency to drop off into the corium. To rephrase my question, is it possible that non-neoplastic melanocytes too may in certain conditions drop off into the corium to change into effete cells filled with lysosomes?

V. SUMMARY AND CONCLUSIONS

1.   There is a masked and potential pigmentation pattern, possibly with some relation to the distribution of melanocytes in the fetal stage, in the normal skin of the palms and soles. In the case of certain pigmented patches in these regions visible to the naked eye, the masked pigmentation pattern become visible with a microscope under certain conditions, while different patterns may be revealed in the nevocytic pigmentation.

2.   Our unitary explanation of the histogenesis of the nevocellular nevi, assuming the outbreak of an abnormal cell in the neural crest, was a milestone leading from the dualistic view of Masson to the coining of the fascinating term "nevoblast."

3.   The lysosomal degradation of melanosomes in melanocytes has been

revealed in various melanoepitheliomas. The findings of melanocytes filled with lysosomes may be similar enough to melanophages to raise the question whether or not the origination of melanophages from epidermal melanocytes should be completely rejected.

REFERENCES

1. Adachi, B., Hautpigment beim Menschen und bei den Affen. *Z. Morph. Anthrop.*, **6**, 1 (1903).
2. Ota, M., Über Pigmentnävi. *Jap. J. Derm.*, **46**, 369 (in Japanese), 99 (in German) (1939).
3. Tanino, H., Über eine in Japan häufig vorkommende Nävusform: "Naevus fusco-caeruleus ophthalmo-maxillaris (Ota)." I. Mitteilung: Beobachtung über Lokalisation, Verfärbung, Anordnung und histologische Veränderung. *Jap. J. Derm.*, **46**, 435 (in Japanese), 107 (in German) (1939).
4. Tanino, H., Über eine in Japan haüfig vorkommende Nävusform: "Naevus fusco-caeruleus ophthalmomaxillaris (Ota)." II. Mitteilung: Beobachtungen über die Augenmelanose als Komplikation dieses Naevus und seine klinische Erscheinung. *Jap. J. Derm.*, **47**, 181 (in Japanese), 51 (in German) (1940).
5. Fitzpatrick, T. B., Zeller, R., Kukita, A., and Kitamura, K., Ocular and dermal melanocytosis. *Arch. Ophthal.*, **56**, 830 (1956).
6. Hidano, A., Nomoto, K., and Mishima, Y., Pigmentation des muqueuses dans le naevus d'Ota. *Bull. Soc. Franc. Derm. Syph.*, **64**, 287 (1957).
7. Hidano, A., Kajima, H., Ikeda, S., Mizutani, H., Miyasato, H. and Niimura, M., Natural history of nevus of Ota. *Arch. Derm.*, **95**, 187 (1967).
8. Hidano, A., The history of the studies of nevus of Ota. *Hifukano-Rinsho*, **10**, 1099 (1968).
9. Ikeda, S., Histological studies on melanin and melanocytes of the normal human skin with special reference to their distribution. *Jap. J. Derm.* Ser. A, **72**, 836 (in Japanese), Ser. B, 866 (in English) (1962).
10. Hayashi, T., Studies on the development of the skin elements originating from the neural crest in the Japanese fetus. I. On melanoblasts and melanocytes. II. On Merkel's tactile cells and nerve endings. *Jap. J. Derm.* Ser. A, **74**, 690 (in Japanese), Ser. B, 298 (in English) (1964).
11. Staricco, R. J. and Pinkus, H., Quantitative and qualitative data on the pigment cells of adult human epidermis. *J. Invest. Derm.*, **28**, 33 (1957).
12. Zimmermann, A. A. and Becker, W. Jr., Melanoblasts and melanocytes in fetal Negro skin. *Illinois Monographs in Medical Sciences*, **6**, no. 3, University of Illinois Press, Urbana (1959).
13. Hagen, E. and Werner, S., Beobachtungen an der Ultrastruktur epidermaler und subepidermaler Nerven vor und nach Röntgenbestrahlung. *Arch. Klin. Exp. Derm.*, **225**, 306, 328 (1966).
14. Kawamura, T., Nishihara, K., and Nakajima, H., On the nevoid spots of palms and soles. *Jap. J. Derm.*, **66**, 75 (in Japanese), 82 (abstract in English) (1956).
15. Mori, M., Congenital pigmented spot on the palms, soles, and lips. *Jap. J. Clin. Derm.*, **2**, 215 (in Japanese) (1948).
16. Kitamura, K., Kojima, R., and Sasagawa, S., Zur Frage des Syndromes von Peutz-Jeghers, Beobachtungen über kleinfleckige Pigmentierungen der Lippenschleimhaut, Handteller und Fussohlen in Kombination mit Darmpolyposis bei den Japanern. *Hautarzt*, **8**, 154 (1957).

17. Kaiserling, C., Naevi, pathologische Anatomie. In *Jadassohns Handbuch der Haut- und Geschlechtskrankheiten*, **12**, no. 2, 600 (1932).

18. Kawamura, T., Über die Herkunft der Naevuszellen und die genetische Verwandtschaft zwischen Pigmentzellnaevus, blauem Naevus und Recklinghausenscher Phakomatose. *Hautarzt*, **7**, 14 (1956).

19. Matsumoto, R., Studies on skin pigment, particularly on vitiligo. *Jap. J. Derm.*, **71**, 57 (in Japanese), 97 (abstract in English).

20. Niebauer, G. and Sekido, N., Über die Dendritenzellen der Epidermis (Eine Studie über die Langerhans-Zellen in der normalen und ekzematösen Haut des Meerschweinchens). *Arch. Klin. Exp. Derm.*, **222**, 23 (1965).

21. Niebauer, G., Dendritic cells of human skin: Dendritic cells in epidermis and their related cells in dermis. In *Cytology, Physiology, and Pathology with Particular Reference to Melanin*, p. 44, S. Karger, Basel-New York (1968).

22. Sekido, N., Epidermal melanocytes of guinea-pigs and their behavior in experimental contact dermatitis. *Jap. J. Derm.* Ser. A, **73**, 37 (in Japanese), Ser. B: 18 (in English) (1963).

23. Kawamura, T., Morioka, S., Kukita, A., Kawada, A., and Taniguchi, K., So-called juvenile melanoma and its allied conditions. *Jap. J. Derm.*, **73**, 67 (in Japanese), 78 (in English) (1963).

24. Plenck, H., Über argyrophile Fasern (Gitterfasern) und ihre Bildungszellen. *Ergebn. Anat. Entwicklungsgesch.*, **27**, 302 (1927).

25. Masson, P., Experimental and spontaneous Schwannomas (peripheral gliomas). *Amer. J. Path.*, **8**, 367. Spontaneous Schwanomas. *Ibid.*, p. 389 (1932).

26. Murray, M. R. and Stout, A. P., Characteristics of human Schwann cells *in vitro.*, *Anat. Rec.*, **84**, 275 (1942).

27. Masson, P., Les naevi pigmentaires, tumeurs nerveuses. *Ann. Anat. path. et Anatom-medico-chirung*, **3**, 417, 657 (1926).

28. Masson, P., My conception of cellular nevi. *Cancer*, **4**, 9 (1951).

29. Kawasakiya, S., Findings in the pigment cell nevi and the other conditions, stained by means of trichromique de Masson and Goldmer's modification of it. *Jap. J. Derm.*, **65**, 83 (in Japanese), 102 (abstract in English) (1956).

30. Soldan, *Arch. f. Klin. Chir.* (1899).

31. John, F., Studien zur Histogenese der Naevi. *Arch. Derm.*, **178**, 608 (1939).

32. Ito, M., Genesis of the nevus cell. *Tohoku J. Exp. Med.*, **52**, 95 (1950).

33. Nishihara, K., Studies on the nature of the Langerhans' cells and findings of the Langerhans staining method on the naevus pigmentosus and others. *Jap. J. Derm.*, **63**, 284 (in Japanese with English abstract).

34. Iriyama, M., Pathogenesis of nevi and phakomatosis based on findings by silver impregnation. *Jap. J. Derm.*, **70**, 395 (in Japanses), 49 (abstract in English) (1955).

35. Feyrter, F., Über die Pathologie der vegetativen nervösen Peripherie und ihrer ganglionären Regulationstätten. Wien, Verlag Wilhelm Maundrich (1951).

36. Feyrter, F., Über den Naevus. *Virchow. Arch.*, **301**, 417 (1938).

37. Terada, M., On the findings of the chromotropic lipoids and lipoproteids in the normal skin, pigment cell nevi and other lesions, demonstrated by means of Feyrter's enclosing staining. *Jap. J. Derm.*, **65**, 203 (in Japanese, with English abstract) (1955).

38. Ishibashi, Y. and Kawamura, T., Über die pigmentierten Naevi bei lepromatösen Leprakranken. *13th Congress Internationalis Dermatologiae*, July 31–Aug. 5, 1967, München (W. Jadassohn and C. G. Schirren, eds.), p. 135, Springer-Verlag, Berlin-Heidelberg-New York (1967).

39. Klingmüller, G. and Ishibashi, Y., Electron microscopic investigation of leprosy, Second Congress of Tropical Dermatology, Aug. 15–20, Kyoto, Japan.

40. Shelley, W. B. and Arthur, R. P., Nerve fibers: A neglected component of intradermal cellular nevi. *J. Invest. Derm.*, **34**, 59 (1960).

41. Delacrétaz, J. and Jaeger, H., Die Melanome, Herkunft der Naevi. *Dermatologie und Venereologie* (H. A. Gottron and W. Schönfeld, eds.), **4**, 559, Georg Thieme Verlag, Stuttgart (1960).

42. Pinkus, H. and Mishima, Y., Benign and precancerous nonnevoid melanocytic tumors. *Ann. N. Y. Acad. Sci.*, **100**, 256 (1961).

43. Lever, W. F., *Histopathology of the Skin* (4th ed.), p. 707, Lippincott, Philadelphia-Toronto (1967).

44. Kawamura, T., The histogenesis of pigmented nevi, nevus fuscocaeruleus ophthalmomasillaris of Ota and melanoepithelioma of Ota. *Proceedings of the Twelfth International Congress of Dermatology*, 9–15 Sept., Washington D. C., p. 134, (1963).

45. Kawamura, T., *G. Ital. Derm.*, **107**, 837 (in English) (1966).

46. Kawamura, T., Mori, S., and Naito, A., Atypical blue nevus (2nd Type) with unique clinical appearance. *Hifu-to Hitsunyo.*, **29**, 1118 (in Japanese) (1967).

47. Kawamura, T., Ikeda, S., Noda, S., Ishizu, S., and Nakajima, T., Mélanoses neurocutanées of Touraine. *13th Congress Internationalis Dermatologiae*, July 31, Aug. 5, 1967, München (W. Jadassohn and C. G. Schirren, eds.), p. 915, Springer-Verlag, New York.

48. Kawamura, T., Nakauchi, Y., and Mori, S., Beitrag zu Pathogenese der Bourneville-Pringlesche Phakomatose. *Hautarzt,* **15**, 476 (1964).

49. Kawamura, T., Kitamura, K., Ohmori, S., Kobori, T., and Noguchi, Y., Certain statistical features of skin carcinoma in Japan. Presented at the Twelfth International Congress of Dermatology, Sept. 9–15, 1962, Washington, D. C.

50. Ota, M., Über Gewebsnaevi, Melanome und Schweissdrüsencarcinome *Jap. J. Derm.*, **47**, 376 (in Japanese), 93 (in German) (1940).

51. Masson, P., l. c. Bloch, B., Über benigne, nicht naevoide Melanoepitheliome der Haut nebst Bemerkungen über das Wasen und die Genese der Dendritenzellen. *Arch. Derm.*, **153**, 20 (1927).

52. Masson, P., Mélanoblastes et Cellules de Langerhans (Réunion Strassbourg), *Bull. Soc. Franc. Derm.*, **42**, 112 (1935).

53. Kamide, I., On the dopa-findings in normal skin, pigment cell nevus, Malpighian proliferation with hyperpigmentation and Recklinghausen spot. *Jap. J. Derm.*, **64**, 125 (1954).

54. Mishima, Y. and Pinkus, H., Benign mixed tumor of melanocytes and Malpighian cells. *Arch. Derm.*, **81**, 539 (1960).

55. Dalton, A. J. and Felix, M. D., Phase contrast and electron micrography of the Cloudman S91 melanoma. In *Pigment Cell Growth* (M. Gordon, ed.), p. 267, Academic Press, New York (1953).

56. Mishima, Y., Macromolecular Changes in Pigmentary Disorders. *Arch. Derm.*, **91**, 519 (1965).

57. Kawamura, T., Ideka, S., Mori, S., and Obata, H., Electron Microscopic findings compatible with those of the lysosome (autophagic vacuole), revealed in the melanocytes in cases of conspicuous pigment blockade. *Jap. J. Derm.* Ser. A, **76**, 713 (in Japanese), Ser. B, 405 (in English) (1966).

58. Obata, H., Electron microscopic studies on the degradation of melanin and certain other findings in the melanoepitheliomas of Ota and blue nevus. *Jap. J. Derm.* Ser. A, **78**, 669 (in Japanese), Ser. B, 414 (in English).

59. Miescher, G., Die Chromatophoren in der Haut des Menschen. Ihr Wesen und die

Herkunft ihres Pigmentes, Ein Beitrag zur Phagocytose der Bindegewebszellen. *Arch. Derm.*, **139**, 133 (1922).

60. Szabó, See ref. 22, above.

61. Furuya, T., Study on incontinentia pigmenti by the use of osmium iodide fixation technique. *Jap. J. Derm.* Ser. A, **74**, 573 (in Japanese), Ser. B, **74**, 345 (in English) (1964).

DISCUSSION

DR. LERNER: When a nevus develops into a malignant melanoma, sometimes repeated wide and deep excisions still are followed by recurrence at the operative site. Is it possible that the malignant potential belongs to the cells in nerves leading to the original nevus, so that no amount of ordinary excision will prevent a recurrence?

DR. KAWAMURA: When the nevus cells degenerate into melanoma cells, they acquire properties other than those of nevus cells. The tendency to recur is one of the prominent properties of melanoma cells.

DR. MISHIMA: I have seen a number of cases developing not only melanomas but even benign nevus cell nevus at the junction layer repeatedly, despite the almost complete removal of the primary lesion. I agree with Dr. Lerner's interpretation of this phenomenon. In fact, I previously proposed the concept of "field potentiality" as an explanation of this phenomenon.

DR. QUEVEDO: Dr. Whimster at Saint Thomas' Hospital in London has proposed that the human integument is a mosaic of differentiated regions maintained by specific cutaneous nerves. In his view, the repeated reappearance of nevus cells following excision may be the result of region-specific stimulation of cell differentiation by the nervous system. This influence, which continues throughout life, may cause repeated differentiation of atypical cells following local excision of skin lesions.

DR. FITZPATRICK: I really enjoyed this, and the idea occurred to me that we may have to extend this Kawamura–Mishima concept a bit to implicate the nevoblast not only in the control of the proliferation of cells but also in a variety of situations seen clinically. For example, in the hypomelanotic macules of tuberous sclerosis, the nevoblast may be programed to make melanocytes with less-melanized melanosomes.

The Mechanism of Normal Human Melanin Pigmentation and of Some Pigmentary Disorders*

Thomas B. Fitzpatrick, Yoshiaki Hori,** Kiyoshi Toda, Shoichi Kinebuchi,*** and George Szabó+

Department of Dermatology, Harvard Medical School, and the Massachusetts General Hospital, Boston, Massachusetts, U.S.A.

I. INTRODUCTION

A large amount of information on the biology of pigment cells has been accumulated in the past two decades, but the study of the pathogenesis of human pigmentary disorders based on this knowledge has only begun. Consideration of human pigmentary diseases can bridge the gap between the biologist and the clinician, inasmuch as there are some genetic pigmentary disorders in man that have animal models.

Pigmentation of the skin and hair, as viewed clinically, is largely related to the content of melanin in the keratinocytes. Although melanin is formed on melanosomes (Fig. 1) in melanocytes, the skin or hair is not visibly pigmented unless the melanosomes have entered the keratinocytes.

The melanocyte in the skin is associated with a group of keratinocytes into which it secretes melanosomes. The melanocyte–keratinocyte pool serves as a structural and functional unit known as the *epidermal melanin unit*. Although the structural basis for the epidermal melanin unit is

---

* This work was supported by Grants 5 RO1 CA-05010-11 and CA-05401 (Dr. Szabó).

** Present Address: Department of Dermatology, University of Tokyo Medical School, Hongo, Bunkyo-ku, Tokyo, Japan.

*** Present Address: Department of Dermatology, Tokyo Teishin Hospital, Fujimi, Chiyoda-ku, Tokyo, Japan.

+ Present Address: Harvard School of Dental Medicine, 188 Longwood Avenue, Boston, Massachusetts, U.S.A.

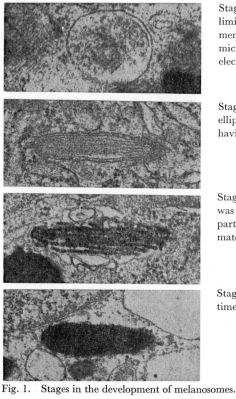

Stage I: large spherical membrane-limited vesicle that shows either a filament with periodicity by electron microscopy or tyrosinase activity by electron-microscope histochemistry.

Stage II: the melanosome becomes ellipsoid and shows numerous filaments having distinct periodicity.

Stage III: the internal structure that was characteristic of Stage II is now partially obscured by electron-dense material.

Stage IV: it is still ellipsoid and by this time is totally electron-dense.

Fig. 1.   Stages in the development of melanosomes.

clearly established, the functional interrelationship of the melanocyte and the keratinocyte is not known at this time. We agree with Breathnach[1] and Quevedo[2] that keratinocytes actively participate in controlling the rate of production and transfer of melanosomes from melanocytes to keratinocytes.

After the melanosomes have been transferred to the keratinocytes, they remain as discrete organelles surrounded by a limiting membrane (as in the skin of Negroids and Australian aborigines) or they aggregate in clusters of three or more surrounded by a limiting membrane. These melanosomes, either isolated or in aggregates, contain acid phosphatase and resemble lysosomes (see later discussion).[3] Within the aggregates, the melanosomes undergo degradation, lose their characteristic internal structure, and finally break up into amorphous melanin particles that then ascend to the upper parts of the epidermis. This phenomenon of melanosome degradation may be one of the important components of the processes that determine the total amount of melanin pigmentation in the skin.

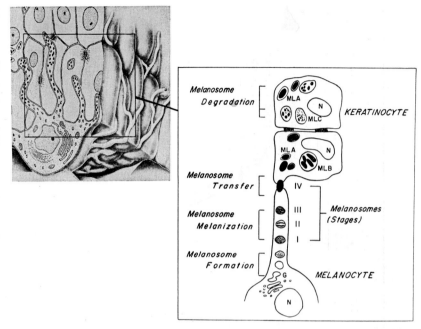

Fig. 2. Four biological processes involved in melanin pigmentation. The diagram at the left depicts a melanocyte supplying melanosomes to the surrounding cluster of keratinocytes (the epidermal melanin unit). The diagram at the right illustrates the four processes involved in melanin pigmentation of the skin. N: nucleus; I–IV: the four stages in the development of the melanosome; MLA: melanosome in the African Negroid and Australian aborigine that is a discrete organelle surrounded by a membrane; MLB: melanosomes in groups surrounded by a membrane (the "melanosome complex"), similar to a lysosome, and present in Caucasoid, Amerindian, and Mongoloid skin; MLC: melanosome complex in which there has been degradation of the individual melanosomes.

Visible pigmentation of the skin and hair is related to four biological processes (Fig. 2):

1. Production of melanosomes in melanocytes.
2. Melanization of melanosomes in melanocytes and, possibly, within keratinocytes.
3. Transfer of melanosomes within lysosome-like organelles in keratinocytes.
4. Degradation of melanosomes within lysosome-like organelles in keratinocytes.

Increased melanin pigmentation (hypermelanosis) or decreased melanin pigmentation (hypomelanosis) results from disturbances in one or more of these four biological processes. In the discussion of pigmentary disturbances that follows, these biological processes will be considered in the

Table 1. Disturbances of human melanin pigmentation.

Causative factors	Classification		
	Hypomelanosis* White	Hypermelanosis* Brown	Hypermelanosi* Gray, slate, or blue[m]
Genetic factors	Piebaldism[a]	Café-au-lait and frecklelike macules in neurofibromatosis[a]	Oculodermal melanocytosis (nevus of Ota)[a,m]
	Waardenburg's syndrome[a]	Melanotic macules in polyostotic fibrous dysplasia (Albright's syndrome)[a]	Dermal melanocytosis (Mongolian spot)[a,m]
	Canities, premature[a]		
	Vitiligo[a,b]	Ephelides (freckles)[a]	Blue melanocytic nevus[a,m]
	Albinism, oculocutaneous[c]:	Lentigines[a]	Incontinentia pigmenti[a,n]
	tyrosinase-positive	Lentigines with cardiac arrhythmias[a]	Franceschetti-Jadassohn syndrome[a,n]
	tyrosinase-negative	Seborrheic keratosis[a]	
	Albinism, ocular	Melanocytic nevus[a]	
	Cross-McKusick-Breen syndrome[c]	Neurocutaneous melanosis[a]	
	Hypomelanotic macules in tuberous sclerosis[a,d]	Xeroderma pigmentosum[a]	
	Nevus depigmentosus[a,d]	Acanthosis nigricans, juvenile type[a]	
	Phenylketonuria[e,f]	Peutz-Jeghers syndrome[a]	
	Fanconi's syndrome[e]		
	Neurofibromatosis		
	Ataxia telangectasia		
Metabolic factors		Hemochromatosis[c]	Hemochromatosis[c]
		Hepatolenticular disease (Wilson's disease)[c]	
		Porphyria (congenital erythropoietic and porphyria variegata and cutanea tarda)[c]	

			Gaucher's disease[j] Niemann-Pick disease[j]
Endocrine factors	Hypopituitarism[c] Addison's disease[a] Hyperthyroidism[a]	ACTH-producing and MSH-producing pituitary and other tumors[c] ACTH therapy[c] Pregnancy[j] Addison's disease[c] Estrogen therapy[k] Melasma[a,l]	Chronic nutritional insufficiency[a]
Nutritional factors	Chronic protein deficiency or loss[e,g]: kwashiorkor nephrosis ulcerative colitis malabsorption syndrome Vitamin-B$_{12}$ deficiency[e]	Kwashiorkor[a] Pellagra[j] Sprue[j] Vitamin-B$_{12}$ deficiency[j]	
Chemical and pharmacologic agents	Monobenzyl ether of hydroquinone[a] Chloroquine and hydroxychloroquine[e] Arsenical intoxication[a]	Arsenical intoxication[c] Busulfan administration[c] Photochemical agents (topical or systemic drugs, tar)[a] Berlock dermatosis[a] Dibromomannitol administration[c]	Fixed (drug) eruption[a,m] Quinacrine toxicity[c] Chlorpromazine administration[j,o]
Physical agents	Burns: thermal, ultraviolet, and ionizing radiation[a,h] Trauma[a,h]	Ultraviolet light (suntanning)[a] Thermal radiation[a] Alpha, beta, and gamma ionizing radiation[a] Trauma (e.g., chronic pruritus)[a]	

Table 1. (continued)

Causative factors	Classification		
	Hypomelanosis* White	Hypermelanosis* Brown	Hypermelanosis* Gray, slate, or blue
Inflammation and infection	Pinta[a,h] Leprosy[a,d] Fungal infections (tinea versicolor)[a,d] Pityriasis alba[a,d] Eczematous dermatitis[a,d] Psoriasis[a] Lupus erythematosus, discoid[a] Postinflammation hypomelanosis[a,d]	Postinflammation melanosis (exanthems, drug eruptions)[a] Lichen planus[a] Lupus erythematosus, discoid[a] Lichen simplex chronicus[a] Atopic dermatitis[j] Psoriasis[a]	Pinta in exposed areas[a] Erythema dyschromicum perstans[a,n]
Neoplasms	In sites of malignant melanoma after disappearance (therapeutic or spontaneous) of tumor[a] Nevus, "halo"[a]	Malignant melanoma[a] Mastocytosis (urticaria pigmentosa)[a] Adenocarcinoma with acanthosis nigricans[a]	Slate-gray dermal pigmentation with metastatic melanoma and melanogenuria[e]
Miscellaneous factors	Alezzandrini's syndrome[a] Vogt-Koyanagi-Harada syndrome[a] Scleroderma, circumscribed or systemic[a] Canities[e]	Scleroderma, systemic[c] Chronic hepatic insufficiency[c] Whipple's syndrome[c] Encephalitis, chronic[a] Lentigo, senile ("liver spots")[a] Cronkhite-Canada syndrome[a]	

Alopecia areata[i]
Horner's syndrome, congenital
and acquired[f]
Hypomelanosis, guttate,
idiopathic[a]

Catatonic schizophrenia[c]

---

* The listing includes the pigmentation disorder itself or the condition with which it is associated.
[a] Pigment change is circumscribed.
[b] Total loss of pigment in the skin and hair may occur.
[c] Pigment change is diffuse, not circumscribed, and there are no identifiable borders.
[d] Loss of pigmentation is usually partial (hypomelanosis); viewed with Wood's lamp, the lesions are not completely devoid of pigment (amelanosis) as in vitiligo.
[e] Pigment is decreased in the hair.
[f] Pigment is decreased in the iris.
[g] Hair is gray or reddish.
[h] There is a loss of melanocytes.
[i] Regrown hair is white.
[j] Pigment change may be diffuse or circumscribed.
[k] Nipples are affected.
[l] Idiopathic or due to progestational agents.
[m] Gray, slate, or blue color results from the presence of *dermal* melanocytes or phagocytized melanin in the dermis.
[n] Areas of brown may be admixed with the slate-gray and blue discoloration.
[o] Pigment has not been definitely identified as melanin.

pathogenesis of only a few of the large number of various pigmentary disorders that occur in man (Table 1).

The pigmentary disorders to be discussed are the result of: (1) change in the number of secretory melanocytes; (2) alterations in the production or structure of melanosomes; (3) alterations in the melanization of melanosomes; (4) disturbances in the transfer of melanosomes to keratinocytes due to structural or pathologic changes in melanocytes or in keratinocytes; (5) alterations in the degradation of melanosomes in melanocytes or in keratinocytes.

II. CELLULAR AND SUBCELLULAR BASIS OF PIGMENTARY DISORDERS

A.   *Change in the number of secretory melanocytes*
1.   Reduction or absence of secretory melanocytes
a.   Vitiligo
Any hypothesis regarding the pathogenesis of vitiligo (see reference 4 for discussion) must take into account that there are no recognizable melanocytes present in vitiligo skin; this is the most striking fact about the disease when the skin is studied by electron microscopy; at least there are no

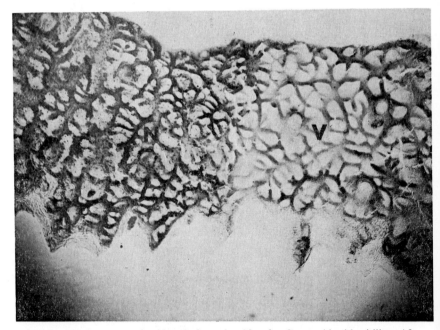

Fig. 3.   Whole mount of epidermis from the skin of a Caucasoid with vitiligo. After dopa incubation, a normal number of melanocytes can be seen in the normal skin (N), but there are practically no melanocytes in the vitiligo area (V).   ×32.

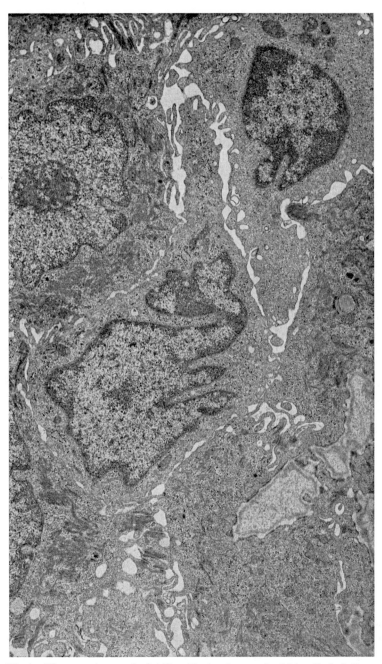

Fig. 4.   Electron micrograph of vitiligo skin showing two dendritic cells. Left: Langerhans cells. Right: unclassified dendritic cell.   ×5,300.

melanocytes that are recognizable as such. The only method for the detection of melanocytes at the present time is electron microscopy, which will reveal the characteristic internal structure of the melanosome in the cytoplasm. In the skin of vitiligo (Fig. 3), we may see round or even dendritic cells at the epidermodermal junction that are in a stage of development in which there are no recognizable organelles (Fig. 4). These cells could be (1) immature keratinocytes without any tonofibrils or desmosomes,[1] or (2) Langerhans cells without the characteristic rod-shaped or racket-shaped organelles, or (3) mature melanocytes in an inactive phase, that is, not producing any identifiable melanosomes, or (4) hypothetical precursors of melanocytes, which one of us, T.B.F., has chosen to call "melanogonia."

We do know that melanocytes can undergo mitosis[5] and also that they can apparently change from an inactive to an active secretory phase, as occurs in patients with vitiligo who are treated with psoralens; in these patients, there is a beginning of pigmentation around the opening of the hair follicle at the sides of the lesions, and, eventually, complete repigmentation may involve the entire lesion. Breathnach et al.[6] observed that the melanocytes at the margins of vitiligo areas have little or no melanogenic activity. This same phenomenon is even more strikingly demonstrated in patients with trichrome vitiligo, in which the normal pigmented skin has normal melanocytes; the tan, or intermediate, zone contains melanocytes with much less tyrosinase activity (as detectable by histochemical techniques); and the white, or central, area does not contain any melanocytes. This gradual diminution until there is a final disappearance of tyrosinase activity and of melanocytes is suggestive of an inhibitor of the enzyme. The occurrence of vitiligo with autoimmune disorders, such as Graves' disease[7] or the syndrome of idiopathic hypoparathyroidism in association with Addison's disease,[8] suggests that vitiligo may be an autoimmune process, as does the striking depigmentation that occurs in the halo nevus that is associated with an inflammatory lymphocytic infiltrate; this same type of depigmentation may occur around primary malignant melanoma.[9]

The notion that vitiligo could be related to neural factors, such as an inhibitor produced by nerve endings, has been championed principally by Lerner.[10] It is certainly interesting that, in some patients, the depigmentation of vitiligo follows a quasidermatomal distribution, which suggests neural control. No substantial proof, however, of a neurogenic factor in the pathogenesis of vitiligo has been demonstrated as yet. There are some well-documented cases of depigmentation in a dermatomal distribution occurring after direct trauma, as reported by Costea et al.[11] Chanco–Turner and Lerner[12] reported diminished sympathetic-nervous system or adrenergic activity (based on measurements of sweat) in the vitiligo areas, and Breathnach et al.[1,6] noted degenerative and regenerative changes of minimal degree in nerves (axons and associated Schwann cells) supplying vitiligo areas. Breathnach[1,6] also suggested that the changes in the nerves and in the skin may be the result of a common lesion of Schwann cells and melanocytes (both of which are derived from the neural crest) that alters the two types of cells in some way during development; their fully differentiated descendants are then susceptible to certain "trigger" factors in postnatal life. It is difficult to explain the absence of any pigment change in the melanocyte of the iris in patients with vitiligo if a neural control mechanism is involved.

Our current belief is that the hypomelanosis in vitiligo is a hereditary defect of parts of the melanocyte system in which melanocytes develop normally but then undergo changes in which the cell stops producing melanosomes; the melanocyte becomes unrecognizable because of the lack of melanosomes. The challenge at present is to develop methods for recognition of inactive melanocytes in order to prove this hypothesis.

b.   Piebaldism of skin and hair

A number of genes control white-spotting patterns in animals.[13-16] These white-spotting patterns are the result of defects in the differentiation of melanoblasts themselves, or, possibly, the result of factors in the environment to which the melanoblasts migrate. A human counterpart of animal white-spotting is *piebaldism,* a circumscribed congenital hypomelanosis of

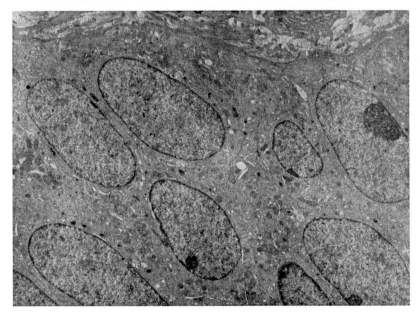

Fig. 5.   Hair bulb from piebald subject, showing absence of recognizable melanocytes. Glutaraldehyde–osmium tetroxide, embedded in Epon. ×5,000.

the skin and hair inherited as an autosomal dominant trait; there is a typical white forelock and there are symmetrical white macules on the trunk and extremities. In this condition, melanocytes are absent from the skin of the white forelock area;[17] identical findings were also reported[18] in guinea pigs in which the spotting was recessively inherited. We[19] established that melanocytes are also absent in the hair bulbs of the white forelock area (Fig. 5). In piebaldism, the absence of melanocytes in the skin and hair bulb could be the result of either (1) a failure of differentiation by melanoblasts that have migrated to the area from the neural crest or (2) a failure of migration of melanoblasts to the area.

c. Hypomelanosis after application of topical depigmenting chemical agents

In the past several years, several chemical agents have been discovered that cause depigmentation of mammalian skin.[20,21] The most effective topical depigmenting agents are 4-isopropylcatechol, hydroquinone, mercaptoethylamine hydrochloride (MEA) and N-(2-mercaptoethyl)-dimethylamine hydrochloride (MEDA); these agents cause a reversible depigmentation of normal skin in pigmented guinea pigs and a reduction of some hypermelanoses (melasma and postinflammation pigmentation) in man. Monobenzyl ether and monomethyl ether of hydroquinone cause an irreversible depigmentation of the normal skin in man.[22] With the use of all the foregoing chemical agents, melanocytes are not recognizable, even by electron microscopy, in the areas of depigmentation. In the depigmented skin, there are undifferentiated cells without tonofibrils and without melanosomes that could be either melanocytes, immature keratinocytes, or Langerhans cells. It is possible that the depigmenting chemical agents could induce an inactive, nonsecretory phase in the melanocytes.

d. Graying or whitening of hair with aging

The graying or whitening of hair with aging appears to result from a gradual diminution of tyrosinase activity in the melanocytes of the hair bulb. In the final stages of this process, when the hair becomes white, the matrix of the hair bulb becomes completely disorganized, and melanocytes cannot be detected.[23] An age-dependent decline in the number of epidermal melanocytes is found in the interfollicular epidermis of the trunk in young pigmented mice[24] (the details of this study are discussed in Professor Quevedo's contribution).

2. Increase of secretory melanocytes

a. Changes in skin after exposure to ultraviolet radiant energy

It has been shown that the melanocyte population in the epidermis in the various human races does not seem to vary in terms of the total number of melanocytes present.[25-27] The response of the human skin to ultraviolet irradiation depends on (1) the basic skin color, (2) the racial background, and (3) whether a single or repeated doses of ultraviolet irradiation are given.

Quevedo et al.[28] were the first to point out that repeated exposure of the skin to ultraviolet radiation caused an increase in the number of dopa-positive melanocytes. A single exposure, however, did not appear to cause an increase in the number of melanocytes, although new melanin formation was observed in the existing population.[29] In a more detailed study of this phenomenon, Szabó et al.[30] reported on the response of melanocytes to irradiation in four racial types: Caucasoid, African Negroid, Amerindian, and Mongoloid. Among these four groups, a macroscopically visible erythema was most prominent in light-skinned Caucasoids on the previously unexposed skin of the buttocks. The skin of persons of the more heavily pigmented races, that is, Negroids, Amerindians, and Mongoloids, showed either no erythema or a hardly perceptible erythema on the buttocks after exposure to artificial ultraviolet light. Visible pigmentation of varying degree was observed, however, on the fifth day after radiation, on the skin of the forearm and the buttock. Light-microscope examination of skin exposed to a single dose of ultraviolet radiant energy did not show change in the melanocyte count. After multiple exposures to ultraviolet light, however, the melanocyte population almost doubled in the areas of the skin that are usually not exposed to sunlight, whereas in the forearm, an area that is exposed to sunlight, there was a much smaller rise in the melanocyte population.

At the subcellular level, there is quite a distinct change in the number and distribution of melanosomes. After single or multiple exposures, it was found that the number of melanosomes increased in the melanocytes and keratinocytes and varied according to the racial background. In very pale Caucasoids with red hair, the melanocytes contain practically no melanosomes in the perikaryon, and the melanocytes are weakly dopa-positive in "split" epidermal preparations. There are very few, if any, Stage III and Stage IV melanosomes in the dendrites, and the keratinocytes contain little or no melanin.

In Caucasoids who are somewhat more pigmented, Stage I, II, and III melanosomes are visible in the perikaryon of the melanocytes, but Stage IV melanosomes are not found; the Stage IV melanosomes are found in groups in the surrounding keratinocytes. In the perikaryon of Mongoloids, there are numerous Stage II, III, and IV melanosomes, whereas, in the perikaryon of Negroids, there are mostly Stage IV melanosomes.

After ultraviolet irradiation, numerous Stage IV melanosomes appear in the melanocytes of Caucasoids, even in those subjects with red hair and a basically very pale skin (Fig. 6); Stage II and III melanosomes and an increased number of Stage IV melanosomes appear in the melanocytes of Negroids.

From these data, it can be concluded that melanosomes become fully melanized (Stage IV) within the melanocytes of Negroids, which indicates

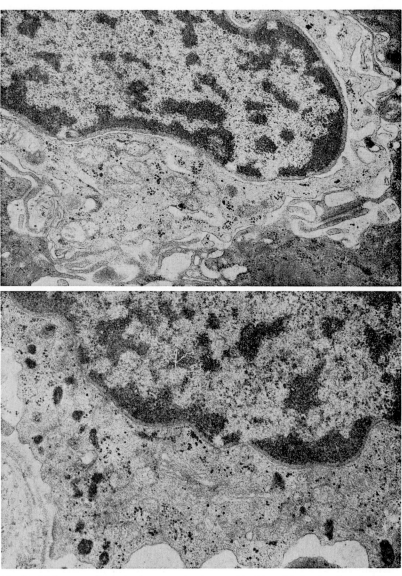

Fig. 6   Melanocyte from the forearm of a Caucasoid with red hair, showing the effects of exposure of the skin to ultraviolet irradiation. ×2,140.

Upper: Before ultraviolet irradiation, only one Stage IV melanosome and several Stage III melanosomes are visible.

Lower: Ten days after a single dose of ultraviolet irradiation, there are a number of Stage IV, Stage III, and Stage II melanosomes in the cytoplasm of the melanocyte.

an important difference between the skin of Negroids and the skin of other racial groups. Inasmuch as it is, as yet, impossible to quantify the increase in numbers of melanosomes in the skin, it is not known whether the increase in the pigmentation of Negroid skin after ultraviolet irradiation is due to an increased production of melanosomes or to fuller melanization of the melanosomes present before irradiation. It is not clear whether this full melanization of melanosomes in Negroids is due to an increase in tyrosinase activity on the melanosomes or to the absence of inhibitors of tyrosinase. We have demonstrated[19] that Stage II melanosomes are produced by melanocytes and transferred to keratinocytes in human albino hair bulb (Fig. 13), and this suggests that the most important factor in the pigmentation of skin after radiation is the degree of melanization of the individual melanosomes rather than the number of organelles produced.

After irradiation with ultraviolet light, an increased number of dopa-positive melanocytes is observed at the light-microscope level. This increased number could represent either a proliferation or an activation of melanocytes that were already present before irradiation but not detectable by the dopa reaction. The reason for the increase in population will not be understood until study of melanocyte turnover is performed with radioactive thymidine and radioactive melanin precursors used simultaneously.

B. *Alterations in production or structure of melanosomes*
1. Increased production of melanosomes
a. Changes in skin after exposure to ultraviolet radiant energy
As stated previously, there is a marked increase in the number of melanocytes after ultraviolet irradiation and also an increase in the number of melanosomes produced.
2. Decreased production of melanosomes
a. Graying or whitening of hair due to aging
As mentioned previously, there is a reduction in the tyrosinase activity in melanocytes and it is associated with a decrease in the production of melanosomes.
3. Structural changes in melanosomes
a. Malignant melanoma
No specific structural abnormality that would characterize a malignant melanocyte has yet been defined. There are just a few reports of the fine structure of melanosomes in malignant melanomas (Table 2). In this small sample of malignant melanomas, changes in the size, shape, and internal structure of melanosomes have been observed, especially in the type classified by Clark et al.[31] as superficial spreading melanoma. In Table 2, we have summarized the reported ultrastructure of melanosomes

Table 2.   Ultrastructure of melanosomes.

Source	Size	Shape	Characteristics of internal structure	Melanization	Other features
Normal epidermal melanocyte (Caucasoid)	Average: 300 × 700 mμ (A, * B) Range: 200–400 × 500–800 mμ (A)	Ellipsoid (A,B,C,D,E,F)	Filaments (A,B,C,D,E,F)	Synchronized (Stages I–IV) (A,E)	
Lentigo-maligna melanoma**	Average: 120 × 240 mμ (C) 200 × 600 mμ (D) "smaller than usual" (E) Range: 150–250 × 450–700 mμ (D)	Ellipsoid (C,E) "Red-cell shape" (D)	Filaments (C,E) "Cross-sectional profiles are abnormal" (E)	Synchronized (E)	Nuclei smaller than normal (C) Large nuclei (E) Melanosome-containing phagosomes in keratinocyte (E) Ribosomes: organized (C)

Superficial spreading melanoma	Average: 400 × 700 mμ Range: 150–700 × 500–1500 mμ (A)	Ring shape (A,E) Irregular dough-nut shape (A,C) Spherical (A,F) Ellipsoid (E)	Less cross-linkage than normal in Stage II (E) Granules (i.e., without filaments (F))	Not syn-chronized (A) Very large nucleolus (A) Incomplete melanization (A,E) Large nucleus (A,E) Normal nucleus (C) Many Stage I melanosomes (E)	Very large nucleolus (A) Large nuclei (A,E) Normal-size nuclei (C) Virus-like particles (A) Melanosome-containing phagosomes in kera-tinocyte (A) Ribosomes: many free (C)

A. Toda, K., unpublished observations.
B. Birbeck, M. S. C. 1963. Electron microscopy of melanocytes: The fine structure of hair-bulb premelanosomes. *Ann. N.Y. Acad. Sci.*, **100**, 540.
C. McGovern, V. J., and Lane Brown, M. M. 1969. The Nature of Melanoma, ed. I. M. Kugelmass. Springfield, Ill.: C. C Thomas.
D. Mishima, Y. 1967. Cellular and subcellular activities in the ontogeny of nevocytic and melanocytic melanomas. *In* The Pigmentary System: 509. (Vol. 8 of Advances in Biology of Skin.) Edited by W. Montagna & F. Hu. Pergamon Press, New York.
E. Clark, W. H., Jr., and R. Bretton. In press. A comparative fine structural study of melanogenesis in normal human epidermal melanocytes and in certain human malignant melanoma cells. *In* The Skin (International Academy of Pathology Monograph). Edited by E. B. Helwig. The Williams & Wilkins Co., Baltimore.
F. Hirone, T. In this symposium.

* The capital letters indicate references; the descriptions are cited therefrom.
** The classification of Mishima and Hirone are different from those of Clark and Bretton and of McGovern and Lane Brown, but the lesion is presumed to be the same, i.e., precancerous circumscribed melanosis of Dubreuilh.

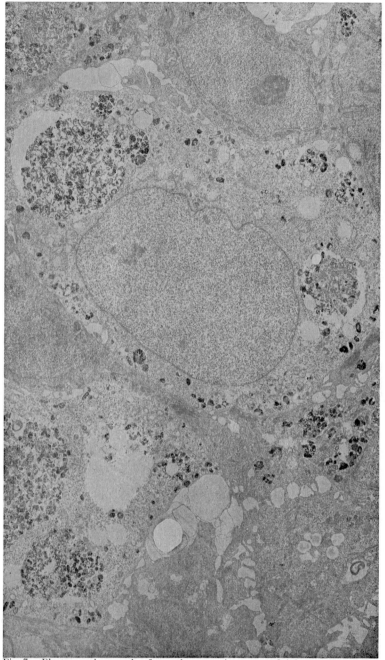

Fig. 7. Electron micrograph of a melanocyte in a superficial spreading melanoma, showing a large autophagosome containing many abnormally melanized melanosomes. ×5,400.

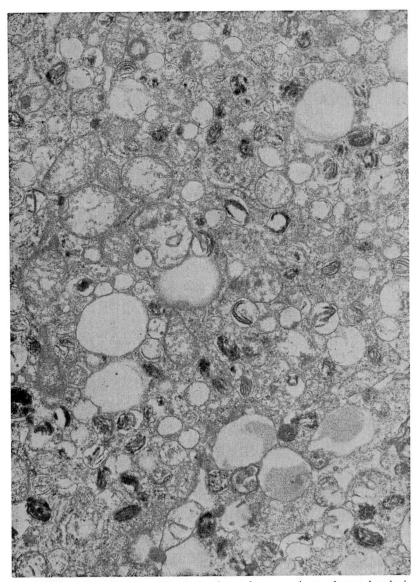

Fig. 8. Higher magnification of Fig. 6, showing melanosomes in an abnormal melanization pattern: doughnut-shape; irregular doughnut-shape; not synchronized (see also Table 2). ×20,000.

in human malignant melanoma. Phagocytic activity is greatly increased in the keratinocytes surrounding malignant melanocytes, and the phagosomes usually contain fully melanized melanosomes. Immature melanosomes are noted in the autophagosomes of the malignant melanocyte

387

(Figs. 7, 8). It is apparent that many more specimens must be studied before any definite statements can be made regarding a specific ultra-structural change in malignant melanoma vis-à-vis benign pigment cell neoplasms.

4. Increase in size of melanosomes

a. Pigmentary dilution of skin, hair, and eyes in Chédiak-Higashi syndrome

Chédiak-Higashi syndrome is a rare autosomal recessive trait character-ized by pigmentary dilution with clinical features suggestive of albinism and the appearance of very large melanosomes and other giant organelles in the lymphocytes in the blood and bone marrow.[32] The pigmentary dilution affects the eyes, skin, and hair in varying degree. The ocular features found in oculocutaneous albinism are present, but the hair color in some patients is not so diluted as it is in most patients with oculocutane-ous albinism. Windhorst et al.[32] suggested that the pigmentary dilution results from a genetic defect that leads to "failure in control of the size of the melanin granules" and related the color defect to the giant granules observed in the leukocytes and their precursors and in the renal tubules, neurons, and other cells in children with Chédiak-Higashi syndrome. A similar type of syndrome is observed in the Aleutian strain of mink (cited by Windhorst et al.).[33] The mechanism of the pigment dilution in

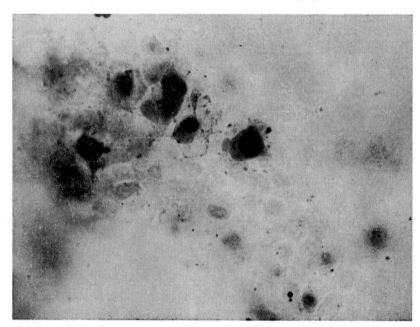

Fig. 9. Café-au-lait macule from a patient with neurofibromatosis. Note the giant pig-ment granules. Split epidermis; dopa. ×900.

Chédiak-Higashi syndrome is not known, but the decrease in pigmentation could be the result of a decrease in the number of melanosomes produced or because the giant melanosomes may be more susceptible to autolysis in the keratinocytes.

b.   Melanotic macules in neurofibromatosis

The pigmentation of neurofibromatosis includes café-au-lait macules and, often, a generally distributed frecklelike pigmentation, especially in the axillae. Light-microscope study of dopa-incubated split-skin preparations showed giant pigment granules in the keratinocytes or in the melanocytes. These pigment granules were present in both the normal skin and the café-au-lait macules of patients with neurofibromatosis (Fig. 9).[34] Electron-microscope studies of these giant organelles have not yet been performed in sufficient quantity for any conclusions to be drawn, and, hence, one can only speculate that these giant particles might represent an increase in the size of melanosomes or might be aggregates of melanosomes either in melanocytes, where they are called autophagosomes, or in keratinocytes, where they are called melanosome complexes. Toda and Fitzpatrick[35] have observed a similar type of giant pigmented particle at the light-microscope level in melanocytic nevus of the skin.

C.   *Alterations in the melanization of melanosomes*

1.   Decreased tyrosinase activity and melanization

a.   Graying or whitening of hair with aging

The changes associated with aging were discussed above, where it was mentioned that there is a reduction in tyrosinase activity and melanization.

b.   Dilution of color of hair in phenylketonuria

A dilution of skin, hair, and eye color is considered typical in patients with phenylketonuria.[36-38] This dilution of color of the hair is most obvious in the Japanese because of the contrast between the normal black and the abnormal dark-brown hair color of the Japanese patient with phenylketonuria.[39] The most likely explanation for the dilution of hair color in phenylketonuria is that there is a competitive inhibition of tyrosinase by the high concentrations of phenylalanine and other metabolites that accumulate in this autosomal recessive disease.[40] Electron microscopic studies of the melanosomes in the hair of Japanese patients with phenylketonuria have not been reported, and, until this is done it is impossible to determine the basis of the hypopigmentation in phenylketonuria at the subcellular level.

c.   Hypomelanotic macules in tuberous sclerosis

Hypomelanotic macules are present in more than 90 percent of the patients with tuberous sclerosis when no other visible signs typical of the disease have yet appeared. The white macules are present at birth but

389

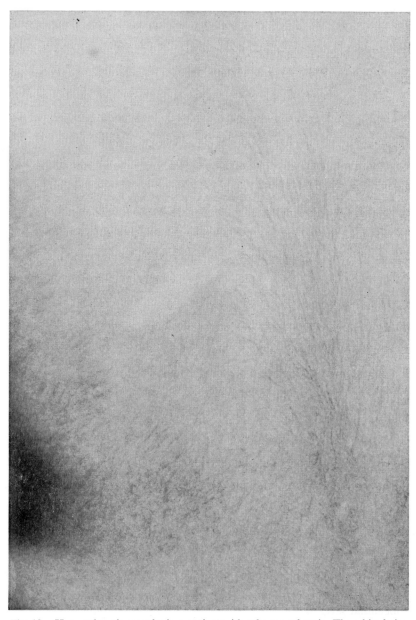

Fig. 10. Hypomelanotic macules in a patient with tuberous sclerosis. The white lesion on the left has a lance-ovate shape (the "ash-leaf" macule). These lesions are present at birth and have been found in a patient 35 years of age. The lesions may not be easily seen in fair-skinned persons unless a black-light (Wood's lamp) is used.

may not be obvious in fair-skinned infants and, therefore, must be searched for with the aid of Wood's lamp. Although these macules have been known for many years as incidental findings in patients with tuberous sclerosis, it was the report of Gold and Freeman in 1965[41] that emphasized that these hypomelanotic lesions may be the earliest sign of tuberous sclerosis. In a study of 31 patients, Fitzpatrick et al.[42] presented evidence that these hypomelanotic macules of tuberous sclerosis were different from those of vitiligo both clinically and histologically. The lesions are distributed all over the body, occurring most commonly on the trunk, especially the buttocks, and only rarely on the face. They are irregularly scattered, isolated, hypomelanotic lesions varying in number from 4 to more than 100. They are usually larger than 1 cm, but on the legs they may be tiny, appearing in a confetti-like pattern. A special feature of these white macules is their shape, which is usually polygonal or sometimes like a "thumb print." A most distinctive profile is the lance-ovate shape, which has the same outline as a leaflet of the mountain-ash tree (Fig. 10). On rare occasions, these white macules may appear in a quasidermatomal distribution.

A black-light (Wood's lamp) examination of the hypomelanotic macules found in patients with either vitiligo or tuberous sclerosis reveals a rather striking difference in the degree of hypomelanosis in the two diseases. In the areas of vitiligo, there is no pigment, whereas in the areas of tuberous sclerosis there is a reduction, but not an absence, of melanin pigment in the white macules. Electron microscopic study of vitiligo skin reveals an absence of detectable melanocytes in the hypomelanotic macules, whereas, in tuberous-sclerosis skin, melanocytes are present in approximately the normal number.[43]

This reduction of pigmentation in the hypomelanotic macules of tuberous sclerosis is related to a decreased melanization of individual melanosomes, as is illustrated in Fig. 11, which shows the normal and hypomelanotic melanocytes. Comparison of these two specimens shows that, in control Negroid skin, there is a marked melanization of the individual melanosomes (Stage IV), whereas, in the abnormal Negroid skin (the hypomelanotic macule of tuberous sclerosis), there is a marked reduction of the melanization in the melanosomes and an apparent diminution in the size of the melanosomes. The decrease in density of melanin deposition on melanosomes in the melanocytes of the white macules of tuberous sclerosis could be the result of a reduction in the concentration of tyrosinase or of the presence of an inhibitor that decreases the total tyrosinase activity and therefore the amount of melanin deposition.

d. Hypomelanotic macules in nevus depigmentosus

It is interesting that a study of three patients with nevus depigmentosus has revealed a pigment defect similar to that present in tuberous sclerosis.[44]

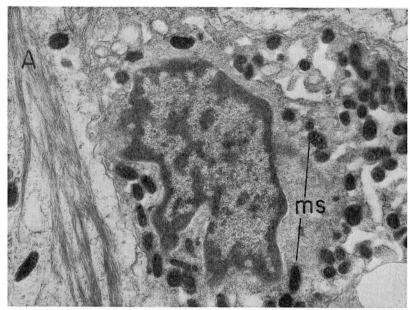

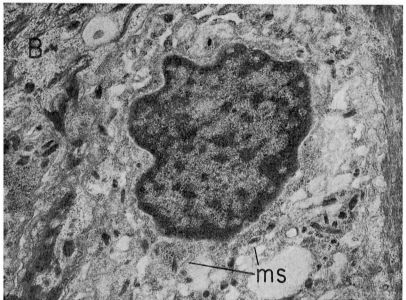

Fig. 11. Specimens from a Negroid patient with tuberous sclerosis. Uranyl acetate and lead citrate. ×16,000 (From Fitzpatrick et al., 1968, *Arch. Derm.*, **98,** 1; with permission).

A: Melanocyte from the normal skin. The melanosomes (ms) are numerous and very dense, and little or no internal structure of the individual melanosomes is seen.

B: Melanocyte from the adjacent hypomelanotic macule on the back. The melanosomes (ms) are smaller, and the internal structure is seen because of the lower amount of melanin deposition.

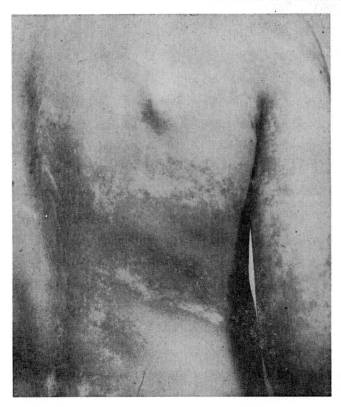

Fig. 12.   Nevus depigmentosus. There is unilateral hypomelanosis in a "nevoid" pattern.

In nevus depigmentosus, not only is there the same degree of hypomelanosis as that seen clinically in tuberous sclerosis, but there is also
electron microscopic evidence of a decreased density of the individual
melanosomes in the white macules. The white macules of tuberous sclerosis
are easily distinguished from those of nevus depigmentosus; in nevus
depigmentosus, the lesions have linear, almost bizarre patterns, and cover
large areas of the trunk and extremities (Fig. 12).

   Congenital circumscribed segmental hypomelanosis can appear in a
variety of conditions, including piebaldism, in which the lesions are usually
bilateral, and nevus depigmentosus, ataxia telangiectasia, and neurofibromatosis, in which the lesions are usually unilateral.

e.   Oculocutaneous albinism—tyrosinase-positive and tyrosinase-negative
Albinism, an anomaly inheritable by man and most species of animals,
occurs only when the lowest alleles in the albino series are present. In its
fullest manifestation, as in albinism in mice, for example, there is a complete absence of tyrosinase activity.[45] In animals, there are several allelo-

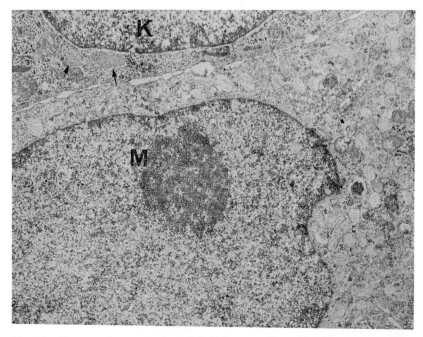

Fig. 13.   Electron micrograph of a hair bulb from a patient with oculocutaneous albinism. The melanocyte (M) contains many immature melanosomes; the keratinocyte (K) contains a few Stage II melanosomes (arrow). Glutaraldehyde-osmium tetroxide; embedded in Epon. ×17,600.

morphs at the c locus, each one of which is correlated with a level of tyrosinase activity. In man, the basic defect in albinism is the failure of melanocytes to synthesize *normal* amounts of tyrosinase. Witkop et al.[46] delineated two forms of oculocutaneous albinism, each based on the ability of the follicular melanocytes of plucked hair to synthesize melanin from tyrosine when they are incubated in tyrosine solutions.[4] It is clear from these studies that these two types of albinism are controlled by separate gene loci and are therefore nonallelic autosomal recessive traits.[47]

In the type of albinism called oculocutaneous tyrosinase-positive albinism, the hair, skin, and eye pigmentation darkens with age, and there are ocular changes (nystagmus and decreased visual acuity), although not severe ones; freckles and pigmented nevi are sometimes present. In this type of albinism, the follicular melanocytes of plucked hair readily form melanin from tyrosine on incubation in tyrosine solutions. Both Stage II and Stage III melanosomes are present in the hair bulbs.

In the tyrosinase-negative type of oculocutaneous albinism, the hair color is lighter and does not darken with age. Photosensitivity and other ocular changes are more severe than they are in the tyrosinase-positive type, and freckles and pigmented nevi are absent. At the subcellular level, only Stage II melanosomes are found in the hair bulbs (Fig. 13).

Applications of tyrosine solutions in the form of wet dressings to the skin of patients with the tyrosinase-positive type of oculocutaneous albinism produce pigmentation, and this suggests that the defect in this type of albinism may be a problem of substrate limitation with or without a deficiency of tyrosinase activity.[47]

D. *Disturbances in the transfer of melanosomes to keratinocytes due to structural or pathologic changes in melanocytes or in keratinocytes*

A functioning epidermal unit implies a melanocyte capable of transferring melanosomes to keratinocytes; this requires not only melanocytes with adequate dendrites but also a "receptive" keratinocyte. Hence, if there are alterations in the melanocyte morphology, such as inadequate dendrites, or if there are pathologic changes in the keratinocytes that do not permit the acceptance of melanosomes after transfer, there will be a reduction in melanin pigmentation. It must be reemphasized that clinically visible melanin pigmentation in the skin requires both the production of melanized melanosomes and the transfer of these melanosomes to keratinocytes.

1. Alterations in melanocyte structure (inadequate dendrites)

The most clear-cut example of inadequately developed dendrites that lead to failure of normal melanin pigmentation is seen in certain mouse genotypes that have genetically determined shapes that do not permit normal pigmentation responses to ultraviolet radiation.[48,49] After irradiation, melanocytes that have "stubby" dendrites supply only a limited amount of melanosomes to the immediate area of the individual epidermal melanin unit. In the epidermis that is fairly uniformly melanized, the melanocytes have well-developed dendrites, and the epidermal melanin units overlap, some of the keratinocytes being supplied with melanosomes by more than one melanocyte. This is especially true for the dilute black $(d/d, B/B)$ mouse. The dendritic processes of melanocytes in the dilute mouse are "shorter" than are the more highly branched processes of melanocytes of the nondilute $(D/D, B/B)$ black mouse.[50,51]

a. Changes in pale skin after chronic exposure to ultraviolet radiant energy

In the pale skin of persons who freckle and who usually have red hair, the melanocytes have been observed to have very much shorter, and therefore inadequate, dendritic processes in comparison with the dendritic processes of the melanocytes in heavily melanized Negroid skin.[52] It has also been noted that, after chronic exposure of the dorsa of the hands, for example, to ultraviolet irradiation, certain individuals who freckle and sunburn quickly will manifest a failure of normal melanosome transfer to keratinocytes; the melanocytes in these areas are stubby or broken off, thereby limiting the degree of transfer of melanosomes.[27]

2.   Increased proliferation of keratinocytes

a.   Psoriasis; verruca vulgaris; condyloma acuminatum

In certain dermatologic conditions, such as psoriasis and viral-induced hyperplasias of the epidermis (verruca vulgaris and condyloma acuminatum), there is a failure of melanosomal transfer, and these lesions are usually nonpigmented, even in the heavily melanized skin of Negroids. It has been assumed that the keratinocytes in this proliferative state are unable to accept the normal complement of melanosomes in transfer from melanocytes, owing either to some intrinsic defect or to the fact that the keratinocytes do not remain in contact with the melanocyte long enough to permit transfer of the normal number of melanosomes.

3.   Pathologic changes (edema; necrosis) in keratinocytes

a.   Eczematous dermatitis

When the epidermis becomes involved by an inflammatory process, such as chronic eczematous dermatitis, the keratinocytes do not usually contain any melanosomes, although the melanocytes are full of melanosomes. It is not known just why melanosome transfer is interrupted, but this failure of transfer does not appear to be related to an inadequate production of melanosomes or to melanocytes with inadequate dendrites; instead, the keratinocyte does not seem to have the capacity to accept melanosomes in transfer. In the epidermis of the normal gingiva, even when keratinocytes appear to be normal in all respects and melanocytes have adequate dendrites, there is a failure of transfer of melanosomes.

E.   *Degradation of melanosomes in melanocytes or in keratinocytes*

It is interesting that the fate of melanosomes in different racial groups is also different (Fig. 14).[30] In all races, melanosomes occur as single organelles surrounded by a unit membrane within melanocytes, but, when they are transferred to keratinocytes, they behave differently. In Caucasoids, Mongoloids, and Amerindians, they aggregate into a group around which a limiting membrane is formed; these aggregates of organelles, which are probably lysosomes, are called "melanosome complexes." In Negroids[30] and Australian aborigines,[27] however, the melanosomes do not aggregate after they are transferred to keratinocytes but usually remain as single, discrete particles, and a membrane forms around each one.

The investigations of Hori et al.[3] showed acid-phosphatase activity in the melanosome complexes, and it is therefore believed that melanosome complexes are similar to lysosomes; this similarity is suggested by (1) the presence of acid-phosphatase activity and (2) observations that melanosomes within the melanosome complexes undergo degradation.

The significance of the single arrangement of lysosome-like melano-

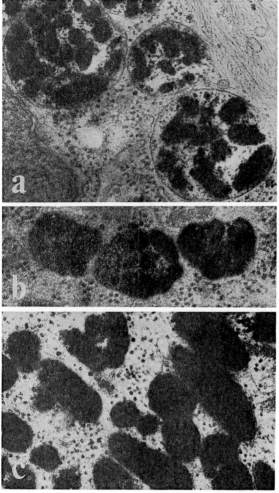

Fig. 14. Melanosomes in the keratinocytes of different races (from Szabó et al., 1969, *Nature*, **222**, 1081; with permission).

a: Melanosome complexes from a keratinocyte of a Caucasoid. The complexes are surrounded by a membrane and contain small particles besides the melanosomes. ×50,000.

b: Melanosome complexes from a keratinocyte of a Mongoloid (Chinese). There is less ground substance between the melanosomes than there is in a. ×50,000.

c: Melanosomes in a keratinocyte of a Negroid. Almost all melanosomes are individually dispersed and are much larger than those of a Caucasoid or Mongoloid. ×50,000.

somes in the skin of Negroids and Australian aborigines and of the grouped arrangement of lysosome-like melanosomes in the keratinocytes in the skin of other racial groups is not known at this time. It is perhaps an oversimplification to regard this type of degradation of melanosomes as a basis for the difference in pigmentation between Negroids and Australian aborigines and people of other races.

1. Malignant melanoma; melanocytic nevus

In Caucasoids (the only persons studied so far), it was found that, in the melanocytes of malignant melanomas and in pigmented nevi, melanosomes aggregated into groups and were surrounded by a membrane;[35] these groups are known as "autophagosomes."

III. CODA

Melanin pigmentation can be regarded as a four-step process: production of melanosomes; melanization of melanosomes; transfer of melanosomes from melanocytes to keratinocytes; and degradation of melanosomes. All four steps appear to be influenced by genetic factors.

Melanin pigmentation as seen clinically in the patient reflects the melanin content of the keratinocytes. Melanin exists in the epidermal melanocytes as fragments of degraded melanosomes in Caucasoids, Mongoloids, and Amerindians, or as intact ellipsoidal particles in the epidermal melanocytes in Negroids and Australian aborigines. The clinically visible pigmentation will also depend on the degree of melanin deposition on the individual melanosomes, for, although melanosomes may be produced by melanocytes and transferred to the keratinocytes, they may contain little or no pigment, as has recently been observed in oculocutaneous albinism (see Fig. 13). Marked reduction in clinically visible pigmentation can occur when the melanosomes are normal but the degree of melanin deposition is "dampened," as in the white macules of tuberous sclerosis and nevus depigmentosus. In these two conditions, melanosomes are elaborated by melanocytes and transferred to keratinocytes, but the reduction of clinically visible pigmentation occurs because of the lesser amount of melanin deposition on the melanosomes in the area of the white macule in comparison with the surrounding normally pigmented skin in which the melanosomes are more heavily melanized.

REFERENCES

1. Breathnach, A. S., Normal and abnormal melanin pigmentation of the skin. In *Pigments in Pathology* (M. Wolman, ed.), p. 353, Academic Press, New York (1969).
2. Quevedo, W. C., Jr., The control of color in mammals. *Amer. Zool.*, **9**, 531 (1969).
3. Hori, Y., Toda, K., Pathak, M. A., Clark, W. H., Jr., and Fitzpatrick, T. B., A fine-structure study of the human epidermal melanosome complex and its acid phosphatase activity. *J. Ultrastruct. Res.*, **25**, 109 (1968).
4. El Mofty, A. M., *Vitiligo and Psoralens*. Pergamon Press, Oxford (1968).
5. Clark, W. H., Jr., Pathak, M. A., Szabó, G., Bretton, R., Fitzpatrick, T. B., and El

Mofty, A. M., The nature of melanin pigmentation induced by furocoumarins (psoralens). In *Abstracts of Proceedings of the Fifth International Congress on Photobiology.* Hanover, New Hampshire, August 26–31, 1968, p. 138.

6. Breathnach, A. S., Bor, S., and Wyllie, L. M., Electron microscopy of peripheral nerve terminals and marginal melanocytes in vitiligo. *J. Invest. Derm.,* **47,** 125 (1966).

7. Ochi, Y. and DeGroot, L. J., Vitiligo in Graves' disease. *Ann. Intern. Med.,* **71,** 935 (1969).

8. Fitzpatrick, T. B., unpublished data.

9. Kopf, A., Morrill, S. D., and Silberberg, I., Broad spectrum of leukoderma acquisitum centrifugum. *Arch. Derm.,* **92,** 14 (1965).

10. Lerner, A. B., Vitiligo, *J. Invest. Derm.,* **32,** 285 (1959).

11. Costea, V., Cusu, Gh., and Matei, P., Plaje de leucodermie, apărute în cursul unei paralizii traumatice a plexului brahial, la un subiect cu canitie insulară. *Dermato-Venerol.* (Bucharest) **6,** 161 (no. 2).

12. Chanco-Turner, M. L. and Lerner, A. B., Physiologic changes in vitiligo. *Arch. Derm.,* **91,** 390 (1965).

13. Silvers, W. K., Genes and the pigment cells of mammals. *Science,* **134,** 368 (1961).

14. Mayer, T. C. and Maltby, E. L. An experimental investigation of pattern development in lethal spotting and belted mouse embryos. *Develop. Biol.,* **9,** 269 (1964).

15. Mayer, T. C., Pigment cell migration in piebald mice. *Develop. Biol.,* **15,** 521 (1967).

16. Mayer, T. C., Temporal skin factors influencing the development of melanoblasts in piebald mice. *J. Exp. Zool.,* **166,** 397 (1967).

17. Breathnach, A. S., Fitzpatrick, T. B., and Wyllie, L. M., Electron microscopy of melanocytes in human piebaldism. *J. Invest. Derm.,* **45,** 28 (1965).

18. Breathnach, A. S. and Goodwin, D. P., Electron microscopy of non-keratinocytes in the basal layer of white epidermis of the recessively spotted guinea-pig. *J. Anat.,* **99,** 377 (1965).

19. Kinebuchi, S., Hori, Y., Toda, K., Fitzpatrick, T. B., and Kobori, T., unpublished observations.

20. Frenk, E., Pathak, M. A., Szabó, G., and Fitzpatrick, T. B., Selective action of mercaptoethylamines on melanocytes in mammalian skin: Experimental depigmentation. *Arch. Derm.,* **97,** 465 (1968).

21. Bleehen, S. S., Pathak, M. A., Hori, Y., and Fitzpatrick, T. B., Depigmentation of skin with 4-isopropylcatechol, mercaptoamines, and other compounds. *J. Invest. Derm.,* **50,** 103 (1968).

22. Fitzpatrick, T. B. and Pathak, M. A., unpublished observations.

23. Fitzpatrick, T. B., Szabó, G., and Mitchell, R. E., Age changes in the human melanocyte system. *Advances in Biology of Skin,* vol. 6, *Aging* (W. Montagna, ed.), p. 35, Pergamon Press, Oxford (1965).

24. Quevedo, W. C., Jr., Youle, M. C., Rovee, D. T., and Bienieki, T. C., The developmental fate of melanocytes in murine skin. In *Structure and Control of the Melanocyte* (G. Della Porta and O. Mühlbock, eds.), p. 228, Springer-Verlag, Berlin (1966).

25. Staricco, R. J., and Pinkus, H., Quantitative and qualitative data on the pigment cells of adult human epidermis. *J. Invest. Derm.,* **28,** 33 (1957).

26. Szabó, G., Quantitative histological investigations on the melanocyte system of the human epidermis. In *Pigment Cell Biology* (M. Gordon, ed.), p. 99, Academic Press, New York (1959).

27. Mitchell, R. E., The effect of prolonged solar radiation on melanocytes of the human epidermis. *J. Invest. Derm.,* **41,** 199 (1963).

28. Quevedo, W. C., Jr., Szabó, G., Virks, J., and Sinesi, S. J., Melanocyte populations in UV-irradiated human skin. *J. Invest. Derm.,* **45**, 295 (1965).
29. Pathak, M. A., Sinesi, S. J., and Szabó, G., The effect of a single dose of ultraviolet radiation on epidermal melanocytes. *J. Invest. Derm.,* **45**, 520 (1965).
30. Szabó, G., Gerald, A. B., Pathak, M. A., and Fitzpatrick, T. B., The ultrastructure of racial color differences in man. *J. Invest. Derm.,* **54**, 98 (1970).
31. Clark, W. H., Jr., From, L., Bernardino, E. A., and Mihm, M. C., Jr., The histogenesis and biologic behavior of primary human malignant melanomas of the skin. *Cancer Res.,* **29**, 705 (1969).
32. Windhorst, D. B., Zelickson, A. S., and Good, R. A., Chédiak-Higashi syndrome: Hereditary gigantism of cytoplasmic organelles. *Science,* **151**, 81 (1966).
33. Windhorst, D. B., Zelickson, A. S., and Good, R. A., A human pigmentary dilution based on a heritable subcellular structural defect—the Chédiak-Higashi syndrome. *J. Invest. Derm.,* **50**, 9 (1968).
34. Benedict, P. H., Szabó, G., Fitzpatrick, T. B., and Sinesi, S. J., Melanotic macules in Albright's syndrome and in neurofibromatosis. *JAMA,* **205**, 618 (1968).
35. Toda, K. and Fitzpatrick, T. B., unpublished observations.
36. Jervis, G. A., Phenylpyruvic oligophrenia: Introductory study of 50 cases of mental deficiency associated with excretion of phenylpyruvic acid. *Arch. Neurol. Psychiat.,* **38**, 944 (1937).
37. Jervis, G. A., Phenylpyruvic oligophrenia (phenylketonuria). *Proc. Assoc. Res. Nerv. Ment. Dis.,* **33**, 259 (1954).
38. Cowie, V. L. A., Phenylpyruvic oligophrenia. *J. Ment. Sci.* **97**, 505 (1951).
39. Shizume, K. and Naruse, H., Dilution of hair colour in Japanese children with phenylketonuria. *J. Ment. Defic. Res.,* **2**, 53 (1958).
40. Miyamoto, M. and Fitzpatrick, T. B., Competitive inhibition of mammalian tyrosinase by phenylalanine and its relationship to hair pigmentation in phenylketonuria. *Nature,* **179**, 199 (1957).
41. Gold, A. P. and Freeman, J. M., Depigmented nevi: The earliest sign of tuberous sclerosis. *Pediatrics,* **35**, 1003 (1965).
42. Fitzpatrick, T. B., Szabó, G., Hori, Y., Simone, A. A., Reed, W. B., and Greenberg, M. H., White leaf-shaped macules: Earliest visible sign of tuberous sclerosis. *Arch. Derm.,* **98**, 1 (1968).
43. Fitzpatrick, T. B. and Szabó, G., unpublished observations.
44. Hori, Y., Fitzpatrick, T. B., and Szabó, G., unpublished observations.
45. Fitzpatrick, T. B. and Quevedo, W. C., Jr., Albinism. In *The Metabolic Basis of Inherited Disease,* 3d edition (J. B. Stanbury, J. B. Wyngaarden, and D. S. Fredrickson, eds.), McGraw-Hill, New York (in press) (1971).
46. Witkop, C. J., Jr., Van Scott, E. J., and Jacoby, G. A., Evidence for two forms of autosomal recessive albinism in man. *Proceedings of the Second International Congress on Human Genetics* (Rome, 1961), 1064 (1963).
47. Witkop, C. J., Jr., Albinism. In *Advances in Human Genetics,* vol. 2 (H. Harris and K. Hirschhorn, eds.), Plenum Press, New York (in press).
48. Quevedo, W. C., Jr., Genetic regulation of melanocyte responses to ultraviolet light. In *Recent Progress in Photobiology* (E. J. Bowen, ed.), p. 383, Blackwell, Oxford (1965).
49. Quevedo, W. C., Jr., Genetics of mammalian pigmentation. In *The Biologic Effects of Ultraviolet Radiation (with Emphasis on the Skin)* (F. Urbach, ed.), p. 315, Pergamon Press, Oxford (1969).
50. Quevedo, W. C., Jr., A review of some recent findings on radiation-induced tanning of mammalian skin. *J. Soc. Cosmetic Chem.,* **14**, 609 (1963).

51. Hadley, M. E. and Quevedo, W. C., Jr., Vertebrate epidermal melanin unit. *Nature,* **209**, 1334 (1966).

52. Szabó, G., Gerald, A. B., Pathak, M. A., and Fitzpatrick, T. B., unpublished observations.

AUTHOR INDEX

404

SUBJECT INDEX

408

410